Taking on Technocracy

Protest, Culture, and Society

General editors:
Kathrin Fahlenbrach, Institute for Media and Communication, University of Hamburg
Martin Klimke, New York University, Abu Dhabi
Joachim Scharloth, Waseda University, Japan

Protest movements have been recognized as significant contributors to processes of political participation and transformations of culture and value systems, as well as to the development of both a national and transnational civil society.

This series brings together the various innovative approaches to phenomena of social change, protest, and dissent which have emerged in recent years, from an interdisciplinary perspective. It contextualizes social protest and cultures of dissent in larger political processes and socio-cultural transformations by examining the influence of historical trajectories and the response of various segments of society, political, and legal institutions on a national and international level. In doing so, the series offers a more comprehensive and multi-dimensional view of historical and cultural change in the twentieth and twenty-first century.

For a full volume listing, please see back matter

Taking on Technocracy

Nuclear Power in Germany, 1945 to the Present

Dolores L. Augustine

berghahn
NEW YORK · OXFORD
www.berghahnbooks.com

First published in 2018 by
Berghahn Books
www.berghahnbooks.com

Library of Congress Cataloging-in-Publication Data
Names: Augustine, Dolores L., author.
Title: Taking on technocracy : nuclear power in Germany, 1945 to the present
/ Dolores L. Augustine.
Description: New York : Berghahn Books, 2018. | Series: Protest, culture and
society ; volume 24 | Includes bibliographical references and index.
Identifiers: LCCN 2017053741 (print) | LCCN 2017056831 (ebook) | ISBN
9781785339042 (ebook) | ISBN 9781785336454 (hardback : alk. paper)
Subjects: LCSH: Antinuclear movement--Germany--History--20th century. |
Nuclear power plants--Germany--History--20th century. | Nuclear
energy--Germany--History--20th century. | Protest
movements--Germany--History--20th century.
Classification: LCC HD9698.G32 (ebook) | LCC HD9698.G32 A94 2018
(print) |
DDC 333.792/40943--dc23
LC record available at https://lccn.loc.gov/2017053741

British Library Cataloguing in Publication Data
A catalogue record for this book is available from the British Library

ISBN 978-1-78533-645-4 hardback
ISBN 978-1-78533-904-2 ebook

This book is dedicated to my husband, Claude LeBrun, and my children, André and Cameron LeBrun

Contents

Illustrations

Illustrations

Figures

Acknowledgments

I thank my husband, Claude LeBrun, and my children, André and Cameron LeBrun—who with their unfailing love and support helped see me through this project and to whom this book is dedicated. My conversations about science with my husband and my mother, Juno Yolanda Augustine (1927–2016), helped shape the direction of this project. I am so lucky to have grown up in a family in which the "two cultures"—represented by my historically interested father and my scientifically inclined mother—intermingled. May the conversation between the sciences and the humanities thrive!

I am very grateful to the German Academic Exchange Service (Deutscher Akademischer Austauschdienst) for providing the funds that enabled me to spend two summers in Germany conducting research. I also thank the Leibniz Association (Germany) and my own university, St. John's University, for funding two further summer research stays in Germany. The Zentrum für Zeithistorische Forschung (Centre for Contemporary History) in Potsdam, Germany provided me with an academic home away from home in Germany. I have benefited greatly from discussions of my research with a number of scholars there, including Thomas Lindenberger, Frank Bösch, Christoph Classen, Jens Giesicke, Melanie Arndt, and Achim Saupe. I am deeply indebted to the German Women's Study Group of New York, which helped me craft this project over the years and gave me invaluable feedback on individual chapters: Bonnie Anderson, Marion Berghahn, Renate Bridenthal, Belinda Davis, Atina Grossmann, Amy Hackett, Maria Höhn, Marion Kaplan, Jan Lambertz, Molly Nolan, K. Molly O'Donnell, Kathy Pence, Nancy Reagin, and Julia Sneeringer.

My conversations with East German anti–nuclear power activist Sebastian Pflugbeil provided crucial insights into the ways opposition to atomic power emerged in the GDR. I learned a great deal from Inge Marszolek, a kind and generous scholar with a deep personal knowledge of the West German anti–nuclear power movement. She passed away in 2016. I am one of the many who will miss her. Dick van Lente spurred me to begin work on the popular culture of the nuclear age in Germany. Together with a fine group of scholars, we put together a comparative volume that gave me a sense of where the two Germanys were similar and different and where they stood in a global perspective. I gained a great deal from discussions with that group of historians: Hans-Joachim Bieber, Christoph Laucht, Sonja Schmid, Hirofumi Utsumi, and Scott Zeman. Dieter Hoffmann's advice was important to this project. I also benefited from the insights of Frank Biess, Alexandra Jaeger, Paul Josephson, Hasso Spode, and Burghard Weiss.

I thank the archivists and staff at the archives I visited, including the Stasi Archive (der Bundesbeauftragten für die Unterlagen des Staatsicherheitsdienstes der ehemaligen DDR), Berlin; Federal German Archives (Bundesarchiv), Berlin; the German Film Archive (Deutsche Kinemathek) of Berlin; the APO-Archive of the Free University Berlin (Freie Universität Berlin); the Green Party Archive (Archiv Grünes Gedächtnis); the National Archives II, College Park, Maryland; the Norddeutscher Rundfunk (Northern German Broadcasting) Archive, Hamburg; the German Broadcasting Archives (Deutsches Rundfunkarchiv), Babelsberg; the Provincial Archives (Landesarchive) of Berlin, Baden-Württemberg (Stuttgart), and Schleswig-Holstein (Schleswig); the State Archive of Hamburg; the Manhattan office of *Stern* magazine; the archive of Westdeutscher Rundfunk (WDR, or West German Broadcasting), Cologne; and the archive of Zweites Deutsches Fernsehen (ZDF) in Mainz, a television consortium. I thank the St. John's University Inter-Library Loan office, as well as the following St. John's students, who helped me find articles and put together my bibliography: Ashley Bozian, Chris Carron, Matthew Halikias, Mark Anthony Lewin, and Pamela Ramirez.

A special word of thanks goes out to Marion Berghahn, whose prodding and encouragement were helpful in equal measure! I am also grateful to Chris Chappell, Amanda Horn, and Caroline Kuhtz at Berghahn Books for the cheerful professionalism with which they helped me prepare my manuscript for publication.

Abbreviations and Acronyms/Glossary

ABCC—US Atomic Bomb Casualty Commission
ADN—Allgemeiner Deutscher Nachrichtendienst, General German News
    Service (East German press agency)
AEC—US Atomic Energy Commission
APO—*Außerparlamentarische Opposition,* extraparliamentary opposition
ARD—Arbeitsgemeinschaft der öffentlich-rechtlichen Rundfunkanstalten
    der Bundesrepublik Deutschland, Consortium of Public Broadcasters
    in Germany (a consortium of public broadcasting stations in West
    Germany)
BBU—Bundesverband Bürgerinitiativen Umweltschutz, Federal
    Association of Citizens' Initiatives on Environmental Protection (a
    West German organization)
BPA—*Bundespolizeiabteilungen,* federal police units (national riot police)
BUU—Bürgerinitiative Umweltschutz Unterelbe, Lower Elbe Citizens'
    Initiative on Environmental Protection
CDU—Christlich Demokratische Union Deutschlands, Christian
    Democratic Union of Germany
COCOM—Coordinating Committee for Multilateral Export Controls
    (Western arms embargo on Eastern Bloc countries)
Cogema—Compagnie Générale des Matières Nucléaires, General
    Company of Nuclear Materials
CSU—Christlich-Soziale Union, Christian Social Union (Bavarian sister
    party to the CDU)
DGB—Deutscher Gewerkschaftsbund, Confederation of German Trade
    Unions
DIVO—Deutsches Institut für Volksumfragen (German Institute for
    Public Opinion)
DKP—Deutsche Kommunistische Partei, German Communist Party
ESG—Evangelische Studentengemeinden, Evangelical student
    congregations
FDJ—Freie Deutsche Jugend, Free German Youth
FDP—Freie Demokratische Partei, Free Democratic Party
GAU—*größter anzunehmender Unfall,* maximum credible accident

GdP—Gewerkschaft der Polizei, Union of Police

GDR—German Democratic Republic (East Germany)

GRS—Gesellschaft für Reaktorsicherheit, Society for Reactor Safety (a West German organization)

HICOG—High Commissioner for Germany (of the US military occupation)

IAEA—International Atomic Energy Agency

ICBM—intercontinental ballistic missile

INES—International Nuclear and Radiological Event Scale

IPCC—Intergovernmental Panel on Climate Change

IRS—Institut für Reaktorsicherheit, Institute for Reactor Safety (a West German institute)

ITER—International Thermonuclear Experimental Reactor

KB—Kommunistischer Bund, Communist League

KBW—Kommunistischer Bund Westdeutschland, Communist League of West Germany

KPD—Kommunistische Partei Deutschlands, Communist Party of Germany

KTA—Kerntechnischer Ausschuss, Nuclear Technology Committee

NATO—North Atlantic Treaty Organization

NBI—*Neue Berliner Illustrierte* (East German illustrated magazine)

NGO—nongovernmental organization

NDR—Norddeutscher Rundfunk, Northern German Broadcasting (a West German public broadcasting network)

NUKEM—Nuklear-Chemie und –Metallurgie GmbH (a West German corporation specializing in nuclear technologies)

RAF—Rote Armee Fraktion, Red Army Faction

Rem—roentgen equivalent in man (measurement of radiation exposure)

RERF—US Radiation Effects Research Foundation

RWE—Rheinisch-Westfälisches Elektrizitätswerk AG, Rhenish-Westphalian Power Corporation

SAAS—Staatliches Amt für Atomsicherheit und Strahlenschutz, State Office for Nuclear Safety and Radiation Protection (of the GDR)

SDI—US Strategic Defense Initiative, or Star Wars

SDR—Süddeutscher Rundfunk, Southern German Broadcasting

SED—Sozialistische Einheitspartei Deutschlands, Socialist Unity Party of Germany (East German Communist Party)

SEK—*Spezialeinsatzkommandos*, special response units (of the West German police)

SPD—Sozialdemokratische Partei Deutschlands, Social Democratic Party of Germany

SWF—Südwestfunk, Southwest Broadcasting

Tageszeitung-taz (Berlin daily newspaper associated with the Green Party)

SZS—Staatliche Zentrale für Strahlenschutz, Central State Office for Radiation Protection

Tennet TSO—Tennet transmission system operator (German subsidiary of a Netherlands energy corporation)

TEPCO—Tokyo Electric Power Company

THTR-300—*Thorium-Hoch-Temperatur-Reaktor*, high-temperature thorium reactor

Tokomak—toroidal chamber with magnetic coils (experimental fusion reactor)

TÜV—Technischer Überwachungsverein, Technical Inspection Association

UCS—Union of Concerned Scientists

Veba—Vereinigte Elektrizitäts- und Bergwerks AG, United Electricity and Mining Works (a West German energy company)

WDR—Westdeutscher Rundfunk, West German Broadcasting

WHO—World Health Organization

WWER—*Wasser-Wasser-Energie-Reaktor*, water-cooled, water-moderated energy reactor (a Soviet nuclear reactor line)

ZDF—Zweites Deutsches Fernsehen, Second German Television Network

Introduction

In February 1975, images flickered across West German television screens of farmers and middle-aged villagers being assaulted with police water cannons and truncheons as they sought to block the construction of an atomic power plant in their village, Wyhl, located in what was then West Germany. Soon, the television station that had dared to transmit these images was decried by politicians and television executives as a communist stronghold.

Thirty-six years later, on 6 June 2011, German chancellor Angela Merkel—a physicist by profession—announced that Germany would abandon nuclear power by 2022. Dissent to this unprecedented decision was muted, coming mainly from the ranks of the leftist, environmentalist Greens, who felt Merkel's timetable unnecessarily delayed the shutdown of nuclear plants.

The slogan, *Atomkraft—Nein, Danke!* (Atomic Power—No Thanks!), once the rallying cry of a marginalized, radical movement, had come to be embraced by an entire society, it would seem. The Chernobyl and Fukushima reactor disasters of 1986 and 2011, respectively, had a far more muted long-term impact in most industrialized nations. In Germany, by contrast, opposition to nuclear power won the upper hand, and environmentalism became central to most Germans' sense of national identity. This was not always the case. In the wake of World War II and the Holocaust, East and West German leaders charted a course that involved a break with the Nazi past and an embrace of technological progress. Nuclear energy[1] was central to this vision. East and West Germans shared futuristic, utopian visions of the Atomic Age as an era of peace and progress for all humankind. With the help of science and technology, they hoped, East and West Germany could leave the Nazi past behind and become modern, forward-looking nations. Within a few years, however, thinking changed dramatically in the Federal

Republic of Germany (West Germany). The younger generation viewed the alliance between the political authorities, nuclear industry, and technical–scientific experts as rooted in the power structures and authoritarian thinking that had made National Socialism possible. West German activists strove to surmount Germany's pariah status by becoming part of the vanguard of transnational, progressive movements. Even in the GDR (the German Democratic Republic, or East Germany), the atomic consensus came to be questioned by a brave few.

Activists across West Germany rose up in what was nothing less than a popular rebellion against the rule of experts. The anti–nuclear power movement was rooted in a breakdown of popular trust in the state, elites, and the scientific establishment.[2] Faced with an almost monolithic nuclear power consensus, very diverse segments of West German society, ranging from Marxist radicals to conservative farmers, banded together to resist government policies. They vehemently rejected plans to turn idyllic regions of southwest and northern Germany into nuclear-powered industrial centers. Public intellectuals provided the foundations for a systemic critique of the West German nuclear power program and the elites that had produced it. In his 1977 book, *Der Atomstaat* (The Atomic State), Austrian journalist Robert Jungk put forth the thesis that reliance on nuclear weapons and nuclear power necessitated security measures that would lead to the reemergence of dictatorship. His biography as a Jew who escaped Austria in the wake of the *Anschluss* (the German takeover of Austria in 1938) lent him particular authority.[3]

Also very influential was sociologist Ulrich Beck's 1986 study, *Risk Society: Towards a New Modernity.* He argued that nuclear power represented a new kind of risk because the occurrence of catastrophic failure was so hard to predict and because its consequences were potentially so great. The government and scientific institutions could not protect the public, he insisted, and in fact resisted public scrutiny. Beck called for a democratization of the decision-making process regarding risky technologies and the application of ethical, philosophical, cultural, and political ways of reasoning to science and technology.[4] Criticism of elites was much more circumscribed in the GDR, yet one East German scientist wrote, "Scientists, doctors, engineers, politicians and military men are, in spite of their expertise, not immune to error, deceit, corruption, carelessness, hunger for power, and vanity."[5]

Ultimately, though, it was not the intellectuals, but the citizenry that forced a fundamental rethinking of the relationship between citizens and the state. The anti–atomic power movement was an unqualified success. Steve Milder has shown in a recent study that anti–nuclear power activism forged a powerful movement out of disparate groups and brought about a deepening of democracy in West Germany.[6] Andrew Tompkins explores the synergy between the West German and French anti–nuclear power movements in another important work.[7] Carol Hager establishes the importance of grassroots mobilization in German environmentalism.[8] The present study places the movement in the larger context of the evolution of the nuclear power issue in Germany, East and West, asking why these activists ultimately triumphed. There is no doubt that the major nuclear power plant accidents in Three Mile Island in 1979, in Chernobyl in 1986, and in Fukushima in 2011 helped turn skepticism into outright opposition. However, Germany was and is a pioneer in attempts to completely phase out nuclear power.

Five factors help explain this German peculiarity: the association in Germans' minds of nuclear war and nuclear power; changes in the media landscape that helped to expand civil society in West Germany; the important role of scientific arguments and counterarguments in the debates concerning atomic energy; a learning process among West German activists that led to an evolution in thinking concerning violence; and the rise of the Green Party and the growing receptiveness of the major political parties to environmentalism. Of these five factors, the role of science in debates proved to be the most surprising, and in some ways the most compelling. In the words of historian Cathryn Carson, "Science is all over the history of the Federal Republic of Germany."[9] However, this dimension of West German history has often been overlooked or underestimated.[10]

Can average citizens weigh in on scientific and technological policies in a meaningful way? This question has gained increasing importance with the tremendous upsurge in scientifically and technologically complex issues since World War II. The nuclear power debate in Germany represents a case in which citizens did, in fact, prove themselves able to grapple with such issues. This is not to say that their interventions were in every case well founded, yet the breadth and depth of popular attempts to understand the ins and outs of nuclear power are impressive. Eventually, public opinion was able to sway Chancellor Angela Merkel, a former physicist.

Recent spectacular examples of popular rejection of science—such as global warming denial and the antivaccination movement—seem to suggest that the public is too irrational, misinformed, or poorly educated to weigh in on scientific issues. However, scholarly research presents a far more complex understanding of the popularization of science than these examples seem to suggest and has pointed to the importance of lay knowledge in the scientific process.[11] Furthermore, the emotional, ideological, and religious commitments of average citizens are not necessarily incompatible with science. Historians have shown that scientific inspiration and competence can be found in what has long been thought to be unlikely places—in medieval Christianity, the Islamic world, astrology, and orthodox Marxist thought.[12]

Marxist and Christian anti–nuclear power activists in West Germany discovered that arguments about nuclear power based on their respective belief systems left them open to criticism. Historian Michael Schüring rightly views Protestant activists' use of a biblical frame of reference as unscientific in nature. However, as he points out, they went on to embrace scientific arguments, drawing on the expertise of pastors with a scientific background, as well as scientific and technical experts who became involved in Church activism.[13] Leftist activists also made serious attempts to understand the technical and scientific fundamentals of nuclear power, as publications from the 1970s and 1980s reveal. This turn to science was crucial to the movement's quest for legitimacy in the eyes of the public, government officials, the courts, and the international community.

While accepting science as a cognitive system, activists were highly critical of scientific institutions, science as a profession, and what they saw as an alliance between science, the state, and the nuclear industry. Activists in West Germany asserted that the supposedly "scientific" consensus behind the atomic power program was in fact highly ideological, politicized, and corrupted by ties to industry. In Communist East Germany, a small group of dissident scientists and their friends agreed in private that "science is a good deal more subjective, corrupt, subservient, and intentionally false than the average citizen might imagine. And in scientific fields related to ecology and nuclear technologies, there is far more science for sale than science that is honest and accurate."[14]

This critical stance emerges out of much older debates and discussions concerning the interpenetration of science, politics, society, the economy, and culture. In the nineteenth and early twentieth centuries, proponents of technocracy claimed that only experts were capable of scientific reasoning or participating in decision making connected with the running of modern societies.[15] The technocracy movement of the 1930s even sought to replace democracy with rule by engineers and scientists.[16] A soft technocratic approach emerged after World War II across the industrialized capitalist world, as many political leaders sought to rationalize and legitimize policymaking through a mutually beneficial alliance with experts. US president Dwight D. Eisenhower warned both of the influence of politics and big money on science and "the equal and opposite danger that public policy could itself become the captive of a scientific-technological elite."[17] Public intellectuals ranging from C.P. Snow to Jürgen Habermas were concerned that such collaboration could undermine democracy.

Scholars disagreed over whether this was likely in the technocracy debate of the 1960s and 1970s. West German antinuclear activists were influenced by these exchanges. According to sociologist Peter Weingart, the nuclear power controversy made the West German public aware for the first time that experts could disagree among themselves concerning major scientific and technological issues and that the political leadership specifically recruited and promoted those experts who supported their political agenda. Weingart asserts that the authority of science was eroded, not only by the "scientification of politics" and the "politicization of science" but also by the "mediatization of science," meaning scientists' efforts to reach and influence the public through the media. In his view, these developments reduced science to the status of one actor among many.[18]

My research confirms that anti–nuclear power activists contested the authority of scientists who entered the fray as experts supporting government nuclear power policies, and in fact they entered into very public and very loud debates with them. However, one of the most important findings of my study is that the anti–nuclear movement was not at all averse to science as a cognitive system and in fact made extensive use of scientific arguments and promoted the popularization of scientific and technical knowledge about nuclear power. They struggled to educate themselves and others, studying the workings of the atom, the health consequences of exposure to radioactivity, technical vulnerabilities of conventional atomic reactors, problems with newer technologies, and the inability of authorities and industry to find safe and acceptable ways of disposing with nuclear waste. They learned a great deal from *Gegenexperten* (counterexperts), the term used in Germany for experts critical of the scientific establishment and its coalition with the state.

What effect did environmentalism have on Germans' approach to public policy and nuclear power in particular? German romanticism and back-to-nature movements of the late nineteenth and early twentieth centuries would seem to suggest that Germans were greater lovers of nature than other peoples. However, David Blackbourn's work on waterways shows that in fact Germans' relationship with the environment was characterized by tremendous tensions between attempts to preserve nature on the one hand and to subjugate nature in the pursuit of progress on the other.[19] The state played a major role in both endeavors. Even the Nazis promoted certain environmentalist policies.[20] During the "Economic Miracle" after World War II, neither policymakers nor the public viewed environmentalist policies such as antipollution measures as undermining the pursuit of prosperity, but rather as a part of the rising standard of living.[21] What changed after 1970 was that environmentalists no longer saw the state and the elites as defenders of the environment. Only then did environmentalism become a protest movement.

New attitudes toward emotions helped legitimize the West German public's participation in debates on nuclear power and other ecological issues. Here, the history of emotions is helpful, providing ways of understanding emotions as a constituent part of human existence and history, rather than merely a disruptive factor in political life and decision making in modern societies.[22] In 1975, supporters of nuclear power asserted that they were representatives of rationality, while tarring opponents of nuclear power as emotional and therefore not competent to weigh in on this issue. Referring to emotional anti–nuclear power demonstrators, television journalist Hans-Gerd Wiegand replied, "It is completely legitimate to show emotions."[23]

Historian Frank Biess argues that the outward-directed emotional regime of the late 1940s and 1950s was supplanted in the 1960s by a greater acceptance of interior emotional life.[24] The rise of environmentalism and the anti–atomic power movement provide good examples of how emotions forced elites to confront scientific and technological problems in new ways and to allow the public to participate in debates and negotiations. In the 1970s and 1980s, many West Germans felt they were living in a world filled with threats. The New Cold War of the 1980s only partially explains the sense of dread so pervasive among the younger generations in that period. Environmentalism grew in tandem with fears of toxic contamination, resource depletion, and forest die-offs. New attitudes toward emotions valorized fears and made it possible to articulate them in public and to turn them into political demands. At the same time, the West German public debate on atomic power in this period gives lie to the notion that science and emotion are intrinsically incompatible.[25]

The West German nuclear power controversy was fueled by the clash of two starkly differing visions of West Germany's future. The major political parties, the state bureaucracies, and the business community advocated joining the capitalist–democratic West through economic modernization, full employment, prosperity, and technological achievement. These priorities became all the more important in their eyes during the energy crises and recession of the 1970s. In addition, West German provinces, called *Länder* (states), provided considerable subsidies and participated in the running of nuclear power plants owned by public utility companies. Thus, the West German political leadership was deeply committed to atomic energy on both ideological and material levels, making retreat almost unthinkable.

Scientists and engineers were among the first advocates of nuclear power, and there were few dissenting voices among them in either Germany. They had long subscribed to an apolitical ethos that made them loyal servants of a state conceived as above politics. This orientation grew out of the struggles of the nineteenth and early twentieth centuries to professionalize, to bolster their status, and in many cases simply to secure employment in their professions.[26] German scientists and engineers displayed little inclination to criticize the state, and they embraced the professional opportunities as well as the ideological benefits of a strategy of economic and technological modernization. Werner Heisenberg, who had played a major role in the Nazi atomic bomb project, became a major spokesman for science as a foundation of modern identity, as Cathryn Carson has shown.[27] West Germans, casting around for a usable past and new foundations for a sense of national identity, were very receptive to Heisenberg's message in the 1950s and 1960s.

Activists of various stripes challenged these values. In the 1960s, young people across the globe led the charge against authority, conformity, and the belief that technological and economic progress should be society's main goal. Intellectuals such as Herbert Marcuse and Antonio Negri spread Marxist thought and inspired a generation of radical students to take up the revolutionary cause. The latter were highly critical of what they saw as Western imperialism in the Third World, and they organized the movement against the Vietnam War. Young West Germans confronted their country's Nazi past. They sought to break down authoritarian structures still in existence in Germany.[28] The state reacted with harsh countermeasures, supported by a large swath of the electorate.

Political and cultural polarization tore at the social fabric. In the long run, however, "1968" contributed to the democratization of West German society. In the short run, West German society was roiled by conflicts regarding the meaning of democracy and citizenship. Conservative professors accused students of behaving like the Hitler Youth. Students responded that the real Nazis were professors who did not want to allow protests. Activists profoundly challenged political authority and law enforcement, disputing the idea that democracy was merely a matter of voting in elections and obeying elected officials.

The anti–atomic power movement emerged as one of the most important "new social movements" of the 1970s. New social movements are generally considered to be popular movements that focus, not on issues revolving around class or socioeconomic status but on quality of life, sometimes described as "postmaterialist values." Competing schools have variously viewed new social movements as irrational reactions to social breakdown; rational attempts to attain concrete goals or political objectives; creators of countercultural milieus that defy social norms and authority, such as Marxist or anarchist groups; or subcultural movements, which have a strong in-group orientation, as in the case of the women's movement and the "alternative" scene. In fact, the anti–nuclear power movement encompassed all these different strands, but sociologists have been reluctant to accept an all-inclusive explanation for the success of new social movements.[29] As Steve Milder has pointed out, new social movement theory fails to recognize the unifying force of the anti–atomic power movement, as well as its profound contribution to the democratization of West Germany.[30]

Diversity gave the movement broad appeal. Not all participants in the anti–nuclear power movement had ties to sixties radicalism. The rural population in particular tended to be wary of outsiders, particularly those leftists whom they perceived as ideologically rigid.[31] Nonetheless, these more conservative elements participated in a mobilization of society from below that challenged the institutions, ethos, and practices of the West Germany system. Local activist and vintner Annemarie Sacherer protested before television cameras against the police handling of demonstrations in Wyhl in 1975, shouting, "This is no longer a democracy!"[32] A viewer who saw a television program about this demonstration wrote in to the TV station, arguing, "The pictures show more clearly than any commentary could, how people trying to exercise their constitutional rights were denounced and treated like heretics and criminals."[33]

The archival records from that period make clear that many who went to demonstrations were not part of any organization. They were motivated by local concerns or felt that as good citizens they should speak up—out of concern for the environment or out of anger against what they saw as an out-of-touch government and elite. To shoehorn all who participated in the movement or who were swayed by it into one category or the other is to miss the great diversity of this mobilization and the deep and broad impact it had on the general population. Gradually, it won over people who simply read a leaflet or watched a television program about opposition to nuclear power. Over time, opinions shifted, creating a fundamental distrust of atomic power that became activated by the Three Mile Island, Chernobyl, and Fukushima disasters.

More enlightening than new social movement theory in trying to explain the successes of the West German anti–nuclear power movement are a body of sociological writings on four dynamics that help make social movements effective: cultural power, organization, negotiation, and disruption. First, movements need to be able to sway public opinion through the dissemination of convincing new ideas. Second, organizational infrastructure is key to mobilizing supporters, providing leadership, and securing resources. The third factor, negotiation, involves politics, engagement with the state, legal action, and bargaining within and outside the activist community.[34] The "strongest weapon of social movements" is the power to physically disrupt through demonstrations, blockades, and occupations.[35] Generally, these involve protesters placing their bodies in harm's way for the purpose of physically impeding the maintenance of the status quo.[36] The West German anti–nuclear power movement was strong in all four areas. But did it have an actual impact on society and the political system?

Activists struggled to win over public support. Initially, nuclear power polarized society. The first major wave of antinuclear activism, in 1975–1977, coincided with a wave of terrorism, spearheaded by the Red Army Faction (RAF, popularly known in the English-speaking world as the "Baader-Meinhof Gang"). Caught up in this fight, state governments countered anti–atomic power protests by building up and deploying massive police and security forces. A significant portion of the West German population thought of opponents of nuclear power as dangerous radicals. This negative image was reinforced by the clashes between demonstrators and police near the planned nuclear power plant construction site in Brokdorf, a small town in the north of West Germany, in 1976–1977.

The anti–nuclear power movement was divided over the advisability of violent confrontations with police during demonstrations. During the campaign to prevent the construction of a nuclear power plant in Brokdorf, demonstrators attempting to enter the construction site engaged in pitched battles with police, while other activists peacefully demonstrated elsewhere. However, as Andrew Tompkins has argued, such distinctions were often not so clear-cut. Some forms of passive resistance and civil disobedience involved physical contact. And many demonstrators who would never have gone on the offensive felt they, along with other protesters, had the right to defend themselves against police.[37] Nonetheless, a turn toward more peaceful methods took place between the mid-1970s and the mid-1980s, accompanied by the rise of the environmentalist Green Party and the increased participation of feminists in the movement. Activists became more tolerant of violence in the wake of Chernobyl but soon turned to blockades and other forms of passive resistance in the fight to stop the delivery of radioactive waste to Gorleben from the 1980s to the early twenty-first century.

The rise of the West German nuclear power movement would not have been possible without the tremendous expansion and transformation of civil society and mass media that took place from the 1960s to the 1980s. The Green Party began as an "anti-party," avoiding the norms of conventional politics and resisting commitment to parliamentary democracy. By the 1990s, it evolved into a conventional party, increasing its electoral support and joining coalition governments, most notably the "red–green" coalition with the Social Democratic Party, which governed Germany from 1998 to 2005.

Profound shifts in the West German media world enabled the anti–
nuclear power movement to bring its case before the court of public opinion.
In the decades after World War II, the media's ability to mobilize popular
opinion against the government was severely constrained. The Federal
Republic's first chancellor, Christian Democrat Konrad Adenauer (in office
1949–1963) saw state domination of the media as a means to democra-
tize West Germany, but also to tamp down opposition. He gave access to
friendly journalists, fed the media official versions of events, bullied editors
and heads of broadcasting networks, and even considered censorship laws.
State governments exercised considerable control over public television and
radio networks, which monopolized the airwaves until the introduction of
commercial television stations in 1981. During the 1950s, politicians largely
succeeded in silencing critical radio commentary.[38]

Fundamental changes and explosive growth in the 1960s transformed
the media landscape, opening it up to points of view not sanctioned by the
federal or state governments. Journalists participated in the seismic cultural
and political shifts of that period. They sought to promote democratization,
help overcome authoritarian cultural patterns, and decisively break with the
Nazi past, as historian Christina von Hodenberg has shown. An ideal of
"engaged journalism" emerged that proposed that journalists could take sides
politically, engage in investigative journalism and social critique, and defend
the downtrodden but that it was not their job to help create and preserve
some sort of cultural or political consensus.

The defenders of this model took on the hierarchies and centers of
authority in the media. Journalists working for *Stern*, a very popular weekly
magazine, were assured that they did not have to write anything that violated
their convictions. On the other hand, Rudolf Augstein, editor in chief of *Der
Spiegel*, a major news magazine, fired "engaged journalists."[39] As the example
of the Nazi rise to power illustrates, expanded media presence in the political
realm does not always bring about liberalization and democratization.[40] In
this particular case, however, it did. Television journalists were no longer
content to allow government and industry spokesmen to drown out other
viewpoints. Over state objections, they gave common citizens the opportunity to speak on TV about why they opposed the building of nuclear power
plants. These journalists were trying to expand both media presence and the
realm of free speech.

Less idealistic forces were at work as well, as historian Bernd Weisbrod
has argued. The tremendous growth of journalism as a profession in the 1960s
created fierce competition among journalists. The rise of television added to
the competitive atmosphere, as magazines struggled to maintain their readership. The struggle for market shares and professional advancement spurred

an expansion of investigative reporting.[41] The *Spiegel* scandal of 1962 made the media into political participants. Defense Minister Franz Josef Strauß accused the news magazine *Der Spiegel* of having published state secrets in compiling investigative reports on the West German military and caused journalists to be arrested. Strauß lost his job as a result. This run-in with the government gave journalists a deeper sense of autonomy and valorized their work as never before. At the same time, the public began to expect the media to take on a more confrontational role vis-à-vis the state.

In the GDR, the SED (the Socialist Unity Party, as the ruling Communist Party was called) put the full force of its power and influence behind the defense of nuclear power. Criticism of atomic energy was long kept in check through state censorship. Suppression of non-Communist organizations precluded the emergence of a mass nuclear power movement. The SED did, however, treat society as an important actor in the quest for "technical–scientific progress." A central aspect of citizenship was mobilization—in factories, schools, and universities—for the technology-driven advancement of socialist society.[42] Moreover, key aspects of professionalism were left intact. Even in an era in which the Soviet Union was supposedly in charge of ensuring the safety of nuclear power plants, East German engineers were at work on obscure but important projects to improve nuclear safety.

In the 1980s, as East Germany's economic and technological problems mounted, so too did political dissent. Emerging peace, human rights, and ecology activism began to create tentative and vulnerable beginnings of a civil society. This enabled a few activists to disseminate criticism of uranium mining and nuclear power.

Although East German media operated under dictatorial conditions, they were not the lifeless tool of the SED. Certainly, the SED did everything in its power to prevent the emergence of a public realm separate from the state and the SED. The SED viewed the unity of state and society as central to the success of socialism and did not like to be contradicted or undermined. However, the GDR was much more than just the SED. Earlier German traditions, institutions, and ways of thinking lived on until at least the late 1960s. By that time, Western youth culture began intruding on what was called "really existing socialism," meaning the imperfect form of socialism then actually in existence. Something like a public sphere was beginning to emerge by the late 1980s, mostly under the stewardship of the Protestant Church. However, I avoid the term "public opinion" because some might object that it implies the existence of a public realm independent of the state. The term "*popular* opinion" is more open-ended in this regard and is more appropriate for nondemocratic societies.[43]

The SED tried to mold popular opinion, but nonapproved views some-times made their way into popular culture.[44] As cultural theorist John Fiske has pointed out, the producers of popular culture must respond to viewer preferences or lose their audience.[45] Popular tastes helped mold East German television programming,[46] as well as the content of illustrated magazines and other down-market publications. Visual culture was not monitored as closely by censors as texts. Under some circumstances, popular opinion came bursting through the usual reserve and conformity, for example after the Chernobyl nuclear power plant explosion in 1986, when citizens inundated the authorities with their concerns.

Additional factors make a comparison between East and West Germany fruitful, starting with their common history and language. During the Cold War, the two countries saw each other as rivals in many realms, including technology, scientific research, culture, and social development. Emulating and reviling each other, both pursued social, economic, and technological modernization. Détente very much promoted East–West contacts.[47]

At the same time, the GDR's ties to the Soviet Union and the Federal Republic's to the United States made them part of different, although not entirely separate, transnational networks. After World War II, the United States tried to induce West German politicians and citizens to follow the American lead in developing nuclear power but not nuclear weapons.[48] Criticism of nuclear power safety came out of the US science community. Disquieting research on radiation and human health was conducted in many countries, but US scientists were the most likely to take the issue to the popular press. West German and US antinuclear activists learned from each other. The occupation of the nuclear power plant construction site in Wyhl, West Germany set an example for the Clamshell Alliance, an activist organization that was fighting the building of an atomic plant in Seabrook, Massachusetts.

The East German leadership relied on the Soviet Union for most of its nuclear energy hardware, but also for expert advice and safety monitoring. Soviet reluctance to fulfill this role, as well as accidents in GDR nuclear power plants, caused the GDR to seek greater technological self-reliance by the 1970s. East German nuclear authorities increasingly adopted Western standards and imitated Western technologies. The GDR also embraced the Soviet glorification of nuclear power as a powerful motor of socialist progress. This dream of socialist technological superiority faded by the 1980s. Mikhail Gorbachev's reformism encouraged expression of discontent over the coun-try's nuclear power regime. The harsh suppression of such dissent—modeled on older Soviet practices—prevented East Germans from mounting a serious challenge to technocratic patterns of decision making.

My first chapter looks at the popular culture of nuclear power from the end of World War II until the early 1970s, focusing on the rallying of public support for nuclear power. The Soviet Union tried to match the American Atoms for Peace program with its own claims to be the biggest proponent of "peaceful uses of the atom" and staunchest opponent of atomic war. Consensus regarding the desirability of nuclear power existed among political, economic, and scientific elites in both Germanys. However, despite bursts of euphoria concerning the cornucopia that nuclear energy ostensibly offered, there were some signs of popular unease regarding radiation and atomic power, particularly in the wake of the Bikini nuclear bomb test of 1954. Nuclear power could not shake its association with "the bomb" in the popular mind. This chapter asks whether West German culture was atypical in this regard, comparing analyses of popular depictions of nuclear power in illustrated magazines across cultures. Television programs relating to nuclear energy and accounts of an "Atoms for Peace" exhibition also provide interesting insights into the place of nuclear power in the popular imagination.

Chapters 2 and 3 turn to science and technology. Chapter 2 compares approaches to safety, risk, human error, and nuclear power accidents in East and West Germany from the 1960s to the 1980s. The engineering community, in tandem with industrial leaders and state regulators, developed "engineering philosophies" that guided fundamental approaches to safety. Initially, there were striking East–West differences. As a result of conflicts within and between institutions involved in nuclear power production and oversight, the GDR eventually adapted itself to Western standards, particularly in the wake of the Chernobyl disaster. This chapter also asks how much the public knew about safety problems. Security concerns kept accidents in the GDR secret, while West German nuclear power plant owners attempted to avoid bad publicity by covering up safety problems. However, the West German public demanded and received better information as time went on.

Chapter 3 looks at the origins and diffusion of the scientific and technological arguments that became central to opposition to atomic power in West Germany and, later, the GDR. US scientists made key contributions to research on the health consequences of radiation exposure as well as to criticism of nuclear reactor safety. Maverick scientists and other "counter-experts" in West Germany took up these arguments against nuclear energy, just as the anti–nuclear power movement was moving onto a national and international stage, giving it a crucial boost. This turn toward science was embraced by West German popularizers, who disseminated and modified these arguments.

The spectacular emergence of the West German anti–atomic power movement is the topic of Chapter 4. The government of Baden-Württemberg, headed by Christian Democrat Hans Filbinger, made the planned nuclear power plant in Wyhl a cornerstone of a technocratic policy of development of Baden, a rural region of West Germany, as well as a step toward overcoming the 1970s oil crisis. Protests ensued, culminating in the occupation of the nuclear power plant construction site in 1975. What began as a regional protest of wine growers and villagers concerned about the impact of the proposed plant on the local economy and on the microclimate grew into a national movement. This movement questioned the equating of nuclear power and progress; the close alliance of the state and nuclear industry; the objectivity of technical and scientific "experts" friendly to the government; and the top-down model of decision making that had prevailed up until that time in West Germany.

Television and the press helped shift power relations between opponents of nuclear power and the state. The Filbinger government engaged in a heated media campaign against the Wyhl protesters and against WDR, a major public broadcasting network. The immediate result was a deeply divided, polarized public realm. Scientific findings played a significant role in these debates. The debates problematized the role of emotion, which was variously interpreted as an impediment to rational, scientific discourse or, conversely, as a gateway to greater public participation.

Conflicts over nuclear power grew into what was nearly a civil war in Brokdorf, the topic of Chapter 5. The government of Gerhard Stoltenberg wanted to construct a nuclear power plant there that would serve both Schleswig-Holstein and Hamburg, in the north of the Federal Republic. Many of the same phenomena observable in the Wyhl case were also present in the Brokdorf conflict: determination on the part of the government to defend this project both as an important component of a modernization program and as a profit-making public utility; government attempts to intimidate the media (in this case, the NDR—Norddeutscher Rundfunk, or Northern German Broadcasting, a public broadcasting network) and discredit the protesters as dangerous radicals; involvement of political activists, some quite radical, in the protests; and polarization of the public. Each side tried to deploy science to defend its case.

While the Wyhl plant was never built, Stoltenberg carried out his plans for Brokdorf with iron determination. More police were deployed than ever before in the history of the Federal Republic. The activists engaged in an intense debate regarding politically motivated violence, leading to a slow but steady decline in violence among anti–nuclear power protesters, lasting until 1986.

The 1986 Chernobyl disaster unleashed a wave of violence among anti–atomic power protesters, the subject of Chapter 6. However, state mishandling of the upsurge of unrest discredited state actions. In Hamburg, hundreds of anti-Brokdorf demonstrators were held for eighteen hours, tightly packed and largely without access to food, water, or bathrooms. This incident was treated as a national scandal. Many people who had never before participated in a demonstration joined in protests after Chernobyl. Ulrich Beck's *Risk Society* spoke to the national mood. The Greens served to consolidate opposition to atomic power, although infighting slowed the rise of the more conciliatory and pacifistic wing of the party.

I return to the GDR in Chapter 7, which traces the rise of anti–nuclear power activism there. It originated with scientists, unlike in the Federal Republic. Sebastian Pflugbeil (a physicist and biomedical researcher at the Academy of Sciences) was one of perhaps five scientists who became interested in nuclear power, its risks, and its possible impact on human health. They quietly conducted their studies for years, reading and analyzing scientific publications. Through these scientists, atomic power became one of the topics of interest to the ecological movement that enjoyed the Protestant Church's protection.

In the 1980s, particularly after the Chernobyl disaster, they began to write samizdat publications (photocopied "publications" not sanctioned for general circulation by the state) and give talks within the framework of the Protestant Church. Chernobyl unleashed an unprecedented wave of appeals to the state for information, assistance, and advice. In Church circles, Chernobyl became the touchstone of a wave of environmentalist activism. This chapter examines the outlooks, politics, and habits of GDR activists. State oppression of the fledgling movement fostered a sense of solidarity among its members. Some of these activists later became involved in the New Forum, which negotiated the transition to a multi-party system.

The book's final chapter discusses debates about atomic power since reunification and asks why Angela Merkel's government decided in 2011 to phase it out. Two quite contradictory tendencies contributed to this "energy turn." The first is professionalization of the Green Party and of environmentalist research as well as the emergence of a vibrant alternative energy sector and its incorporation into the capitalist economy. The second is the continued militancy of the anti–atomic power movement, which was focused on the disposal of nuclear waste in Gorleben. The Fukushima disaster sounded the death knell of atomic energy in Germany. Or did it? In light of climate change, the parameters of the debate concerning nuclear power have shifted considerably, and the future remains uncertain. I agree with historian Frank Uekötter's view that the rise of environmentalism has been historically contingent and is reversible.[49]

Notes

1. In German, the term *Atomkraft* (atomic power) came to be used by its opponents, while *Kernkraft* (nuclear power) or *Kernenergie* (nuclear energy) were used by its proponents. This distinction dwindled by the mid-1980s. Matthias Jung, *Öffentlichkeit und Sprachwandel: Zur Geschichte des Diskurses um die Atomenergie* (Wiesbaden: Springer Fachmedien, 1994), 43–46, 60–62, 82–89, 134–36, 194–95. The difference between atomic power, atomic energy, nuclear power, and nuclear energy is not as pronounced in English. To avoid tedium, I use them interchangeably throughout this study.
2. Albrecht Weisker, "Expertenvertrauen gegen Zukunftsangst: Zur Risikowahrnehmung der Kernenergie," in *Vertrauen: Historische Annäherungen*, ed. Ute Frevert (Göttingen: Vandenhoeck & Ruprecht, 2003), 394–421.
3. Robert Jungk, *Der Atom-Staat: Vom Fortschritt in die Unmenschlichkeit* (Munich: Kindler, 1977). English translation: Robert Jungk, *The New Tyranny: How Nuclear Power Enslaves Us*, trans. Christopher Trump (New York: F. Jordan Books/Grosset & Dunlap, 1979).
4. Ulrich Beck, *Risikogesellschaft: Auf dem Weg in eine andere Moderne* (Frankfurt am Main: Suhrkamp Verlag, 1986), 37 (on the last point). First English edition: Ulrich Beck, *Risk Society: Towards a New Modernity*, trans. Mark Ritter (London and Newbury Park, CA: Sage, 1992). The democratization theme comes out more clearly in Ulrich Beck, *Gegengifte: Die organisierte Unverantwortlichkeit* (Frankfurt am Main: Suhrkamp Verlag, 1988), 277, 288.
5. Sebastian Pflugbeil, preface to Michael Beleites, "Pechblende: Der Uranbergbau in der DDR und seine Folgen," Samizdat publication, unnumbered pages, BStU, HA XVIII, Nr. 18237, 84.
6. Stephen Milder, *Greening Democracy: The Anti-Nuclear Movement and Political Environmentalism in West Germany and Beyond, 1968–1983* (Cambridge, UK and New York: Cambridge University Press, 2017).
7. Andrew Tompkins, *Better Active Than Radioactive! Anti-Nuclear Protest in 1970s France and West Germany* (Oxford: Oxford University Press, 2016).
8. Carol Hager, *Technological Democracy: Bureaucracy and Citizenry in the German Energy Debate* (Ann Arbor, MI: University of Michigan Press, 1995).
9. Cathryn Carson, *Heisenberg in the Atomic Age: Science and the Public Sphere* (Cambridge, UK and New York: Cambridge University Press, 2010), 3.
10. The main problem is the divide between historians of technology and science on the one hand and those who study society, culture, and politics on the other.
11. Brian Wynne argues that while lay knowledge is socially constructed, so too is science in "May the Sheep Safely Graze? A Reflexive View of the Expert-Lay Knowledge Divide," in *Risk, Environment and Modernity: Towards a New Ecology*, ed. Scott Lash, Bronislaw Szerszynski, and Brian Wynne (London: Sage, 1996), 44–83. By contrast, Steven Shapin takes a realist stance, positing in *Never Pure: Historical Studies of Science* (Baltimore, MD: Johns Hopkins Press, 2010) that science is an imperfect, human enterprise, influenced by social and historical context, yet universalistic and fundamental to the technological foundations of the modern world. Roger Cooter

and Stephen Pumphrey also emphasize this relationship between the scientific community and the public in the making of science in "Separate Spheres and Public Places: Reflections on the History of Science Popularization and Science in Popular Culture," *History of Science* 32, no. 3 (1994): 237–67.

12. Edward Grant, *The Foundations of Modern Science in the Middle Ages: Their Religious, Institutional and Intellectual Contexts* (Cambridge, UK and New York: Cambridge University Press, 1996); George Saliba, *A History of Arabic Astronomy: Planetary Theories during the Golden Age of Islam* (New York: New York University Press, 1994); Lawrence Principe, "Alchemy Restored," *Isis* 102, no. 2 (2011): 305–12; Loren Graham, *Science in Russia and the Soviet Union* (Cambridge, UK and New York: Cambridge University Press, 1993), 112–16.

13. Michael Schüring, *'Bekennen gegen den Atomstaat': Die evangelischen Kirchen in der Bundesrepublik Deutschland und die Konflikte um die Atomenergie 1970–1990* (Göttingen: Wallstein Verlag, 2015), esp. 181–97.

14. Email from Sebastian Pflugbeil to Dolores Augustine, 31 March 2015.

15. These include the utopian socialist Henri de Saint-Simon, sociologist and economist Thorstein Veblen, and engineer Frederick W. Taylor, the founder of "scientific management," or Taylorism.

16. Edwin Layton, *The Revolt of the Engineers: Social Responsibility and the American Engineering Profession* (Cleveland, OH and London: Press of Case Western Reserve University, 1971).

17. Dwight D. Eisenhower, "Farewell Radio and Television Address to the American People, January 17th, 1961," Eisenhower Archives, retrieved 20 October 2017 from https://www.eisenhower.archives.gov/all_about_ike/speeches/farewell_address. pdf. See also Jeff Hughes, *The Manhattan Project: Big Science and the Atom Bomb* (Cambridge, UK: Icon Books, 2002), 128–29.

18. Peter Weingart, *Die Stunde der Wahrheit? Zum Verhältnis der Wissenschaft zu Politik, Wirtschaft und Medien in der Wissensgesellschaft* (Weilerswist, Germany: Velbrück, 2001), 133–39, 143; Peter Weingart, "Verwissenschaftlichung der Gesellschaft—Politisierung der Wissenschaft," *Zeitschrift für Soziologie* 12 no. 3 (1983): 225–41.

19. David Blackbourn, *The Conquest of Nature: Water, Landscape, and the Making of Modern Germany* (New York: Norton, 2006).

20. Joachim Radkau and Frank Uekötter, eds., *Naturschutz und Nationalsozialismus* (Frankfurt and New York: Campus Verlag, 2003).

21. Frank Uekötter, *Deutschland in Grün: Eine zwiespältige Erfolgsgeschichte* (Göttingen: Vandenhoeck & Ruprecht, 2015), chs. 2–4.

22. For a useful approach to emotions on the radical left, see Joachim Häberlen and Jake Smith, "Struggling for Feelings: The Politics of Emotions in the Radical New Left in West Germany, c.1968–84," *Contemporary European History* 23, no. 4 (2014): 615–37. On "fear" and the quest for "security" in the German debates over nuclear disarmament, see Holger Nehring and Benjamin Ziemann, "Führen alle Wege nach Moskau? Der NATO-Doppelbeschluss und die Friedensbewegung—eine Kritik," *Vierteljahrshefte für Zeitgeschichte* 59, no. 1 (2011): 81–100. For an overview of the history of emotions, see Jan Plamper, *The History of Emotions: An Introduction,* trans. Keith Tribe (New York: Oxford University Press, 2015).

23. WDR Historisches Archiv, Signatur 12523, transcript of "Glashaus—TV Intern," dated 1 April 1975. My translations throughout.

24. Frank Biess, "Feelings in the Aftermath: Toward a History of Postwar Emotions," in *Histories of the Aftermath: The Legacies of the Second World War in Europe,* ed. Frank Biess and Robert Moeller (New York: Berghahn Books, 2010), 30–48. Also Frank Biess, *German Angst? Fear and Democracy in Postwar West Germany* (New York: Oxford University Press, forthcoming).

25. For example Spencer Weart, *Nuclear Fear: A History of Images* (Cambridge, MA: Harvard University Press, 1989). He sees popular fears regarding nuclear power as emotional and therefore a rejection of science.

26. Dolores Augustine, *Red Prometheus: Engineering and Dictatorship in East Germany, 1945–1990* (Cambridge, MA: MIT Press, 2007), 22–27.

27. Carson, Heisenberg, chs. 5 and 6.

28. Timothy Brown, *West Germany and the Global Sixties: The Anti-Authoritarian Revolt, 1962–1978* (Cambridge, UK and New York: Cambridge University Press, 2013), esp. 4–12, 81–84.

29. Ruud Koopmans, *Democracy from Below: New Social Movements and the Political System in West Germany* (Boulder, CO: Westview Press, 1995), 7–37 and 229–36, and literature cited therein. On the "alternative scene," see Sven Reichardt, *Authentizität und Gemeinschaft: Linksalternatives Leben in den sieziger und frühen achtziger Jahren* (Berlin: Suhrkamp Verlag, 2014).

30. Milder, *Greening Democracy.*

31. Tompkins, *Better Active,* 25, 41–50.

32. WDR Historisches Archiv, Archive number 0165239. Ulrich Eith, "Nai hämmer gsait! Stilbildender ziviler Widerstand in Wyhl am Kaiserstuhl," in *Aufbruch, Protest und Provokation: Die bewegten 70er- und 80er-Jahre in Baden-Württemberg,* ed. Reinhold Weber (Darmstadt: Theiss Verlag, 2013), 52; article on 35–54.

33. WDR Historisches Archiv, 7210, letter dated 26 February 1975.

34. Kenneth Andrews, *Freedom Is a Constant Struggle: The Mississippi Civil Rights Movement and Its Legacy* (Chicago, IL: University of Chicago Press, 2004), 2–6, 198–200.

35. Sidney Tarrow, *Power in Movement: Social Movements and Contentious Politics,* 3rd rev. ed. (Cambridge, UK and New York: Cambridge University Press, 2011), 103–4; see 25, 124–29, 231.

36. Tiya Miles, "Fighting Racism Is Not Just a War of Words," *New York Times* (21 October 2017). The roots of physical resistance are many, ranging from Mahatma Gandhi's conception of passive resistance to long-standing practices of rebellion and revolution.

37. Tompkins, *Better Active,* ch. 5; Michael Sturm, "Polizei und Friedensbewegung," in *"Entrüstet Euch!" Nuklearkrise, NATO-Doppelbeschluss und Friedensbewegung,* ed. Christoph Becker-Schaum, Philipp Gassert, Martin Klimke, Wilfried Mausbach, and Marianne Zepp (Paderborn: Ferdinand Schöningh, 2012), 277–93.

38. Matthias Weiss, "Öffentlichkeit als Therapie: Die Medien- und Informationspolitik der Regierung Adenauer zwischen Propaganda und kritischer Aufklärung," in *Medialisierung und Demokratie im 20: Jahrhundert,* ed. Frank Bösch und Norbert Frei (Göttingen: Wallstein Verlag, 2006), 73–120; Christina von Hodenberg,

Konsens und Krise: Eine Geschichte der westdeutschen Medienöffentlichkeit, 1945–1973 (Göttingen: Wallstein Verlag, 2006), 152–215.

39. Von Hodenberg, *Konsens und Krise*, 245–439, esp. 420–33.

40. Frank Bösch und Norbert Frei, "Die Ambivalenz der Medialisierung: Eine Einführung," in *Medialisierung und Demokratie im 20: Jahrhundert*, ed. Frank Bösch und Norbert Frei (Göttingen: Wallstein Verlag, 2006), 7– 23.

41. Bernd Weisbrod, ed., *Die Politik der Öffentlichkeit—die Öffentlichkeit der Politik: Politische Medialisierung in der Geschichte der Bundesrepublik* (Göttingen: Wallstein Verlag, 2003), esp. the introduction and articles by Frank Bösch and Willibald Steinmetz.

42. Andrew Port, *Conflict and Stability in the German Democratic Republic* (Cambridge, UK and New York: Cambridge University Press, 2007); Jörg Roesler, "Die Produktionsbrigaden in der Industrie der DDR. Zentrum der Arbeitswelt?" in *Sozialgeschichte der DDR*, ed. Hartmut Kaelble, Jürgen Kocka, and Hartmut Zwahr (Stuttgart: Klett-Cotta Verlag, 1994), 144–70. On state citizenship, see Jan Palmowski, "Citizenship, Identity and Community in the German Democratic Republic," in *Citizenship and National Identity in Twentieth-Century Germany*, ed. Geoff Eley and Jan Palmowski (Stanford, CA: Stanford University Press, 2008), 73–93. He interprets the GDR conception of citizenship as encompassing economic, social, and cultural participation in the socialist community, although he does not focus on the workplace.

43. Usage in Ian Kershaw, *Popular Opinion and Political Dissent in the Third Reich, Bavaria 1933–1945* (Oxford: Oxford University Press, 1983), 4.

44. An example is the glorification of (East) German technology and the downplaying of Soviet technological achievements in a book handed out to eighth graders who completed their *Jugendweihe*. See Alfred Kosing, Diedrich Wattenberg, and Rolf Dörge, *Weltall Erde Mensch: Ein Sammelwerk zur Entwicklungsgeschichte von Natur und Gesellschaft*, 13th ed. (Berlin: Verlag Neues Leben, 1965), unnumbered pages 336–37, 352–53, 393, 400–401, 404, and 415–49. See also my analysis in Augustine, *Red Prometheus*, 223–24.

45. John Fiske, "The Popular Economy," in *Cultural Theory and Popular Culture: A Reader*, ed. John Storey, 3rd ed. (Harlow, UK: Pearson Education Ltd., 2006), 537–53. Also see comments and application of Fiske in Thomas Lindenberger, "Einleitung," in *Massenmedien im Kalten Krieg: Akteure, Bilder, Resonanzen* (Cologne: Böhlau Verlag, 2006), 16–17 and introduction on 9–23.

46. Heather Gumbert, *Envisioning Socialism: Television and the Cold War in the German Democratic Republic* (Ann Arbor, MI: University of Michigan Press, 2014), 13. For a broad and deep analysis of popular entertainment programs, see Rüdiger Steinmetz and Reinhold Viehoff, eds., *Deutsches Fernsehen Ost: Eine Programmgeschichte des DDR-Fernsehens* (Berlin: Verlag für Berlin-Brandenburg, 2008). For East–West German comparative studies on television, see Lindenberger, *Massenmedien im Kalten Krieg*, esp. Uta Schwarz, "Der blockübergreifende Charme dokumentarischer Filme," 203–34 and Thomas Heimann, "Television in Zeiten des Kalten Krieges: Zum Programmaustausch des DDR-Fernsehens in den sechziger Jahren," 235–61. Schwarz finds parallels in developments in East and West Germany.

47. Frank Bösch, "Geteilt und verbunden: Perspektiven auf die deutsche Geschichte seit den 1970er Jahren," in *Geteilte Geschichte: Ost- und Westdeutschland 1970–2000* (Göttingen: Vandenhoeck & Ruprecht, 2015), 7–38.
48. On American attempts to mold science policies in postwar Europe, see John Krige, *American Hegemony and the Postwar Reconstruction of Science in Europe* (Cambridge, MA: MIT Press, 2006).
49. Uekötter, *Deutschland in Grün,* 137.

Chapter 1

Nuclear Dreams and Radioactive Nightmares

Popular Culture and the Quest for Nuclear Consensus in East and West Germany, 1945–1970

As Germany dug out of the rubble in 1945–1946, the power of the atom in its military and peaceful forms became a major theme in the press. The Munich-based *Süddeutsche Zeitung,* the first newspaper given permission to publish in the American zone of occupation, ran an article in 1946 that asserted, "The bombing of Hiroshima and Nagasaki ushered in a new era," in which "it is vital to transform the power of the atom into a blessing." Atomic energy could bring forth "wonders" in the areas of medicine, biology, transportation, and public welfare, according to the article. But it also raised the terrifying possibility that it would eventually be possible to develop nuclear bombs that could smash apart the earth or suck all the oxygen out of the atmosphere.[1]

A similar narrative was to be found in a 1946 article in the illustrated magazine *Neue Berliner Illustrierte,* published in the Soviet-occupied sector of Berlin. It showed a frightening artist's depiction of the 1 July 1946 US atomic bomb test at Bikini Atoll, along with actual photos from the test. Should atomic power be used this way?, it asked. "Up until now, atomic energy has only been used and tested for destructive ends. This development must be brought to a complete halt and prohibited by international law because it could eventually mean the end of human culture. By contrast, the use of atomic energy in industry will provide the entire world with untold possibilities for peaceful progress."[2]

The atomic bomb and atomic power were inextricably intertwined in popular narratives of this period. Media across the globe depicted humankind as standing "at the crossroads," confronted with a choice between the annihilation of the human race through nuclear war and the creation of a better world through nuclear-fueled prosperity and progress.[3] This was not a new idea. In his 1913 novel, *The World Set Free,* H.G. Wells first formulated the idea that radiation either could provide huge amounts of energy, thus providing prosperity for all, or, if used in a bomb, could spell the destruction of humanity. Wells invented both the concepts and the terms "atomic bomb" and "Atomic Age."[4]

According to historian David Nye, the destructive power of the bomb evoked religious feelings, a sense of awe and profound fear among Americans, giving way to "a somber feeling" and a new sense of the vulnerability of life on earth. Paul Boyer shows that Americans were initially gripped with an overwhelming sense of terror. However, he argues that within a couple of years, government, the media, and corporate America succeeded in convincing the US public that the "peaceful atom" somehow outweighed and even cancelled out the threat of destruction through nuclear war.[5]

The American and Soviet military administrations in Germany and, later, German authorities made major attempts to mold popular perceptions of the unfolding Atomic Age. Germans were receptive because, to them, nuclear technologies represented a modern world that would help them escape from the past.[6] The theme of redemption through the turn away from war and the building of peace and prosperity was fundamental to the American Atoms for Peace policy, as well as its Soviet equivalent. West and East Germans sought security, defined in three ways: peace and protection as allies of the United States and the Soviet Union, economic security, and progress through science. Nuclear power was widely held to be compatible with and even essential to all three. More importantly, by being welcomed into the new nuclear regime of their respective blocs, East and West Germany gained new respectability that might allow them to shed their Nazi past.

However, the association of atomic power with the atom bomb never wore off. In the decades after World War II, political leaders and elites who promoted nuclear power certainly viewed these misgivings as irrational and attempted to build a nuclear consensus from the bottom up by educating the public, promoting technical safety, and providing a legal framework for the safe use of nuclear power. However, an undercurrent of fear lived on. This chapter looks at conflicting tendencies in depictions of nuclear power and at the relationship between science and emotion in East and West German popular culture from the 1940s to the 1960s, placing this analysis in the context of wartime experiences, the Cold War, and the development of a dictatorial socialist system in the GDR and a capitalist–democratic system in the Federal Republic. The media did not just filter and transmit attitudes and ideas about nuclear power but also helped to mold them.

Media and Popular Opinion in the Atomic Age

Shattered and subjugated during the Nazi period, West German media rapidly recovered in the postwar period, in many cases returning to Weimar era patterns. The elite upheld the idea of Germany as a *Kulturnation* (cultured nation) and disdained popular entertainment. The educated classes held certain newspapers and newsmagazines to be the most dependable purveyors of political news. These included the Center-Right daily *Frankfurter Allgemeine Zeitung* (founded in 1949), the Center-Left daily *Süddeutsche Zeitung,* and the Center-Left weekly news magazine *Der Spiegel* (founded in 1947). However, tabloids and illustrated magazines far surpassed the "serious" press in terms of circulation.[7]

Public opinion polls carried out by or commissioned by the US occupation authorities provided a corrective to overly elitist views of the West German press.[8] Illustrated magazines such as *Stern* regained their pre-1933 popularity after the war. In 1950, about 16.3 percent of the adult population in West Germany read *Stern,* according to a US survey. This translates into a readership of about eight million.[9] *Stern* had a circulation of 1.2 million in 1960, but *Stern* estimates that it had over ten million readers at that time. Thus, about one in five West Germans read it, making it as popular a magazine in West Germany as *Life* was in the United States.[10] Many read the magazine in a beauty salon, a barbershop, or a waiting room, or they read a friend's, relative's or colleague's copy. The readership of magazines cut across class lines. Professionals, businessmen, and office workers were the most avid readers of magazines and illustrated magazines, according to a US survey of 1955.[11]

Gradually, television became the dominant popular medium. Most Germans did not watch television in the 1950s. Only 24 percent of West German households and 17 percent of East German households owned a television set in 1960.[12] By contrast, 60 percent of West Germans surveyed in 1955 read magazines.[13] By 1966, however, over half of all households in the Federal Republic and the GDR had a TV.[14]

The driving force behind popular media was the desire to appeal to the broad masses. In the early days, *Stern* published a good many articles about scandals, stars, and members of royal families, but by the mid-1950s, it was doing a good deal of investigative journalism and featuring pointedly critical articles on world affairs. Its editors saw it as part of the "fourth power" or "fourth estate," that is, a watchdog overseeing the three branches of government. Its Center-Right leanings up into the mid-1960s gave way to a more Center-Left orientation by the 1970s.[15]

The East German press was state or party owned and was subject to strict censorship. The SED's party newspaper, *Neues Deutschland,* closely hewed to the party line. Journalists were carefully monitored by the authorities and were expected to adhere to guidelines issued by the "agitation/propaganda" division of the SED and by bodies such as the German Peace Council, a GDR affiliate of the Soviet-dominated World Peace Council.[16]

Although never criticizing the Communist system, some publications sought to appeal to a popular audience. One example is *Neue Berliner Illustrierte,* an illustrated magazine published since 1945 in the Soviet sector of Berlin and later the GDR. With a circulation of 700,000 in a country with a population of about seventeen million, NBI was the most popular magazine in the country around 1970, although the authorities restricted the number of copies printed, and it was sold from "under the counter."[17] From its inception, it was a lively magazine that sought, not just to promote a socialist consciousness but also to draw in readers with arresting images, drama, and teaser headlines. Its articles, although never critical of the SED or the Soviet Union, were sometimes illustrated with photos rich in ambiguity and Westernized imagery, particularly after Joseph Stalin's death in 1953.

Internal documents from the Berliner Verlag (Berlin Publishing House), the publisher of NBI and the tabloid BZA (*Berliner Zeitung am Abend,* the evening edition of the *Berliner Zeitung,* or Berlin News), make clear that journalists were under great pressure to fulfill SED mandates. It was evidently rare for a journalist to push back, yet I was able to find one such instance. At a 1962 meeting behind closed doors, an SED party official objected to the headline, "He Does Not Want to Commit Suicide," placed above a photo of a Western demonstrator carrying a banner that read, "War is our greatest enemy." This item, published in BZA, "contains pacifist tendencies," he contended. (This was problematic because it might call into question Soviet "self-defense.") A journalist countered that such scruples stood in the way of making BZA *massenwirksam,* meaning "appealing to the masses."[18] This desire to connect with the reader opened mass media in the GDR to currents in popular culture that did not align perfectly with official goals.

The first part of this chapter places the popular culture of the Atomic Age in the context of German history and the Cold War, drawing on my quantitative content analysis of all issues of *Stern* and NBI published between 1945 and 1965 (about 2,000 in number—none digitized). This study was compiled as part of a larger project on popular culture depictions of nuclear technologies in eight countries.[19] In keeping with Germany's position on the potential front lines of any nuclear war, military technologies overshadowed civilian technologies in articles on the Atomic Age from the late 1940s to 1965. Of 270 articles on nuclear technologies published in

Stern between 1948 and 1965, only thirty-five discussed atomic power or other nonmilitary technologies (such as medical uses). NBI published forty-seven articles about civilian nuclear technologies out of a total of 186 articles with nuclear themes in the same period.[20] The following sections discuss some of the most important tropes, beginning with one whose significance is sometimes overlooked.

Scientists as Heroes, 1945–1953

Popular attitudes were molded, not just by the peaceful atom/destructive atom narrative but also by the stature of science and scientists. Across the globe, science was presented as an objective anchor in an uncertain world and a return to rationality after the irrationality of the ideological extremism of wartime. In Germany, whose scientific tradition was the source of considerable pride, journalists, authorities, and the people could construe science as representing a "better Germany." Such retrospection invoked a redeemable German past that ostensibly stood above politics, making it useful in both the Soviet and Western zones of occupation. The first article about nuclear technology published after the war by NBI focused on Wilhelm Röntgen (1845–1923), a German scientist responsible for the discoveries that led to the development of modern radiology. It extolled not only his scientific contribution but also his ethos: "In a selfless manner, he made his beneficial discovery available to all of humankind."[21] Similarly, a 1947 article in *Der Spiegel* about the death of Max Planck recounted that the physicist, who was pushed out as head of the Kaiser-Wilhelm Society in 1936, had asked in vain to speak to Hitler to protest against the persecution of Jews.[22]

In reality, the place of scientists in German society, both during and after the Nazi era, was rather different. According to historian Dieter Hoffmann, Planck had in fact displayed a marked lack of solidarity with Jewish physicists such as Albert Einstein.[23] Physicist Erwin Schrödinger was one of the rare non-Jewish scientists who fled Nazi Germany.[24] Most scientists were only too happy to work on projects with military applications during the war.[25] They included a project to build an atomic bomb. The reasons for its failure have been the subject of a good deal of academic debate. German scientists ended up working on a wartime project to build an atomic reactor that used heavy water as a moderator.[26] Few scholars believe Werner Heisenberg's claim that as head of an important research team working on the bomb, he deliberately sabotaged the project.[27] Forced to flee from Germany, Albert Einstein and Leo Szilard induced President Franklin D. Roosevelt to initiate the Manhattan Project.

In a mad rush in 1945–1946, the United States, Great Britain, and the Soviet Union seized scientists and technical specialists that they feared might go to work for the other side. Those captured by American forces under the ALSOS Mission and Operation Paperclip were quickly released and found they had good professional prospects in the United States or the Western zones of occupation. German atomic scientists in Soviet custody, many of whom worked on the Soviet atomic bomb program, were not released until the mid-1950s.[28] East and West Germany competed to recruit these and other scientists, especially recipients of the Nobel Prize, not only because of their potential contribution to science but also because of their prestige value. Each side was anxious to burnish its reputation as heir to the German tradition of excellent in science.[29]

Worldwide trends helped German scientists leave their tainted past behind. Science and scientists enjoyed great prestige across the globe after the war. Atomic physicists such as Albert Einstein and Niels Bohr emerged from the war as ambassadors of international understanding and world peace. Albert Schweitzer and Leo Szilard gained a reputation for "speaking truth to power." Scientists "were seen as a kind of international brotherhood, hampered in their exchange of views by the barriers of the Cold War, which therefore impeded progress."[30] Werner Heisenberg was able to take advantage of the moral stature of leading members of the international scientific community. Quickly rehabilitated and in 1946 named director of what became the Max Planck Institute for Physics, he became an advocate of state funding for nuclear power.[31] According to historian Cathryn Carson, he was anxious to head off public concerns regarding nuclear research. In the late 1940s, he became a respected public figure. In public speeches, he addressed the modern sense of alienation and the struggle to find meaning, suggesting that science provided orientation in a disorienting world.[32]

Journalists, scientists, policymakers, and occupational authorities took advantage of the prestige of science and scientists when they developed the narrative of the peaceful atom to promote agendas of various sorts.

"Atoms for Peace" in a Warlike Era, 1954–1957

Nuclear power was developed in the shadow of the Cold War and the arms race. The US nuclear monopoly ended when the Soviet Union detonated an atomic bomb on 29 August 1949. The United States tested its first hydrogen bomb—a far more powerful weapon that was not, yet, a deliverable weapon—on 1 November 1952. The Soviet Union shocked the world in

August 1953 by exploding an H-bomb that, although smaller, could be dropped from an airplane. Germany, straddling the East–West divide, felt acutely exposed. Soviet forces tried to force Western Allied forces out of Berlin during the Blockade of 1948–1949, but the Western Allies were able to keep the city supplied in the Berlin Airlift, and West Berlin became part of the Federal Republic of Germany, which was founded 23 May 1949. On 7 October 1949, the GDR was established in the East.

Anticipating the H-bomb, whose development had just been announced by US president Harry Truman, a 1950 *Stern* article featured an artist's rendition of an H-bomb explosion over a map of Essen, a large city in West Germany. It explained the levels of destruction in zones around the ground zero of such an attack. Only six such bombs were needed to completely destroy Germany, the article asserted.[33] Similar articles ran in illustrated magazines in the United States, Great Britain, and the Netherlands around this time. However, *Stern* and NBI ran articles on the threat of nuclear war more consistently and in greater number, at least up until 1965 (when my quantitative data series ends).[34]

Combating a popular sense of dread, political leaders, opinion makers, and scientists tried to spread a message of optimism regarding the possibilities for peace and prosperity in a world where "the atom" was put to use to better humanity rather than to destroy it. That was the main message of President Dwight D. Eisenhower's "Atoms for Peace" speech, delivered before the United Nations General Assembly on 8 December 1953. In addition, he called for the creation of a U.N. atomic energy agency that would give nations across the globe access to the fissionable materials and technology needed to start up nuclear power programs. Some historians have argued that this speech and the worldwide campaign that followed were primarily propaganda aimed at concealing or justifying the growing US nuclear arsenal and mitigating the aggressive impression it made.[35] However, historians John Krige and Mara Drogan have demonstrated that the spread of nuclear power was a major aim of the Atoms for Peace program.

Access to the technologies and materials needed to initiate atomic energy programs had been severely restricted up to that point by Allied agreement. Krige emphasizes that the US Atoms for Peace program was crucial to the founding of Euratom, which US policymakers saw as the lynchpin of nuclear nonproliferation policy in Western Europe. The timing of Eisenhower's launching of this policy initiative also had to do with international events. The death of Stalin and the end of the Korean War opened the prospect of greater cooperation with the Soviet Union. On the other hand, Soviet proposals for total nuclear disarmament and pledges never to launch a "first-strike" nuclear offensive put the United States, which did not want to renounce either, on the defensive. Eisenhower hoped to mobilize world public opinion in favor of US leadership in the Cold War.[36]

The Soviet Union had long been trying to promote its own leadership role with its own version of the peaceful atom narrative. Science was central to the self-understanding of leaders and loyal citizens in the Soviet Union and the GDR.[37] Following the example of Soviet publications, NBI proclaimed that nuclear technologies in the hands of capitalist nations led to the development of weapons and the spread of aggression, imperialism, and, eventually, war. Socialism, by contrast, was said to promote peaceful uses of the atom, not only for power production but also for medicine, transportation, and food production.

An NBI article from the late 1940s featured an atomic-powered spacecraft that could reach the moon in "3 hours, 27 minutes." The caption of a fantastical drawing of the dramatic nighttime launch, the creation of the great German press illustrator Helmuth Ellgaard (1913–1980),[38] read:

> It is a few minutes before midnight. All eyes on earth are turned to this first United Nations airport for inter-planetary travel. The elegant, shiny, metallic body of the spacecraft lies on a mighty, rotating launching pad ... A pull on a lever unleashes the subdued atomic power ... For the first time, a spacecraft leaves our planet.[39]

An imitative article ran in *Stern* three years later, although the nuclear-powered craft was a freight and mail rocket, and the name stamped on its side—RAK—made it clear that it was inspired by stunt vehicles manufactured by the Opel Automobile Company in the 1920s.[40]

Futuristic daydreaming had a purpose, as can be seen in the 1947 NBI article "Bomb or Philosopher's Stone?" While the West continued to develop and test nuclear weapons, the article asserted, the Soviet Union and the socialist world offered peaceful uses of the same nuclear technology. Nuclear fission could be used to fuel space travel, combat cancer, reverse human aging, and create artificial food in times of poor harvests, the article claimed. However, the author believed that the use of atomic power to generate electricity was impractical because of its high cost.[41] Soviet authorities and the East German press made no mention of the testing of the first Soviet atomic bomb on 29 August 1949. NBI first mentioned Soviet nuclear capabilities in December of that year, unveiling a proposal to change the course of two Siberian rivers by detonating a nuclear bomb.[42] In trying to conceal or rationalize nuclear weapons programs, the superpowers provided arguments for latecomer nations trying to skirt nonproliferation agreements years later.

At the Geneva Conference of 1955, the Soviet Union joined the United States in providing what had, until then, been classified information on atomic reactor designs and the prospect of helping many nations to develop nuclear power. This conference unleashed "messianic hopes" around the world.[43] In West Germany, it unleashed a wave of "nuclear euphoria," particularly on the political left.[44] H.G. Wells, the first writer to embrace the idea of nuclear power as an answer to the danger of nuclear weapons, was a Fabian socialist. In a work written shortly after the war but not published until 1955, prominent Marxist philosopher Ernst Bloch asserted that the betterment of humankind through climate change was the true mission of nuclear fission: "A few hundred pounds of uranium and thorium would suffice to make the Sahara and Gobi Deserts disappear and to turn Siberia, North America, Greenland and Antarctica into the Riviera."[45] Social Democrat Leo Brandt made similar fantasy-filled statements regarding the peaceful atom in a high-profile speech in 1956.[46] Many nonsocialists joined in the paeans to this "Atomic Age," a term that in Germany had a utopian ring to it.

In the GDR, atomic-powered fantasies equated the advance of socialism, "scientific–technical progress," and technological spectacles. These were inspired in part by the start-up of nuclear research in the GDR.[47] NBI, as well as the comic book series *Mosaik,* featured pictures of nuclear-powered jet aircraft. In artists' renditions, passengers sat in the front, far from the reactor, which was located in the tail section.[48] A popular East German illustrated book on nuclear technologies featured atomic trains that pulled into atomic train stations.[49] The source of inspiration for this technological daydreaming was probably a 1946 American study titled *Applied Atomic Power.*[50] The first nuclear-powered Soviet icebreaker, the *Lenin,* forced its way through a mighty sea of ice in another NBI article: "In numerous battles, the Soviet icebreakers have proven themselves stronger that the polar ice's powers of nature."[51] Here, the caption writer, Lothar Hitziger (an author of popular books on science and technology), was invoking several narratives common in Soviet thinking about technology: the struggle to master technology as military engagement, socialism as the technologically superior system, and the taming of nature as central to progress.[52]

The US government also made major attempts to mold popular perceptions of nuclear power by sending traveling "Atoms for Peace" exhibitions around the globe. Although to my knowledge no pictures or materials from a 1955 exhibition in Frankfurt have survived, a survey commissioned by the HICOG (High Commissioner for Germany) research staff gives an impression of its layout and impact on the public.[53] Utopian visions found no place here. Instead, the exhibition was "sober," "factual," and well organized, according to visitors. When asked what their favorite part was, 42 percent of those surveyed mentioned science demonstrations involving, for example, Geiger counters and measuring devices.

A "young lady" demonstrated how "magic hands" were used to manipulate radioactive materials. Other visitors particularly enjoyed the films (the high point for 16 percent) and "lectures and explanations" (favored by 12 percent). Visitors were first shown a film, which provided a "very instructive" and "easily understandable" overview of the subject. They then walked around the exhibition hall, which contained large models illustrating, for example, nuclear fission and the functioning of an atomic reactor. They viewed Otto Hahn's workbench, which was on display.

The comments revealed curiosity, a desire to learn, and respect for science. One attendee offered, "I was impressed by the medical angle, that atomic energy can be used for the benefit of mankind, since I am interested in cancer therapy." Another had high hopes that these new medical breakthroughs would benefit all of society, not just the well-off, believing that it would be "possible in all probability to provide low cost cancer treatment, above all, for the lower income brackets." Other visitors were highly enthusiastic about the exhibition because it contained a great deal of information about the potential benefits of atomic power overall: "Everybody should see it! Atomic energy is a means by which to achieve prosperity throughout the world."

Participants in the survey saw themselves as living in a society in which citizens were expected to have at least a basic understanding of atomic power: "This has become part of the overall knowledge everybody is expected to have. Since it is of current interest, you are expected to know something about it." For them, research in this area was not something esoteric and impossible to understand, and they were grateful to have the opportunity to inform themselves. One attendee, when asked what he or she understood the purpose of the exhibition to be, responded, "To enlighten people about the atom and to popularize the latest findings in atomic research." Another answered, "To make people understand what the aims of atomic research are. That people who live in the 20th century just have to concern themselves with atomic energy." These were not necessarily highly educated people. Of the 400 visitors who participated in the survey, 21.5 percent had only completed elementary school and 42.5 percent had some high school education but had not graduated; only 36 percent had at least a high school diploma. However, the lesser educated felt they should make a real effort to understand the exhibition: "For dumb amateurs things were a little too difficult to grasp, although the staff tried hard to make everything quite clear. One should see the exhibition a number of times."

One man thought that women should not visit the exhibition: "Women can be depended upon to ask terribly stupid questions, that's why I would advise only men to visit this exhibition." However, about 20 percent of the attendees were women. They were slightly more inclined than men to deem the exhibition "excellent." One woman who visited the exhibition when it traveled to Berlin did not feel that it was not intended for women: "This exhibition was arranged mainly because we women know almost nothing about these things." Another was enthusiastic about the "wonderfully comprehensible demonstration of the process of nuclear fission. Being a woman I know very little about all that, but this was really excellent."[54]

The visitors felt the exhibition was highly effective. In Frankfurt, 77 percent found it "excellent" or "very good." Almost all indicated they had learned something. Few saw the exhibition as propagandistic in nature. It was a resounding success for the United States, promoting good feelings and the conviction that US intentions were benevolent. Visitors to the same exhibition in Berlin extolled its objectivity ("The whole way of presentation is so purely factual that I lack the words to praise it properly") and its logical presentation ("I have never before seen such a well-organized and ingeniously contrived exhibition"). German pride expressed itself when one of those surveyed complained "that they don't attach enough importance to the fact that this thing actually originated in Germany. Merely showing Professor Hahn's table isn't enough."[55]

"America Houses" throughout West Germany distributed American booklets on nuclear technologies that were also sober and scientific and lacked a utopian dimension. One such booklet contained small, unspectacular pictures of scientific and technical personnel at work, equipment, and results, such as potatoes that had been irradiated and others that had not.[56]

In Berlin, one attendee struck a critical note: "I noticed the tendency to divert attention from the military to the civilian sector. They didn't show the destruction atom bombs caused in Japan." Indeed, "Atoms for Peace" was only one side of the story.

Dystopian Visions in an Era of Military Threat, 1954–1960

Images of a world of peace and plenty spread around the world at the same time as dystopian visions of death and destruction spurred by East–West tension and, particularly, the Castle Bravo hydrogen bomb test conducted by the United States at Bikini on 1 March 1954. The sheer force of the explosion and the four-mile–wide fireball it produced symbolized for many the terrible destructiveness of this new weapon. Inhabitants of the Marshall Islands, who were not evacuated at first, suffered from radiation sickness.

NBI coverage marked a shift in focus from a Stalin-era focus on capitalist conspiracies to humanitarian concerns. A drawing depicted a Pacific Islander mother trying to protect the toddler in her arms from the terrible blast. A grass skirt and lei identified her and her child as part of the indigenous population—a defenseless victim of Western imperialism.[57] The caption expressed outrage: "The U.S.A. won new enemies, and the World Peace Council, whose demand for the prohibition of all atomic and H-bomb experiments was also heard on the Pacific islands, won new friends and comrades in the struggle for peace and the happiness of humanity."[58]

Stern employed sarcasm to condemn Castle Bravo: "When the head of American civil defense, O'Brien, saw the first pictures of the explosion of the hydrogen bomb on the Marshall Islands, he said, 'All of that is so fantastic! We don't know where to begin!' He must mean, where will it all end[?]"[59] Scholar Ilona Stölken-Fitschen has argued that, thanks to the highly restrictive information policies of the Western occupation powers during the period of occupation, Germans did not fully absorb the great peril posed by nuclear weapons until this American H-bomb test at Bikini in 1954.[60]

Moreover, that test changed the fundamental perception of the nature of the threat posed by nuclear weapons, shifting focus from the impact of the blast to radiation, which became a worldwide obsession. The fallout of the hydrogen bomb explosion fell on a Japanese fishing vessel, the *Lucky Dragon*. The irony of the ship's name and its Japanese identity came to endow this event with great symbolic significance. Six months after the test, the Japanese crew member Aikichi Kuboyama died as a result of exposure to the fallout. An article about this event in *Stern* featured a photograph of Kuboyama, lying dead or dying in the hospital, his body emaciated and maimed. The autopsy revealed how deeply the radiation had penetrated and damaged his body: "Radiation had destroyed all his internal organs, leaving them beyond all recognition."[61] The article contrasted images of Japanese women as angelic caregivers and Americans as destroyers of life.

In 1954 physicist Pascual Jordan warned against "heedless optimism" regarding the triumph of peaceful uses of atomic technologies, writing, "No one can know what clouds of horror we may have to march through or crawl through." Nonetheless, he believed the development of atomic power to be a worthwhile endeavor.[62] Rejection of nuclear technologies were often framed in emotional and nonrational terms. A 1956 article expressed the new understanding of the significance of Hiroshima in metaphysical terms: "In reality, the fireball over Hiroshima set in motion the destruction of the balance of fundamental forces in the godly economy of nature." However, the article also included a brief discussion of genetic mutations caused by radiation as well as the problems associated with the disposal of radioactive waste from atomic power plants.[63]

Stern became far more critical of US defense policies in the mid-1950s. A 1956 article on intercontinental ballistic missiles (ICBMs) made clear that both the United States and the Soviet Union bore responsibility for the arms race, depicted here in frightening terms: "If one of the world powers wins the race, the pendulum of the world clock will point to doom."[64] By contrast, NBI articles in 1957–1958 presented Soviet success in developing ICBMs and in expanding and upgrading its nuclear arsenal as the best guarantee for peace.[65]

Articles in both magazines reflect shifting sensibilities resulting from the Berlin Crisis of 1958–1961, when the Soviet Union attempted to dislodge Western forces from West Berlin. During this period, East and West Germans contemplated the possibility of war on a daily basis. Unlike earlier articles on the arms race, a 1960 NBI article on an ICBM tested by the Soviet Union did not trumpet this as a success for the socialist world. Instead, the piece assured readers that the target area was far from islands, fishing grounds, and shipping routes and that there was no nuclear warhead on the missile. It also claimed that its purpose was not military, but rather related to the Soviet space program.[66] It is not known how East Germans viewed this cover-up.

This was the era of *On the Beach,* a US film released in December 1959 that depicted the end of the human race owing to a nuclear war. *Stern* was quite taken by the film's realism, asserting that is was "not a sermon, not an appeal, [and] not a warning," but rather "a realistic picture of the day after tomorrow, when it is too late." It was a "beacon of hope" that might bring political leaders who saw it, such as Eisenhower and Khrushchev, to their senses.[67] A review in the East German newspaper *Neues Deutschland* had kind words for the film: "A film with such a perspective is something new for American [film] production. One can clearly see in it the spirit of détente developing between the two camps."[68]

Articles on Hiroshima and Nagasaki reflected a new awareness of the bleak horror of nuclear war. At the height of the Berlin Crisis, *Stern* published a serialized novel featuring Claude Eatherly, the pilot of a small scouting plane that accompanied the *Enola Gay* on its mission to drop an atomic bomb on Hiroshima on 6 August 1945. Eatherly was the subject of intense fascination because it was thought that feelings of guilt had caused him to suffer a mental breakdown.[69] This novelization described the sufferings of the people of Hiroshima in graphic detail. The author, Hans Herlin, a pilot who deserted from the *Luftwaffe* in 1944,[70] explored themes of guilt, remorse, and victimization in ambiguous ways. Germans could either identify with the American pilot, who, although he was responsible for many deaths, was ennobled by his mental suffering. Or they could identify with the Japanese victims of the bombing. Side-by-side photos of Berlin and Hiroshima in August 1945 compared the two cities' fate. The message appeared to be that the atomic bomb could have been dropped on Berlin if the historical circumstances had been a bit different.[71] This very effectively brought home the danger nuclear war posed to Germany.

NBI also ran articles about Hiroshima and Nagasaki around this time, but there were no hints of soul searching as in the novel published in *Stern*. NBI viewed Hiroshima as an example of imperialist warmongering, which NBI said had continued in the postwar era and was the cause of East–West tensions. However, like the *Stern* serialized novel, the NBI articles put a human face on wartime suffering and reinforced feelings of utter horror and revulsion toward nuclear war. The GDR's first science fiction film, *The Silent Star*, released in 1960, depicted an international crew's visits to Venus, whose inhabitants wanted to annihilate earth's population, but instead were themselves wiped out by their own weapons. One member of the crew was a Japanese survivor of the dropping of the atomic bomb on Hiroshima.[72]

In "The New Sun," part of the comic book series *Mosaik*, the central characters of the series discovered a planet destroyed by atomic war. To make it inhabitable again, they replaced its double sun—nearly burned out—with a new, nuclear-powered sun.[73] Works by Stanislav Lem and other Eastern European science fiction writers served as an inspiration for this story.[74]

Only slightly less utopian was a 1961 NBI article that imagined a better world that could be created if the arms race ended. In Africa, the money saved could be used for advanced medical care, modern housing complexes, and atomic power. "Out of the billions (in savings) from disarmament, one or several atomic reactors could be built for every African country."[75] At that time, the GDR was still not recognized as a sovereign nation by most countries and was trying to court favor with African nations.[76] A drawing (Illustration 1.1) accompanying the NBI article did not show the European experts and the indigenous population as equals, however. The socialist planners and builders of African atomic reactors had brought their instruments, blueprints, and expertise to the jungle. They were wearing colonial gear—pith helmets and khakis. A native passively looked on holding a bamboo pole, perhaps waiting for orders. The picture showed discernable racial and cultural hierarchies. It did not appear to be a manifestation of specifically socialist racism so much as an unintentional revival of much older imperialist fantasies linked to an ideology of socialist modernization.[77]

This socialist version of the "Atoms for Peace" narrative also took on more pragmatic aspects at times. A 1958 article held up research at the Central Institute for Atomic Physics in Rossendorf as proof that the GDR

Illustration 1.1 East German fantasy of bringing nuclear power to Africa. "Wohin mit dem Geld?" NBI 4, 1961. Courtesy of Verkehrsmuseum Dresden, GmbH.

was a peace-loving nation, committed to the development of the peaceful atom, unlike West Germany, which was pursuing military uses of nuclear technologies. Journalist Lothar Hitziger enthused, "The atom will become part of our daily life to an extent that today we wouldn't dare to imagine in even our boldest dreams."[78] This, along with articles from the early 1960s, emphasized the cleanness and safety of this and other East German nuclear facilities as well as the authority of male scientists.[79] Photographs featured healthy, attractive young women working as laboratory technicians in a safe, socialist environment.[80] Here, the healthy female body was depicted as enjoying the protection of male scientists, the state, and socialism.

No such sense of security was found in articles on radiation in *Stern*. That magazine began reporting on nuclear accidents in 1957 with an article on an incident at a nuclear laboratory in Houston, Texas. The director of the laboratory had unwittingly tracked radioactive material into his automobile and his home. He was photographed in a hospital, where he was being examined for radiation poisoning. His young son's friends were forbidden to play with the child: "He doesn't understand what it means when the parents of his friends tell their children, 'Don't play with that radioactive [boy].'" According to the title of the article, radioactivity was "Worse than the Plague."[81]

Stern also reported on a 1958 accident at a Yugoslav nuclear power plant that exposed six plant employees to high doses of radiation. The article focused on the lone woman in the group, Rosa Ristic. She was photographed lying in bed, holding a hand to her forehead and looking weak and vulnerable.[82] The cover of another 1958 *Stern* issue featured a nubile young woman in her underwear. "Through naked force to the x-ray machine?" asked the title. The article was more than an excuse to titillate, although a picture in the inside article of young women wearing nothing but sheets or towels wrapped around themselves probably did just that. The women in the photograph were subject to compulsory mass x-ray screenings for tuberculosis. The message was that no one could shield the vulnerable female body from radiation.

In fact, the state, embodied by a male judge, was trying to force citizens to expose themselves to this danger. The article cited a U.N. study that warned against unnecessary exposure to radiation. According to a chart, the average person was exposed to more radiation through radiological examinations than as a result of environmental pollution caused by nuclear weapons testing or nuclear power plants.[83]

Such sensitivity to unseen poisons was probably heightened by the Contergan tragedy. Known as thalidomide in the United States, Contergan was a pharmaceutical developed in West Germany and approved for use to combat morning sickness in pregnant women. In 1957–1961, many thousands of babies were born with severe birth defects because they had been exposed in utero to this medication. Initially, it was feared that nuclear weapons testing was to blame.[84] Although this fear proved unfounded, the tragedy intensified fears that the creations of modern science could maim and kill.

Although GDR press coverage of nuclear power was generally quite positive, fears concerning radiation did make their way into NBI. A 1958 article picked up on transnational discussions on the health impact of radioactive fallout from atmospheric nuclear weapons testing. Of particular concern was Strontium-90, which got into food and water supplies and could cause birth defects and cancer, particularly in children. Here, the vulnerability of the human body was illustrated with a picture of an infant. The need to protect vulnerable innocents from the dangers posed by radioactive contamination, blamed here on *Western* atomic tests, was manifested in another 1958 article, accompanied by a photo of a father and his children happily running in the rain, oblivious to the potential danger.[85] Despite the anti-Western slant, these two articles contained a message about the dangers of radiation that could potentially undermine trust in the ability of the socialist system to protect the population from harm.

Stern contained virtually no positive coverage of nuclear power. *Stern* did publish one utopian article on energy—not fission-based nuclear power, but fusion power, a form of power whose development has eluded researchers to the present day.[86] In a "city of the future," this "radiation-free" form of energy would provide enough power for ten million people, according to the article.[87] A 1959 Allensbach Institute survey indicated that the overwhelming majority of West Germans had reservations against nuclear energy.[88] These misgivings derived from the association of atomic power with nuclear weapons, testing, radiation, and the very real danger of nuclear war. The GDR was also affected by such concerns.

Winning Popular Support through Education and Safety?

The state, industry, and scientists were determined to educate the public concerning nuclear power, an effort that took on particular importance with the start-up of nuclear power programs in East and West Germany. The first East and West German research reactors went online in 1957. West Germany completed the first nuclear power plant, in Kahl, which began producing power in 1961, while the first East German atomic power plant, in Rheinsberg, went critical in 1966. In these years, state and industry hoped to establish the foundations on which both safety and trust were based.

The West German Atomic Law (*Atomgesetz*) of December 1959 addressed well-known safety issues in ways specific to the West German system. It did so within the framework of US hegemony and European integration. Allied prohibition of nuclear research in Germany officially ended in 1955 and was superseded by the new legal framework of the Atomic Law. The United States sought to ensure the long-term pacification of West Germany through binding international agreements, notably the Rome Treaties of 1957, which created the Common Market and Euratom. Political leaders in West Germany and elsewhere successfully resisted handing authority over the development of nuclear power to a supranational organization—a role that US leaders had hoped to entrust to Euratom.

Instead, the West German model allowed the private sector to play a central role in the development of nuclear power, while the federal (i.e., national) government took over primary responsibility for overseeing nuclear safety. The West German constitution, the *Grundgesetz,* had to be modified to give the state this power. State (in the sense of provincial) governments were in charge of licensing nuclear facilities. However, there was no central regulatory agency in West Germany. The institutions overseeing nuclear safety were comparatively weak. On the other hand, bureaucratic procedures were rather complex. The approval process was split among various offices and courts, making it much more complicated than in other countries. The courts had to determine questions of safety without adequate technical knowledge. Initially, however, this bureaucratic and judicial oversight was rather lax. Technical supervision became more rigorous by the late 1960s.[89]

The early history of nuclear power in the Federal Republic was punctuated by conflicts among elites, institutions, and experts, as historian Joachim Radkau has shown, but only rarely did these spill over into the public realm. Once industry became aware of the tremendous costs involved in developing nuclear power, it turned to the state for support. Although the Atomic Ministry was created under Adenauer, the Christian Democrats were very skeptical about a large government role in the development of atomic power, mainly because of their free market orientation. Rather, it was left-leaning Social Democrats who spearheaded efforts to institute state subsidies. It was not until the mid-1960s that earlier frugality was cast aside and the state became heavily involved in atomic power. Most plants were public utilities partially owned and run by state governments.[90]

The GDR took the opposite course. After a period of state support for nuclear research and reactor development, the SED surprised its scientists by largely shutting down the East German nuclear power program in 1962–1965. Historians Mike Reichert and Burghard Weiss believe that the high production cost of atomic power and the inability to produce equipment in the GDR were the primary reasons for this about-face. However, one East German nuclear scientist—Heinz Barwich—claimed after his defection to the West that the Soviet Union wanted to stop all nuclear power research and atomic power plant construction in its satellite states so as to be able to take over this lucrative industry itself. And in fact, this is essentially what happened, making the GDR highly dependent on the Soviet Union, which built East Germany's nuclear power plants, supplied most of the needed equipment, and initially established the parameters of nuclear safety.[91]

The GDR continued to rely heavily on brown coal as an energy source. It is ironic, as Burghard Weiss noted, that in capitalist West Germany the state came to play a dominant role in the development of atomic power, whereas in socialist East German, the state largely withdrew from involvement in atomic power.[92]

There were no public protests in the Federal Republic against the first generation of nuclear power plants. Throughout the 1950s and 1960s, antiwar activists saw no connection between nuclear weapons and nuclear power. According to historian Holger Nehring, this was true of the Campaign against Atomic Death, launched in 1958 by the Social Democratic Party (SPD) and labor unions, as well as of the Easter Marches, yearly peace marches that began in 1960.[93] Environmentalists had not yet discovered nuclear power as an issue.[94] However, water authorities made the press and the public aware of their concern that the use of rivers to cool reactors would overheat rivers. Such issues generated distrust and a certain amount of local resistance to the construction of atomic power plants.[95]

In the GDR such criticism of nuclear power would never have made it into the press. However, consensus was achieved, not just through dictatorial control but also through popular participation in technology. The SED's model of citizenship was rooted in participation in the building of a modern economy—in factories, on farms, and on construction sites. Atomic energy was depicted as part of this collective socialist effort in a 1967 educational television program, evidently aimed at primary and secondary school students. Nuclear power was part of a comprehensive plan in the GDR to achieve peace and prosperity and to make the GDR one of the most modern industrial states in the world, according to the program. Upbeat, jazzy music underlined the theme of industrial modernization. Young people were exhorted to participate in this collective effort, following the example of the popular campaign to build the International Seaport at Rostock.[96] This megaproject, completed in 1960, had mobilized the East German population, which contributed money and even stones.[97]

The socialist world's technical and scientific expertise was the most important argument used on television to win public approval for nuclear power. East German TV programs often highlighted the Soviet Union's leading role in developing nuclear power and ensuring the safety of East German atomic power plants.[98] Programs from around the time that the Rheinsberg nuclear power plant went into operation also contained pedagogical elements, explaining to viewers the splitting of the atom, the way nuclear reactors worked, and safety systems. These programs tried to make science appear as appealing as possible. The development of nuclear power was, for example, depicted as an "adventure" involving the "drama of daring lines of thought."[99]

But under socialism, technological progress was also compatible with preservation of the cultural legacy of the past—cleansed, of course, of Nazi and capitalist elements. One television program assured viewers that the Rheinsberg nuclear power plant would not diminish the charm, cultural significance, or historical value of the town of Rheinsberg. "Much is like back then," according to a voice-over that narrated scenes evoking the past: a horse-drawn wagon going along a cobblestone street, the Rheinsberg castle, chestnut trees, the lake, people out in rowboats. It was "a city that attracted writers and poets." The narrator quoted the nineteenth-century German writer Theodor Fontane, who had described just such scenes in an account of a trip that brought him to Rheinsberg.

The program emphasized how well that traditional German world harmonized with the modernity of socialist society. The castle had been turned into a diabetes clinic. Life went on, and citizens were safe and secure. The atomic power plant was producing energy in accordance with the state's economic plan, and it had a better than average safety record, according to the voice-over. This scene featuring the Rheinsberg nuclear power plant was squeezed in between a segment on vacationing in Rheinsberg and one on the inhabitants' quality of life. A family consisting of a man, a woman, and a child walked in the sunshine past shops, holding hands, as modern, optimistic music played. Many healthy, happy children were featured, the offspring of young families that had moved to Rheinsberg to work at the nuclear power plant. A woman was shown walking with three small children and pushing a baby carriage. "When Rheinsberg began to grow, the babies came." Diapers and children's clothing were drying on a clothesline. "Diapers waving in the breeze tell of life, fecundity, and love."[100]

A decade later, *Our Sandman*, a universally popular children's show that children across the GDR watched at bedtime, referred to the new Greifswald nuclear power plant in somewhat similar terms. The evening's short film took children on a tour of Greifswald. As cheerful music played, the camera panned the town from the air. Children were shown at play. The brief film then cut from street scenes to a glimpse of the nuclear power plant, which was under construction. Immediately after came scenes of housing and children on a playground. A reference to Caspar David Friedrich, a major nineteenth-century artist who had painted the market square, established the town's cultural continuity. Technological modernity did not disturb the environment either, the final scene suggested, featuring boats on the river, where children were swimming.[101] This, like the 1967 program on Rheinsberg, sought to embed atomic energy in a narrative about progress, health, and the welfare of the population. In my sample, however, this theme appeared far less often than the attempt to legitimize and normalize the use of nuclear power through science.

Conclusion

Science was an important anchor of the Western Atoms for Peace narrative and its Eastern equivalent. A postwar sense of disorientation, disillusionment with ideology, and desire to overcome the odium of the Nazi past encouraged many Germans to look to science and scientists for guidance. Even those without an advanced education respected science and felt that they should engage with it. In the GDR, science and engineering provided a career path for those who wished to avoid onerous party and ideological commitments.[102] The long German tradition of scientific excellence was a supposedly unsullied aspect of the German past that Germans could identify with. At the same time, the internationalism and public roles of scientists promoted a belief in the rationality and humane qualities of science. This image of science endowed nuclear power with reassuring emotional associations.

Hopes that it would be possible to create a better world through science also sprang from American and Soviet culture. Americans had long found science and technology deeply inspiring, even "sublime."[103] In the Soviet Union, science and technology were understood to be the foundations of the socialist project.[104] Under the aegis of dialectic materialism, East German ideology portrayed science as a replacement for religion. For example, all young people completing a secular substitute for confirmation, the *Jugendweihe,* were given a book that focused largely on science and technology. It began with this sentence: "This book is the book of truth,"[105] an echo of the "book of truth" out of which the Angel Gabriel reads to Daniel in the Old Testament.[106]

Nuclear power occupied a place of honor in the Soviet Union as a force helping to build a socialist society. A Moscow exhibition on nuclear power that opened in 1956 aimed to educate, enlighten, and inspire Soviet citizens.[107] The 1955 American "Atoms for Peace" Exhibition also aimed to educate, but in a more pragmatic, less emotional manner. At that exhibition, science was presented as a lingua franca and a bridge between the two societies. The intention was to gain West German acceptance of American leadership. Actual transfer of nuclear technology served to cement that bond—as it did in the case of the Soviet Union and the GDR.

The Berlin Crisis, the arms race, and aboveground bomb testing disrupted and undermined the narrative of the "peaceful atom." The Bikini H-bomb test of 1954 and the ensuing *Lucky Dragon* incident shifted the focus of popular fears from the explosive power of the bomb to radioactive fallout. In West Germany, *Stern* became increasingly critical of US policies and weapons testing and at the same time began to highlight nuclear power plant accidents and concerns about exposure to radiation. Criticism of the Soviet Union was impossible in the GDR, but articles in NBI about the nuclear arms race lost their swagger, while a couple of articles about the dangers of radiation made their way into the magazine. The latter was the kind of topic that might catch the attention of readers in both countries—an older equivalent of "click bait." In NBI, the West was blamed for the problem. Nonetheless, popular anxieties were awakened.

Whatever small anti–nuclear power protests occurred in West Germany in the 1950s and 1960s were small and not widely known. There was no network of environmentalist or political sympathizers. On the whole, science appeared to be on the side of atomic energy's proponents. The narrative of progress through science prevailed in East German television programs about nuclear power, reflecting the SED's ability to maintain a nuclear consensus by force, but also through persuasion. In West Germany, trust in science was declining, reflected in works such as *Die Physiker* (The Physicists), a 1962 play by Swiss writer Friedrich Dürrenmatt that dealt with the responsibility of scientists for the development of nuclear weapons.

What ultimately was to lead to the emergence of a full-blown anti–nuclear power movement in West Germany was the confluence of science and a questioning, skeptical spirit nourished by the late 1960s protest movement. Nuclear power experts hoped that they would be left solely in control of nuclear power plant safety. The next chapter shows how they conceptualized safety and implemented safety regimes and how these safety philosophies were discussed and disrupted in West and East Germany. As will be seen in Chapter 3, dissenting scientists and activists challenged this technocratic approach in West Germany. Since this sort of fundamental challenge to technical expertise arose later in the GDR and in the context of a different sort of system, the emergence of scientifically motivated activism in the GDR will not be discussed until Chapter 7.

Notes

1. Max von Brück, "Völkerrecht im Zeichen des Atoms," *Süddeutsche Zeitung* (13 August 1946), 3. Even in the international scientific community, there was some uncertainty as to whether such horrors might one day be technically possible. See, for example, Anthony Leviero, "Bomb vs. Big Ships Posed to Senators," *New York Times* (6 December 1945); William Laurence, "Langmuir Urges Atom Pact, Says War Might Strip Earth," *New York Times* (17 November 1945). On Manhattan Project discussion as to whether the explosion of an atomic bomb could ignite the atmosphere, see Spencer Weart, *Nuclear Fear: A History of Images* (Cambridge, MA: Harvard University Press, 1989), 100–101.
2. No author, "Atomkraft so oder so?" NBI (*Neue Berliner Illustrierte*) 28, 1946.
3. Dolores Augustine and Dick van Lente, "Conclusion: One World, Two Worlds, Many Worlds?" in *The Nuclear Age in Popular Media: A Transnational History,*

1945–1965, ed. Dick van Lente (New York and Houndmills, UK: Palgrave Macmillan, 2012), 234; ch. on 233–48.

4. H.G. Wells, *World Set Free* (London: Macmillan, 1914), 47; W. Warren Wagar, "H.G. Wells and the Scientific Imagination," *The Virginia Quarterly Review* 65, no. 3 (1989): 390–400; Justin Busch, *The Utopian Vision of H.G. Wells* (Jefferson, NC and London: McFarland, 2009), 118, fn. 113.

5. Paul Boyer, *By the Bomb's Early Light: American Thought and Culture at the Dawn of the Atomic Age* (New York: Pantheon Books, 1985), esp. 294–302; David Nye, *American Technological Sublime* (Cambridge, MA: MIT Press, 1994), 228–29.

6. Joachim Radkau, *Aufstieg und Krise der deutschen Atomwirtschaft 1945–1975: Verdrängte Alternativen in der Kerntechnik und der Ursprung der nuklearen Kontroverse* (Reinbek bei Hamburg: Rowohlt Taschenbücher, 1983), 92.

7. On popular media such as tabloids and illustrated magazines, see Corey Ross, *Media and the Making of Modern Germany: Mass Communications, Society, and Politics from the Empire to the Third Reich* (Oxford and New York: Oxford University Press, 2008); Frank Bösch, *Mass Media and Historical Change: Germany in International Perspective, 1400 to the Present,* trans. Freya Buechter (New York: Berghahn Books, 2015), 148–49; Wolfgang Pensold, *Eine Geschichte des Fotojournalismus* (Wiesbaden: Springer VS, 2015).

8. Polling was introduced into US-occupied regions of Germany by the Opinion Survey Section of the Information Control Division of the Office of Military Government in late 1945. After the founding of the Federal Republic of Germany in 1949, the US High Commission for Germany (HICOG) took over the conducting of surveys, some in parts of Germany outside of the American zone of occupation. In 1950, the Deutsches Institut für Volksumfragen (DIVO) took over the fieldwork. See Anna Merrit and Richard Merrit, *Public Opinion in Semisovereign Germany: The HICOG Surveys, 1949–1955* (Urbana, IL: University of Illinois Press, 1980), 5, retrieved 24 September 2016 from https://archive.org/details/publicopinionins00merr. Also Anna Merrit and Richard Merrit, *Public Opinion in Occupied Germany: The OMGUS Surveys, 1945–1949* (Urbana, IL: University of Illinois Press, 1970).

9. In 1961, the total population of the Federal Republic was about 56.6 million. Thus, it passed through the hands of about one in five West German inhabitants. On 1950: NARA USIA RG 306 Records of the United States Information Agency, Office of Research, Reports and Related Studies 1948–53, Germany, West, Box 29, "Germany: Media Impact Study I," 63.

10. "Basisdaten des *stern,*" *Gruner + Jahr, Media-Forschung und -Service, Stern* Archives.

11. NARA USIA RG306/250/67/04/02-04/Box 7, report 214, "Written Media in West Germany: A Study of Public Reactions and Extent of Penetration," 62. According to this US study, 70 percent of professionals, 75 percent of business professionals, and 76 percent of office workers read "magazines and/or illustrated magazines." By contrast, only 49 percent of semiskilled workers and 33 percent of farmers and farmhands read magazines.

12. Michael Meyen and William Hillman, "Communication Needs and Media Change: The Introduction of Television in East and West Germany," *European Journal of Communication* 18, no. 4 (2003): 465; article on 455–76.

13. NARA USIA RG306/250/67/04/02-04/Box 7, report 214, "Written Media in West Germany: A Study of Public Reactions and Extent of Penetration," 59, 61, 65, 77.

14. Sixty-one percent of West German and 54 percent of East German households had a television set in 1966. Meyen and Hillman, "Communication Needs," 465.

15. Otto Haseloff, *Stern: Strategie und Krise einer Publikumszeitschrift* (Mainz: v. Hase & Koehler, 1977), 453, 893, 896; Daniela Münkel, *Willy Brandt und die 'Vierte Gewalt': Politik und Massenmedien in den 50er bis 70er Jahren* (Frankfurt am Main: Campus Verlag, 2005).

16. Simone Barck, Martina Langermann, and Siegfried Lokatis, *"Jedes Buch ein Abenteuer" Zensur-System und literarische Öffentlichkeiten in der DDR bis Ende der Sechziger Jahre* (Berlin: Akademie-Verlag, 1997); Julia Martin, "Der Berufsverband der Journalisten in der DDR (VDJ)," in *Journalisten und Journalismus in der DDR,* ed. Jürgen Wilke (Cologne: Böhlau Verlag, 2007). Example in SAPMO/Barch (*Stiftung Archiv der Parteien und Massenorganisationen der DDR im Bundesarchiv,* Foundation of the Archive of the Parties and Mass Organization of the GDR in the Federal German Archives) DY 30/IV 2/9.02 37, Krüger (of the SED division for security questions) to Sindermann (of the SED division for Agitation/Propaganda), dated 1 December 1960. On cooperation with Soviet counterparts, see SAPMO/ Barch DY 30/IV 2/9.02 58, 128–29. On the World Peace Council guidelines, see SAPMO/Barch DY 30/IV 2/9.02 119, "Hinweise für Presse, Rundfunk und Fernsehfunk... ," dated 3 May 1962.

17. Michael Meyen, *Denver Clan und Neues Deutschland* (Berlin: Christoph Links Verlag, 2003), 138.

18. Landesarchiv Berlin (Provincial Archive of Berlin), C Rep. 904-089, Nr. 16 Bestand GO Berliner Verlag, Aktentitel Protokolle der Leitungssitzungen der ZPL, Bd. 7 (1962), "Protokoll der ZPL-Sitzung vom 21.3.62."

19. Dick van Lente, ed. *The Nuclear Age in Popular Media: A Transnational History, 1945–1965* (New York and Houndmills, UK: Palgrave Macmillan, 2012). Augustine, "Learning from War," 79–116. None of these issues were digitized.

20. Here, I have double counted the six articles that covered both civilian and military technologies. On the other hand, I have not included eight articles on "other" nuclear themes, which often centered on personalities or used the concept of "nuclear" as a metaphor or metonym. There were 201 articles on nuclear themes.

21. "Das Strahlenwunder. Röntgen," NBI 8, 1945.

22. No author, "Bergknappe der Physik," *Der Spiegel* 41, (1947).

23. Dieter Hoffmann, *Max Planck: Die Entstehung der modernen Physik* (Munich: Verlag C.H. Beck, 2008), 84–104. As head of the Kaiser Wilhelm Society, Planck did not show solidarity with Albert Einstein's protests against the Nazis, then a member of the society. In fact, he allowed the Kaiser Wilhelm Society to issue a letter in his name criticizing Einstein and calling on him to resign.

24. On the circumstances, see Walter Moore, *Schrödinger: Life and Thought* (Cambridge, UK and New York: Cambridge University Press, 1989), chs. 7–9.

25. Comprehensive research has been done on scientific research in the Kaiser Wilhelm Society. See the book series "Geschichte der Kaiser-Wilhelm-Gesellschaft im Nationalsozialismus," edited by Reinhard Rürup und Wolfgang Schieder on behalf

of the Max-Planck Society. For a list of the individual studies, see http://www. mpiwg-berlin.mpg.de/KWG/publications.htm#B%C3%BCcher (retrieved 25 October 2017). For a brief English-language analysis, see Margit Szöllösi-Janze, "National Socialism and the Sciences," in *Science in the Third Reich* ed. Margit Szöllösi-Janze (Oxford and New York: Berg, 2001), 1–35.

26. Mark Walker, *German National Socialism and the Quest for Nuclear Power 1939–1949* (Cambridge, UK: Cambridge University Press, 1989); Charles Frank, *Operation Epsilon: The Farm Hall Transcripts* (Berkeley, CA: University of California Press, 1993); Manfred Popp, "Misinterpreted Documents and Ignored Physical Facts: The History of 'Hitler's Atomic Bomb' Needs to be Corrected," *Berichte zur Wissenschaftsgeschichte* 39, no. 3 (2016): 265–282. DOI: 10.1002/bewi.201601794. Mark Walker, "Physics, History, and the German Atomic Bomb," *Berichte zur Wissenschaftsgeschichte* 40, no. 3 (2017): 271–288. DOI: 10.1002/bewi.201701817.

27. For a solid interpretation, see David C. Cassidy, *Uncertainty: The Life and Science of Werner Heisenberg* (New York: Freeman, 1992).

28. Christoph Mick, *Forschen für Stalin: Deutsche Fachleute in der sowjetischen Rüstungsindustrie 1945–1958* (Munich and Vienna: R. Oldenbourg Verlag, 2000); Tom Bower, *The Paperclip Conspiracy: The Hunt for the Nazi Scientists* (Boston: Little, Brown, 1987); Dolores Augustine, *Red Prometheus: Engineering and Dictatorship in East Germany, 1945–1990* (Cambridge, MA: MIT Press, 2007), ch. 1; Samuel Goudsmit, *Also: The Failure of German Science* (New York: Schuman, 1947).

29. Agnes Tandler, "Geplante Zukunft: Wissenschaftler und Wissenschaftspolitik in der DDR 1955–1971" (Diss. Europäisches Hochschulinstitut, Florence, Italy, 1997), 50–54.

30. Augustine and van Lente, "Conclusion," 236.

31. Radkau, *Aufstieg und Krise*, 36, 39. Werner Heisenberg, held only briefly by the British, was able to begin organizing postwar German atomic research in 1946. Atomic research was officially prohibited until 1955, so he had to maintain the appearance of doing only theoretical work.

32. Cathryn Carson, *Heisenberg in the Atomic Age: Science and the Public Sphere* (Cambridge, UK and New York: Cambridge University Press, 2010), 103–13, 225–26.

33. "6 davon vernichten Deutschland," *Stern* 22, 28 May 1950.

34. Appendices I and II, in van Lente, *Nuclear Age,* 249–50, 262–63.

35. Kenneth Alan Osgood, *Total Cold War: Eisenhower's Secret Propaganda Battle at Home and Abroad* (Lawrence, KS: University of Kansas, 2006).

36. Mara Drogan, "Atoms for Peace: U.S. Foreign Policy and the Globalization of Nuclear Technology: 1953–1960" (PhD diss., SUNY Albany, 2011), iii, 1, 57–58, 73, 85–86, 121; John Krige, *Sharing Knowledge, Shaping Europe: U.S. Technological Collaboration and Nonproliferation* (Cambridge, MA: MIT Press, 2016), 1–7, 20–26.

37. On the Soviet Union, see Alvin W. Gouldner, *The Two Marxisms* (New York: Seabury Press, 1980), esp. 42–43, 73, 269–75, 385–86; Jonathan Coopersmith, *The Electrification of Russia, 1880–1926* (Ithaca, NY: Cornell University Press,

1992); Kendall Bailes, *Technology and Society under Lenin and Stalin: Origins of the Soviet Technical Intelligentsia, 1917–1941* (Princeton: Princeton University Press, 1978), 64–120; Loren R. Graham, *The Ghost of the Executed Engineer: Technology and the Fall of the Soviet Union* (Cambridge, MA and London: Harvard University Press, 1993). On the GDR, see Augustine, *Red Prometheus* and literature cited therein.

38. Ellgaard moved from Berlin to West Germany in 1949.
39. "In 3 Stunden 27 Minuten—zum Mond! Erster Teil," NBI 1, 1946.
40. Walter Heise, "Luftbrücke von Morgen," *Stern* 7, 1949.
41. "Bombe oder Stein der Weisen?" NBI 1, 1947.
42. "Ein Erdteil verändert sein Klima," NBI 51, 1949. Under the name Operation Plowshare, the United States also tried to develop techniques for using nuclear bombs for peaceful construction projects. Scott Kirsch, *Proving Grounds: Project Plowshare and the Unrealized Dream of Nuclear Earthmoving* (New Brunswick: Rutgers University Press, 2005). On Soviet technological gigantomania, see Paul R. Josephson, "'Projects of the Century' in Soviet History: Large-Scale Technologies from Lenin to Gorbachev," *Technology and Culture* 36, no. 3 (1995): 519–59.
43. Joachim Radkau, *Nature and Power: A Global History of the Environment*, trans. Thomas Dunlap (New York: Cambridge University Press, 2008), 256.
44. Ilona Stölken-Fitschen, *Atombombe und Geistesgeschichte: Eine Studie der Fünfziger Jahre aus deutscher Sicht* (Baden-Baden: Nomos Verlagsgesellschaft, 1995), 166–79; Radkau, *Aufstieg und Krise*, 78–88.
45. Ernst Bloch, *Das Prinzip Hoffnung*, vol. 2 (Frankfurt am Main: Suhrkamp, 1973), 775; 1st ed. Berlin [GDR]: Aufbau-Verlag, 1955). Bloch wrote this three-volume work in 1938–1947 in the United States.
46. Bernd-A. Rusinek, "'Kernenergie, Schöner Götterfunken!' Die 'umgekehrte Demontage': Zur Kontextgeschichte der Atomeuphorie," *Kultur und Technik* 4 (1993): 15–21; quotation on 19.
47. Under Soviet tutelage, the GDR established a number of institutes devoted to the development of nuclear power in the mid-1950s. In 1957, preparatory work for reactor building in the GDR started. See Johannes Abele and Eckhard Hamper, "Kernenergiepolitik der DDR," in *Zur Geschichte der Kernenergie in der DDR*, ed. Peter Liewers, Johannes Abele, and Gerhard Barkleit (Frankfurt am Main: Peter Lang, 2000), 31–39; ch. on 29–89.
48. "Start frei für A1," NBI 49, 1956. *Mosaik von Hannes Hegen: Geheimsache Digedanium*, reprint (Berlin: Buchverlag Junge Welt, 2003), 28.
49. Karl Böhm and Rolf Dörge, *Gigant Atom* (Berlin: Verlag Neues Leben, 1957), 272–73.
50. Edward S.C. Smith, A.H. Fox, R. Tom Sawyer, and H.R. Austin, *Applied Atomic Power* (New York: Prentice-Hall, Inc., 1946), frontispiece. The proposal to use atomic power for trains, planes and ships was endorsed in this work by H.D. Smyth (author of a major US government report on the Manhattan Project) and Leslie Groves (head of the Manhattan Project).
51. "Atomkraft bricht Eismacht," NBI 42, 1957.
52. Josephson, "Projects," 519–59; Susanne Schattenberg, *Stalins Ingenieure* (Munich: Oldenbourg Verlag, 2002).

53. NARA USIA RG 306, 250/67/04/02-04, Box 7, Report 208, "Frankfurt Visitors Appraise the Atomic Energy Exhibit 'Atoms for Peace,'" 15 February 1955.
54. NARA USIA RG 306, 250/67/04/02-04, Box 6, Report 205.
55. Ibid.
56. "Atomenergie für den Frieden" (Bad Godesberg: USIA, 1957).
57. However, grass skirts and leis were not everyday attire in the Marshall Islands in this period.
58. "Es war, als ob zehn Sonnen aufgingen," in "Zeitgeschehen im Bild," NBI 25, 1954.
59. "Wehe wenn sie losgelassen," Stern 15, (11 April 1954).
60. Ilona Stölken-Fitschen, "Der verspätete Schock: Hiroshima und der Beginn des atomaren Zeitalters," in Moderne Zeiten: Technik und Zeitgeist im 19. und 20. Jahrhundert, ed. Michael Salewski and Ilona Stölken-Fitschen (Stuttgart: Franz Steiner Verlag, 1994), 144; article on 139–55. However, US authorities did turn a blind eye when, in September 1946, the Neue Zeitung published a German translation of John Hershey's Hiroshima without official permission. The Neue Zeitung was published under the auspices of the Information Control Division of OMGUS, but its German reporters and editors won a large measure of independence. See Jessica Gienow-Hecht, Transmission Impossible: American Journalism as Cultural Diplomacy in Post-War Germany (Baton Rouge, LA: Louisiana State University Press, 1999), 92.
61. "Nur außen noch ein Mensch," Stern 41 (10 October 1954).
62. Pascual Jordan, Atomkraft: Drohung und Versprechen (Munich: Wilhelm Heyne, 1954), 60, 61.
63. "Die radioaktive Sintflut ist schon im Steigen," Stern 27 (7 July 1956).
64. "Da hilft nur beten!" Stern 22 (2 June 1956).
65. "Sensation um eine Gurke," NBI 36, 1957; "Verteidiger der Luft," NBI 40, 1957; "Vom Meeresgrund in die Stratosphäre," NBI 8, 1958.
66. L. Hitziger, "Raketen, Politik und Wasserwüste," NBI 5, 1960.
67. "Die letzten Tage unseres Lebens," Stern 2 (8 January 1960).
68. "Jeder stirbt für sich allein?" Neues Deutschland (10 January 1960). From the collection of the Film Museum (Filmmuseum), Potsdam, Germany. According to this archive, the film was never shown in the GDR; however, some East Germans may have seen it in West Berlin since the film was released in 1959, two years before the building of the Wall.
69. Hans Herlin, "Kain, Wo ist Dein Bruder Abel?" Stern 12 (19 March 1960); 18 (30 April 1960); 19 (7 May 1960).
70. "Vor 10 Jahren gestorben: Hans Herlin (1925–1994): Schriftsteller und Journalist aus Stadtlohn," Dokument des Monats [Stadtarchiv Stadtlohn], December 2004, 1–2, retrieved 6 January 2018 from http://www.stadtlohn.de/pics/medien/1_1190134091/Dokument_des_Monats_2004.pdf; "Hans Herlin gestorben," Frankfurter Rundschau, 23 December 1994.
71. Hans Herlin, "Kain, Wo ist Dein Bruder Abel?" Stern 19 (7 May 1960).
72. Burghard Ciesla, "Droht der Menschheit Vernichtung? Der Schweigende Stern—First Spaceship on Venus: Ein Vergleich," in Apropos: Film 2002: Das Jahrbuch der DEFA-Stiftung (2002): 124, 132–33.

73. "Die Neue Sonne," in *Mosaik von Hannes Hegen: Die Reise ins All*, reprint (Berlin: Buchverlag Junge Welt, 2007), 54–75; first published as Hannes Hegen and Lothar Dräger, *Die Neue Sonne: Mosaik* 27 (1959).

74. Thomas Kramer, *Micky, Marx und Manitu* (Berlin: Weidler Buchverlag, 2002), 218–21.

75. "Wohin mit dem Geld?" NBI 4, 1961.

76. Under the Hallstein Doctrine, in force until 1972, the Federal Republic vowed to break off diplomatic relations with any country that recognized the GDR. On competition in Africa, see Katherine Pence, "Showcasing Cold War Germany in Cairo: 1954 and 1957 Industrial Exhibitions and the Competition for Arab Partners," *Journal of Contemporary History* 47, no. 1 (2011): 69–95.

77. Susanne Zantop, *Colonial Fantasies: Conquest, Family and Nation in Pre-Colonial Germany, 1770–1870* (Lincoln, NE: University of Nebraska Press, 1990).

78. "Revolution hinter Türen aus Stahl und Paraffin," NBI 45, 1958.

79. "Krebs," NBI 26, 1962; "Wissenschaft am seidenen Faden," NBI 35, 1962; "Atomreaktor aus dem Baukasten," NBI 5, 1963; "Die grosse Koalition," NBI 46, 1964; "Die Wissenschaft von der Wissenschaft," NBI 49, 1964; "Ein Leben für die Physik," NBI 52, 1964.

80. "Phönix aus der Asche," NBI 7, 1960; "Isotopen-Labor," NBI 15, 1960.

81. "Schlimmer als die Pest," *Stern* 38 (21 September 1957).

82. "Erst wenn der Tod kommt, kann das Leben siegen," *Stern* 50 (13 December 1958).

83. "Mit Gewalt vor den Röntgenschirm?" *Stern* 34 (23 August 1958).

84. Willibald Steinmetz, "Ungewollte Politisierung durch die Medien: Die Contergan-Affäre," in *Die Politik der Öffentlichkeit—die Öffentlichkeit der Politik*, ed. Bernd Weisbrod (Göttingen: Wallstein Verlag, 2003), 195–228.

85. "Ja zum Leben," NBI 43, 1958; "Angst vor Regen?" NBI 32, 1958.

86. While fission involves splitting the atom, fusion involves fusing atoms of hydrogen, the process at work in the sun.

87. "Ein Glas Wasser genügt für die Stadt von morgen," *Stern* 18 (30 April 1960).

88. Radkau, *Aufstieg und Krise*, 89.

89. On Allied prohibition of nuclear research in Germany, see John Krige, *American Hegemony and the Postwar Reconstruction of Science in Europe* (Cambridge, MA: MIT Press, 2006), 47–50. On Euratom and West German resistance to this supranational model, see John Krige, "The Peaceful Atom as Political Weapon: Euratom and American Foreign Policy in the Late1950s," *Historical Studies in the Natural Sciences* 38, no. 1 (2008): 20–22; article on 5–44. On the system that emerged out of the Atomic Law, see Radkau, *Aufstieg und Krise*, 398–408.

90. Radkau, *Aufstieg und Krise*, 20–27, 37–45, 54–56, 58, 71, 90–91, 96, 103–5, 118–19, 138–47, 196–98, 259–64, 278–89.

91. In the GDR, nuclear safety was overseen by the Central State Office for Radiation Protection (Staatliche Zentrale für Strahlenschutz, or SZS), later the State Office for Nuclear Safety and Radiation Protection (Staatliches Amt für Atomsicherheit und Strahlenschutz, or SAAS).The SED shut down the School of Nuclear Physics of the Technical University of Dresden in 1962, the Office for Nuclear Research and Nuclear Technology in 1963, and the Scientific Council for the Peaceful Use

of Atomic Energy in 1966. Nonetheless, nuclear research continued at the Central Institute for Nuclear Research (Zentralinstitut für Kernforschung) in Rossendorf, which became part of the Academy of Sciences of the GDR. Also important for the training of nuclear power plant personnel was the Rheinsberg nuclear power plant, which not only generated power for the grid but also served as a research and training facility. Some research was conducted by the nuclear industry. East Germans also participated in research at an international center of atomic research, which was established in 1956 and was located in Dubna in the Soviet Union; at that time it was the site of the world's largest particles accelerator. However, Heinz Barwich, a leading nuclear researcher until his defection to the West in 1964, found during research stays in Dubna that he and other non-Soviet scientists were excluded from the most interesting research conducted there.

92. Mike Reichert, *Kernenergiewirtschaft in der DDR* (St. Katharinen: Scripta Mercaturae Verlag, 1999); Burghard Weiss, "Nuclear Research and Technology in Comparative Perspective," in *Science under Socialism: East Germany in Comparative Perspective,* ed. Kristie Macrakis and Dieter Hoffmann (Cambridge, MA and London: Harvard University Press, 1999), 217–20; article on 212–229. My sources on Barwich: NARA FBI (65/230/86/15/07/105), File #134040, CIA teletype dated 15 October 1964; CIA report dated 6 November 1964; CIA report dated 8 February 1965; CIA report dated 16 June 1965; CIA report dated 20 October 1965; United States Mission Berlin, Dispatch No. 119, dated 15 September 1961. See my interpretation in Augustine, *Red Prometheus,* ch. 4.

93. Holger Nehring, "Cold War, Apocalypse and Peaceful Atoms: Interpretations of Nuclear Energy in the British and West German Anti-Nuclear Weapons Movements, 1955–1964," *Historical Social Research* 29, no. 3 (2004): 150–70.

94. Uekötter, *Deutschland in Grün,* ch. 4, esp. 82.

95. Radkau, *Aufstieg und Krise,* 397. He cites resistance to the building of a plant in Biblis in 1969. For examples, see R. Gerwin, "Schwieriges Ja zum ersten deutschen Atommeiler," *Zeitung für Kommunalwirtschaft* (October 1955). This article claimed that there was a "major propaganda campaign" against the building of the Karlsruhe reactor. Also (on escape of radioactive gases from the French nuclear power plant Marcoule), see "Durch diesen Atom-Schornstein," *Der Tag* (1 July 1956).

96. Deutsches Rundfunkarchiv (German Broadcasting Archives), Babelsberg (abbreviated throughout as DRA) DRA OVC 6442, "Schülerprogramm: Geographie. Beweise unserer Kraft," first broadcast on 22 June 1967.

97. Rolf Geffken, *Arbeit und Arbeitskampf im Hafen: Zur Geschichte der Hafenarbeit und der Hafenarbeitergewerkschaft* (Bremen: Edition Falkenberg, 2015), 101.

98. Programs highlighting the Soviet Union as the protector of the GDR as far as nuclear safety was concerned, while also emphasizing the GDR's scientific and technological prowess: DRA OVC 1465, "Im Blickpunkt in Aktuelle Kamera: VEB Atomkraftwerk Rheinsberg," first shown on 9 May 1966; DRA 058210, C7968, "Physik, Klasse 9–10: Energie aus dem Atom—Physik als Produktivkraft"; DRA AC 17901, "Kämpfer und Sieger T: 15," first broadcast on 30 June 1967; DRA OVC 2102, "Im Blickpunkt in Aktuelle Kamera: Besuch der Partei- und Regierungsdelegation der DDR in Nowo-Woronesch," first broadcast on

10 November 1967; DRA AC 17909, "Atomzeitalter," first broadcast on 20 March 1968; DRA AC 18816, "Flamme des Atoms: Professor Kurtschatows Entscheidung," first shown on 30 June 1971; DRA OVC 0134, "Strom fließt aus der Heide ... auf der Großbaustelle Kernkraftwerk-Nord erlebt und notiert," first shown on 3 February 1974; DRA AC 5680, "Bericht der neuen Fernseh-Urania: Löst das Atom unser Energieproblem?," first broadcast on 12 March 1975; DRA 004015, "Einsatz Atomstation: Dialog mit einem Mann am Don," first broadcast on 9 June 1976; DRA, 003998, "Am Tag, als der Sputnik kam: 25 Jahre nach einem Jahrhundertereignis," first shown on 4 October 1982.

99. DRA 002700, "Atomstrom aus Rheinsberg," first broadcast on 7 June 1966.

100. DRA OVC 2908/1, "Im Blickpunkt in Aktuelle Kamera. Rheinsberg," first broadcast on 29 July 1967.

101. DRA FP3973, "Unser Sandmännchen: Städtebilder: Greifswald," first shown on 2 July 1976.

102. Augustine, *Red Prometheus*, 64–66, 284–87.

103. David Nye, *American Technological Sublime* (Cambridge, MA: MIT Press, 1994).

104. Jonathan Coopersmith, *The Electrification of Russia, 1880–1926* (Ithaca, NY: Cornell University Press, 1992); Susanne Schattenberg, *Stalins Ingenieure: Lebenswelten zwichen Technik und Terror in den 1930er Jahren* (Munich: R. Oldenbourg Verlag, 2002), 70–107; Paul Josephson, *Red Atom: Russia's Nuclear Power Program from Stalin to Today* (New York: Freeman, 2000), esp. 7–19.

105. Alfred Kosing et al., *Weltall Erde Mensch: Ein Sammelwerk zur Entwicklungsgeschichte von Natur und Gesellschaft*, 13th ed. (Berlin: Verlag Neues Leben, 1965), 5.

106. Daniel 10:21.

107. Sonja D. Schmid, "Celebrating Tomorrow Today: The Peaceful Atom on Display in the Soviet Union," *Social Studies of Science* 36, no. 3 (2006): 331–65.

Chapter 2

On the Brink of Disaster?

Safety Regimes and Nuclear Accidents in the Two Germanys

Profound disagreements concerning risk were fundamental to the nuclear power debates across the globe. In West Germany, the nuclear industry and other proponents of nuclear power tried to define nuclear safety as a purely technical issue, following the lead of German engineers. This stance was deeply rooted in the history of engineering and science in Germany. In the nineteenth century, German engineers and scientists fended off professional rivals and attained respect and status by staking claims to unique expert knowledge and objectivity. Resisting the interference of laymen in issues defined as scientific or technical, they claimed (disingenuously) to stand above politics in promoting the good of the nation.[1] In the postwar period, scientists and engineers involved in the development of nuclear power laid claim to this tradition. Historian Michael Schüring points to the importance of the concept of *Sachzwang*—the idea that the facts dictated a particular technical or scientific approach, which should not be diluted or disputed with ethical or political considerations.[2]

Critics contested the validity of this concept and the self-conception of German nuclear physicists and engineers. In his magisterial study on nuclear energy, historian of technology Joachim Radkau depicts the process of evaluation, selection, and deployment of atomic power in West Germany as so politicized, corrupted, and illogical as to be irredeemable. "The origins of the [anti–nuclear power] protest movement lie in the development of nuclear energy," he argues.[3] By contrast, German scholars who view atomic energy in a positive light, such as Paul Laufs and Wolfgang Müller, employ classic arguments regarding the competence of experts and the effectiveness of the process of constant questioning and improvement. Any mistakes made in the past leading to nuclear accidents can be corrected, they believe.

Laufs, an expert with a career in nuclear industry and politics behind him, describes a process of ongoing inquiry and examination of weaknesses in reactor and nuclear power plant design in the Federal Republic; this is overseen by the state as well as by institutions that combine to various degrees state oversight, scientific and engineering expertise, and industrial self-monitoring.[4] In 1981, Marion Dönhoff, publisher of the widely read West German weekly *Die Zeit,* took a position between those of Radkau and Laufs, arguing that anti–nuclear power activism had led to "many safety regulations that make our reactors safer than those of other countries."[5] These three viewpoints reflect different understandings of the evolution of safety cultures in West Germany.

The development of nuclear power in the GDR has generally been treated as a failure, particularly in light of the GDR's abandonment of an independent nuclear power program in the mid-1960s.[6] However, a volume of essays by scholars and former East German engineers published in 2000 discusses systematic attempts to improve nuclear power safety in the GDR in the 1970s and 1980s.[7] This chapter looks at the construction of expert knowledge regarding nuclear power safety and at the interaction between engineers, the political leadership, nuclear safety regulators, and nuclear industry management in both East and West Germany. Chapter 3 then discusses public discussion and criticism of technical approaches to safety and risk.

Safety had been the stepchild of nuclear power development in its early stages in the two Germanys. Nuclear power, growing out of military technologies and euphoric hopes for an endless energy supply, was developed in a cultural context that downplayed safety issues. Greater safety awareness grew out of a complex web of factors, which included an economic interest in continuity of the electricity supply, the influence of US and Soviet safety regimes, and the postwar reemergence of German technological systems, involving the rebuilding and expansion of research infrastructure and competition among politico-technological institutions. Growing public concern regarding the risks involved in nuclear power and intensifying activism in West Germany pushed the nuclear industry and the research establishment to greatly expand efforts to improve atomic power plant safety. "It has always been clear to all involved that a serious nuclear power plant accident that had a detrimental impact on the population would bring about the immediate termination of this technology in Germany," writes Laufs.[8]

But what impetus to improve safety existed under real socialism? It has become abundantly clear that Communist East Germany, like the Soviet Union, was able to produce first-rate technologies in high-priority areas, at least up through the 1960s.[9] It is less clear what role safety could and did play in an advanced Communist society. Given the dictatorial nature of Communist rule in the Eastern Bloc, public pressure to improve nuclear safety was much weaker than in the West. However, there are striking convergences in approaches to nuclear power plant safety in the GDR and the Federal Republic, although major differences remained.

Accidents in nuclear power plants were treated differently in the two Germanys. Disaster research tells us that catastrophes and, under some circumstances, more garden-variety accidents can be catalysts to change, bringing about safety improvements and, occasionally, deeper social changes. They also make manifest aspects of industrial and political reality that are normally difficult to discern.[10] Thomas Lindenberger sees a preoccupation with safety and security (hard to distinguish in German because both are termed *Sicherheit*) as central to Communist rule in the GDR. His work on industrial disasters shows that in their attempts to prevent a weakening of SED rule, industrial management and the secret police blocked safety improvements and social change.[11]

Nuclear accidents, outside of the scope of Lindenberger's work, were a category unto themselves. In East and West, accidents connected with military nuclear research were treated as state secrets during the Cold War.[12] In the GDR, as in the Soviet Union, any accidents that occurred in civilian nuclear power production were also treated as highly confidential matters. This chapter asks whether they nonetheless generated internal debate in the GDR, while also asking whether nuclear accidents and lesser mishaps were handled in a fundamentally different way in West Germany.

Safety Philosophies in East and West

"Safety philosophies," an Americanism that came into widespread use in West German industry and from there spread to the GDR, were overarching concepts that helped engineers formulate norms, rules, and procedures for use in planning, building, and operating nuclear power plants. Historian of technology Joachim Radkau points to four that played a major role in West Germany. The first emphasized "experience." The idea was that the safest technologies were those with which industry had the most experience. The light water reactors (specifically, boiling water and pressurized water reactors) that became the energy industry standard across most of the world in the 1960s and 1970s and are still in use in most nuclear power plants evolved from reactors used on nuclear submarines. One problem, pointed to by Radkau, was that such tried-and-true technologies were the oldest and, given the military origins of these technologies, were often not the safest. Nor could they necessarily be safely scaled up.

A second philosophy looked to spatial location (i.e., distance from population centers) for safety. A third philosophy was that of "inherent safety," arising out of a search for technologies, for example reactor designs, that were inherently safer than others. According to Radkau, however, fundamental decisions, such as the adoption of the US-made light water reactors for use in commercial nuclear power plants, were based on politics, not scientific rationality. By the 1970s, the US safety philosophy known as "engineered safeguards" triumphed, not only in West Germany but also in the GDR and across the globe. This fourth approach involved the use of monitoring devices and "fail-safe" systems that minimize harm in case of equipment failure or that interrupt operations if malfunctions are detected, such as overheating caused by human error or equipment failure.[13]

Engineers and regulators in West Germany and the United States long clung to a "deterministic" approach to safety involving the cause-and-effect analysis of particular safety issues and vulnerabilities. This went hand in hand with the development of engineered safeguards. The US Atomic Energy Commission found this piecemeal approach lacking because it did not provide a way to estimate overall risk, leaving the public in a state of apprehension. According to historian Thomas Wellock, it was the 1975 Rasmussen Report, also known as WASH-1400, that ushered in a paradigm shift to a probabilistic method that made quantitative risk assessment possible. The methodology was based on the use of "event trees" or "fault trees," which allowed the calculation of the probability of the occurrence of a series of events that could lead to a nuclear plant disaster. This method was severely criticized by many engineers and bureaucrats.

From a purely technical point of view, this method of risk calculation was deficient because probability can be used to calculate risk, but not uncertainty (that is, an open-ended process in which the number of possible combinations of events is incalculable). Moreover, as sociologist Charles Perrow argued in his 1984 study *Normal Accidents,* efforts to prevent accidents had become a major *cause* of accidents. Faced with an unfolding nuclear power plant crisis, human operators had great difficulty in diagnosing the fundamental problem because multitiered redundant safety systems were so complex.[14] However, probabilistic methods did help improve safety by identifying problematic areas. Eventually, a fruitful synthesis of probabilistic and deterministic methods was achieved. The Rasmussen Report did not, however, reach a second goal of the US Atomic Energy Commission, which was to allay public fears of nuclear power accidents.[15] These quandaries and disputes became known in West Germany, leading to a questioning of nuclear power safety there, as will be discussed in Chapter 3.

West Germany was much influenced by the United States with regard to standards, approaches, and artifacts that were part of the technological system of nuclear power, at least up into the 1970s. The Federal Republic adopted the American light water reactor as its standard reactor for commercial nuclear power plants, evidently because of confidence in US technological prowess, as well as a desire to please its most important ally.[16] West Germany also adopted and improved on technologies associated with nuclear power production developed in the United States, such as airtight containment domes, automated monitoring systems, and improvements aimed at preventing the explosion of reactor pressure vessels (which contain the reactor core). By the 1970s, an internationalization of the West German nuclear safety regime was under way, as institutions such as the IAEA (the International Atomic Energy Agency) attained new stature and a multinational exchange of technological knowledge became the norm.[17]

The GDR became completely dependent on the Soviet Union after the SED shut down most of East Germany's nuclear power program in 1962–1965. The USSR supplied nuclear power plants for energy production to the GDR. The SED leadership simply assumed that the Soviet Union would ensure the safe operation of the Rheinsberg nuclear power plant (which went critical in 1966), as well as that in Lubmin, near the town of Greifswald (also known as Nord or "Bruno Leuschner," but referred to in the present study as "Greifswald"). This did not go well, however. In 1965, the GDR complained about Soviet disengagement: a lack of technical assistance, poor customer service, and the unavailability of replacement parts for Soviet equipment.[18]

When, however, Karl Rambusch (then general director of the socialist combine for nuclear energy) tried to put together a coordinated research and development (R&D) program that would address safety issues in 1967, the party leadership accused him of harboring anti-Soviet attitudes. During the building of unit 3 of the Greifswald plant engineers identified safety issues, but in 1975 representatives of the East German government initially refused to allow alterations to the Soviet design that East German experts deemed necessary.[19]

A series of events, including a "serious incident" at the Greifswald nuclear power plant in 1975 and the far graver Three Mile Island (Harrisburg) nuclear power plant accident of 1979, brought about a rethinking of nuclear safety on the part of the SED leadership. Intense worldwide preoccupation with nuclear safety put the GDR under international pressure to improve the safety of its atomic power plants.[20] The GDR nuclear power industry transitioned to the Western paradigm of "engineered safeguards"—implicitly if not explicitly. In July 1979, four months after Three Mile Island, East German authorities introduced new regulations for the licensing of nuclear power plants. The central regulatory agency of the GDR, the State Office for Nuclear Safety and Radiation Protection (Staatliches Amt für Atomsicherheit und Strahlenschutz, or SAAS), took on a more active role.[21]

The SED leadership could not let go—politically, psychologically, or in practical terms—of its fixation on the Soviet Union, but technical experts were increasingly critical of weaknesses in these Soviet technological systems. The Soviet Union did not export intrinsically risky graphite-moderated reactors such as the one responsible for the Chernobyl catastrophe to the GDR, but rather pressurized water reactors of its WWER[22] line. The Soviet authorities were motivated, not by safety concerns, but rather by total unwillingness to give the GDR a reactor line that produced plutonium. They also did not provide the GDR with the latest models, which incorporated technological improvements, notably containment structures. (Only the later-generation reactor planned for Stendal in the GDR was to have containment, but it never went into operation.[23])

The Soviet Union did make up for this to a certain extent by helping to rebuild units 1–4 of the Greifswald nuclear power plant in the 1980s.[24] On the whole, Soviet safety systems for nuclear power plants relied far more on human intervention than on automated systems, such as those prevalent in the West.[25] East German researchers attempted to develop nuclear power plant monitoring systems that could compensate for deficits in Soviet equipment, most notably *Rauschdiagnostik,* or noise diagnostics, a technique developed at the Central Institute for Nuclear Research at Rossendorf to deal with dangerous vibrations of Soviet fuel rods.[26]

A third phase, characterized by the adoption of international standards, was ushered in by the Chernobyl disaster. In 1986, both the GDR and the Soviet Union joined the IAEA's newly established early warning system.[27] Chernobyl and Gorbachev's accession to power also unleashed open criticism of safety of Soviet nuclear power plants by GDR officials.[28] The GDR began to measure itself by international standards, as can be seen in SAAS reports, which will be discussed in a later section of this chapter.

In the final phase of GDR history, the slow collapse of the GDR economy and SED rule exacerbated long-term problems and made it very difficult to follow through with an internationalization of the safety regime. Both on the shop floor level and among management, discipline was breaking down. The physical and psychological decay of East German industry in the late 1980s had a very negative effect on nuclear power plant safety. In addition, the Soviet Union stopped shipments in the Gorbachev era. In some ways, however, decay and internationalization converged: By 1987, the GDR was importing energy from West Germany and other Western countries.

In 1989, the Soviet Union suggested—and the GDR considered—importing a nuclear power plant from the West and seeking Western help in repairing the Greifswald and Stendal plants.[29] The SED leadership decided in June 1989 to shut down Rheinsberg in 1992. But after German reunification, all former East German nuclear power plants were shut down. Whatever convergence had taken place was insufficient in the eyes of Federal German policymakers after reunification. Many East German experts saw this as the outgrowth of the West German energy industry's desire to rid itself of unwanted competition, but Western experts insisted that they based their evaluation on normative safety standards.

Interpreting Nuclear Power Safety Problems in West Germany

In mid-February 1979, members of the West German environmental organization, the Federal Association of Citizens' Initiatives on Environmental Protection (Bundesverband Bürgerinitiativen Umweltschutz, or BBU), broke into the Hamburg offices of the Society for Reactor Safety (Gesellschaft für Reaktorsicherheit, or GRS)[30] and stole reports on mishaps and accidents in West German nuclear power plants dating back to the mid-1960s.[31] Until that time, these reports (compiled on behalf of the West German government) had not been released to the public, but this act of civil disobedience and the publication that resulted from it forced open the records on nuclear plant safety to public scrutiny.[32]

Figure 2.1 shows that between 1979 and 1990, Category B[33] incidents decreased, while less serious Category C events became more common.[34] The most severe sorts of incidents, Category A, were rare, yet there were five in 1980 and one each in 1979 and 1990. The overall number of accidents and incidents reached a high of 334 in 1986—an unfortunate coincidence in the year of the Chernobyl disaster—and remained around 300 per year through the end of the decade. The report for 1986 attributed this to start-up difficulties with the newest atomic power plants and reactors.[35]

In its report on incidents and accidents at West German nuclear power plants in 1977–78, the Society for Reactor Safety asserted, "The great majority of the incidents can be attributed to technical deficiencies; only in a few cases did managerial negligence or human error play a role."[36] The data, shown on Figure 2.2, bear out this interpretation. Component failures were faulted for a third to half of all of nuclear power plant accidents in

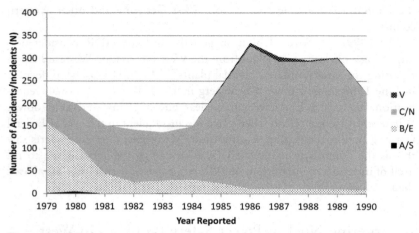

Figure 2.1 Nuclear power plant accidents and incidents, Federal Republic, 1979–1990.

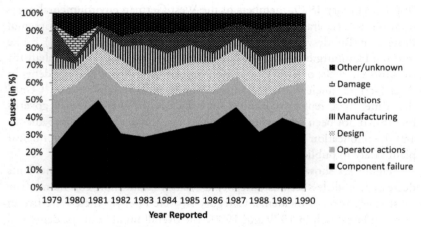

Figure 2.2 Causes of incidents, Federal Republic, 1979–1990.

1980–1990.[37] Manufacturing errors were blamed in a further 3–17 percent of cases in those years. Problems arising from design and layout of the plants were thought to play the main role in 9–17 percent of the accidents. [38] "Human error" was not given much prominence: Operator errors and repair and installation errors[39] were cited as the primary cause in only 18–27 percent of the cases in 1980–1990. Conditions in the specific plants[40] were viewed as the main culprit in 3–16 percent of the incidents. This last factor was probably understood to encompass errors made by plant managers.

Illustration 2.1 Town seal of Gundremmingen, West Germany, with atomic symbol. Courtesy of Gemeinde Gundremmingen. Wiki Commons.

Thus, nuclear incidents and reactions to them were framed more in terms of control over technology than over people.

The most serious nuclear incident in the history of the Federal Republic took place on January 13, 1977 in the first commercial nuclear power–generating plant in the Federal Republic, at Gundremmingen, which had opened in 1966. (See Illustration 2.1 for town seal with atomic symbol.) While the 1975 "serious incident" in Greifswald, East Germany (to be discussed below) was eventually given a rating of 3 on the INES scale,[41] the 1977 incident in West Germany "only" rated a 2.[42] (Like the Richter scale for earthquakes, the INES scale is exponential, so that each step represents a jump by a factor of ten.) However, this incident is emblematic of perennial problems with this, the original Gundremmingen reactor, built by General Electric in partnership with the German AEG corporation.

Historian Joachim Radkau is highly critical of the way the fundamental decisions regarding the Gundremmingen A plant (so called because two other units, Gundremmingen B and C, were built in 1976–1985) were arrived at. The beneficiary of West German, US, and Euratom subsidies, it was considered too big to fail. According to Radkau, it was built with very serious safety defects, omitting safety features used by General Electric in the United States. This was done despite warnings from TÜV (Technischer Überwachungsverein, or Technical Inspection Association), a semiofficial German organization with wide-ranging powers to inspect machines, vehicles, products, and amusement park rides and to develop new technological concepts.[43]

On 13 January 1977, the Gundremmingen A plant was "totaled" by the most serious accident to ever take place in the Federal Republic. It was, like the Fukushima reactor, a boiling water reactor. Due to extreme cold and ice, the electrical line transmitting power for the grid shorted out, necessitating an emergency shutdown of power generation at the Gundremmingen plant. The reactor's emergency shutdown system worked. However, due to a malfunctioning relay switch, the turbine's shutoff valve did not work. Water pressure in the main loop (loop A[44]) fell. The automated system interpreted this as a leak. The emergency cooling water system went on, pumping water into the main loop. The hot, radioactive water flooded the reactor building. In response to the rapidly rising temperature, the emergency sprinkler system went on, adding to the flood. After the accident, it was also discovered that many of the pipes in the reactor core were corroded and cracked.[45] RWE (Rheinisch-Westfälisches Elektrizitätswerk AG, or Rhenish-Westphalian Power Plant) never put the plant back into service after the accident.

Gundremmingen A had been plagued by a series of technological breakdowns, mishaps, and accidents in the decade in which it was in operation (1966–1977). Leaks and material fatigue were widespread in the plant. In 1975, two workers died while trying to repair a leaky valve on a pipe that carried steam from the reactor to the turbine.[46] Remarkably, although the 1975 accident immediately became known to the public, the full significance of the 1977 accident did not become clear until years later. The 1977 accident took place during a period of tremendous turmoil over nuclear power in West Germany. The authorities evidently colluded with plant officials to keep the story quiet. Der Spiegel and Die Zeit mentioned the accident in passing, but there was little indication of how potentially serious it had been—because no one outside of top government and industrial circles understood this.[47]

In the Federal Republic, a new generation of reactors promised greater inherent safety. Most of the second-generation commercial reactors in West Germany were pressurized water reactors. They were considered safer than boiling water reactors because radioactive water never left the reactor building. In this kind of reactor, water is not converted to steam, thus reducing corrosion and cracking of pipes—a perennial problem in boiling water reactors. Boron mixed into the water in loop A of the pressurized water reactor does promote corrosion in carbon steel, although not in stainless steel. However, pressurized water reactors have safety-enhancing fail-safe design features, such as control rods located above the reactor core (instead of below it as in the boiling water reactor). In case of loss of electrical power, gravity causes the control rods to fall into the core, thus putting the brakes on the nuclear reaction. An additional feature, known as negative temperature coefficient, is that a rise in temperature causes the core to automatically cut back output.[48]

The thorium reactor was also considered as an alternative to "light water reactors," a category to which boiling water and pressurized water reactors belong. Proponents of the liquid fluoride thorium reactor argue that it is safer, among other reasons, because of its negative temperature coefficient and its low-pressure operation (making it far less likely that the reactor will blow up or leak) and because it produces far less radioactive waste than light water reactors.[49] However, the high-temperature thorium reactor, the THTR-300, developed by West German physicist Rudolf Schulten, had a different design and serious safety issues. Particularly problematic was its fuel delivery system, which took the form of uranium and thorium embedded in graphite balls, called "pebbles."

In 1986, a paradigmatic "normal accident" of the sort predicted by Charles Perrow took place at the THTR-300. According to *Der Spiegel,* the automated system needed to convey exactly sixty pebbles into the reactor at a time. For whatever reason, only forty-one were available. This led to a pebble getting stuck in the pipe leading into the reactor. In the process of trying to dislodge the pebble by blowing gas, engineers accidently released radioactive helium into the atmosphere. According to *Der Spiegel,* this came to light because an employee tipped off the authorities anonymously and because an environmentalist group was keeping close track of radioactivity readings around the plant. Otherwise, the release of radioactive dust might not have been noticeable such a short time after Chernobyl. The utility company that owned and ran THTR-300 initially denied that a mishap had occurred.[50] This incident and the ensuing cover-up contributed to the decision to shut down THTR-300, making it one of the most expensive failed projects of all times in West Germany. Thus, in West Germany, technocratic approaches to safety failed not only on technical grounds but also on political and ethical grounds.

The contention that safety should come first in the development of nuclear power was not accepted by all West German politicians and nuclear power experts. The fast-breeder reactor had many proponents worldwide, particularly in the 1960s, because it produces more fuel than it consumes. However, it also produces plutonium, which can be used to produce nuclear weapons and which is highly dangerous to humans, particularly since its deadly radiation accumulates in bones. The use of sodium as a coolant was also considered a major safety concern since sodium can explode upon contact with water.[51] Huge amounts of taxpayers' money were poured into the building of the Kalkar fast-breeder reactor, but it never went into service.

Another major problem for the West German nuclear power program was how to safely dispose of or recycle nuclear waste. A 1976 amendment to the West German Atomic Law required utility companies seeking permission to build a new nuclear power plant to draw up a plan that would ensure the safe disposal of spent nuclear fuel and radioactive waste. Following American advice, West Germany explored the possibility of using Asse, an abandoned salt mine, as a permanent repository for nuclear waste starting in the 1960s. Water eventually filled the mine, making it unusable. Another possibility was to reprocess nuclear fuel, extracting uranium and plutonium from the spent fuel. In 1977, the minister-president of the West German state of Lower Saxony, Ernst Albrecht, proposed that the salt deposits near the town of Gorleben become a permanent storage site. Originally, a nuclear reprocessing plant was supposed to be built there as well, but in the face of protests Albrecht gave up on that project.

At the time, too little was known about the salt deposits and geological characteristics of the area to make a firm decision on permanent storage of nuclear material. Exploratory work on the salt domes lasted for years and faced considerable protests. In the 1980s, it became clear that ground water was seeping into the salt deposits, threatening to leach radioactive materials into the soil and the groundwater.[52] Plans for a nuclear reprocessing plant in the town of Wackersdorf also faced stiff resistance, mainly because of the high toxicity of plutonium, the danger of nuclear proliferation, and the suspicion that the West German government wanted to use plutonium to develop a nuclear weapons program of its own as well as because plutonium could be used to fuel fast-breeder reactors.[53] The project was never completed. The German nuclear industry never really solved the perennial problem of nuclear waste, settling in the 1990s on a medium-term strategy of nuclear reprocessing in France and temporary storage in Germany.

Conflicting Understandings of Accidents in the GDR

The East German safety record was lackluster at best. According to secret police reports, there were 290 events between January and October 1974 at the Greifswald nuclear power plant, 234 in 1978, 272 in 1979, and 275 in the first eleven months of 1980.[54] This seems like an extraordinary number of disruptions—nearly six per week on average—even for a plant with four units, each with its own reactor. Exact comparisons with West Germany are impossible. What is clear is that, first, no public discussion of nuclear power plant safety could take place in the GDR. Second, understandings of nuclear power safety and accidents differed greatly in East and West Germany. Third, there was considerable disagreement among East German institutions and elites regarding the nature of nuclear safety and accidents.

In the GDR, industrial authorities, the party, and the secret police helped to construct the *Havarie,* or industrial accident, as a distinctive socialist phenomenon. The ways in which *Havarien* in nuclear power plants were dealt with were tied to particular safety regimes and claims about socialist society. Ironically, the focus was on assigning blame to a guilty party. Sonja Schmid has shown that in the Soviet Union, the authorities wanted to deflect blame from technological failures, notably in the aftermath of Chernobyl. To question Soviet domination of science and technology was to question the foundations of Communist rule—and that had to be avoided at all costs.[55] Something similar was going on in the GDR. On a more practical level, the state and industry hoped to educate, motivate, and, if necessary, compel employees to better performance.

However, individuals held responsible for a nuclear power plant accident were *not* accused of sabotage. Under communism, sabotage was understood to be a political crime against the state and socialist society. A charge leveled mainly in the early GDR, its absence from any discussion of accidents in the atomic energy sector signals a reduction of political intrusion into this high-tech area. By the late 1980s, the GDR adopted a Western understanding of sabotage, considering the possibility of a terrorist attack on a nuclear power plant in much the same terms used in the Federal Republic and the United States.[56]

Essentially, there were two kinds of *Havarien:* those involving the escape of radioactive materials and those that were considered nonnuclear. In the latter case, factory and industrial regulations were the basis of a quasi-judicial hearing presided over by secret police and industrial administers and conducted within the walls of the atomic power plant. During an in-depth investigation of the accident, employees who had played some role in the relevant events were interviewed. Investigators identified the culprits, generally fairly low-level employees; for example, in a 1974 case, it was two young mechanists who had ended their apprenticeships only a month before and a foreman who had not properly supervised them to make sure that the proper procedures were carried out.

It was left to the plant management to punish the guilty, but the plant had to report to the Stasi how the employees had been punished. The non-nuclear *Havarie* was also discussed in meetings with plant employees, giving the procedure a pedagogical component.[57] It is noteworthy that the secret police played a major role in the proceedings and that their goal was to assign personal blame to specific individuals. The possibility that immediate problems—such as the reliance on inexperienced workmen and the lackadaisical attitude of the foreman in the 1974 case—were only symptoms of systemic problems was not explored.

In cases of *Havarien* that endangered the nuclear integrity of the plant, the procedures differed in important points. Whereas conventional accidents were discussed with the work collectives, accidents involving the escape of radioactive materials were only evaluated and discussed on a need-to-know basis.[58] True nuclear accidents (as opposed to garden-variety mechanical and electrical accidents) were shrouded in secrecy. In addition, criminal charges were often brought against the person or persons whose negligence had led to the *Havarie.* One of the most frightening and serious accidents occurred in 1975. It was later classified by the IAEA as an INES-3 event, making it the most serious accident ever to have taken place in a German nuclear power plant.[59] Knowledge of this accident was kept from the public and was not revealed until after the fall of the Berlin Wall.

A 27-year-old electrician was trying to show a 19-year-old helper how equipment worked, when they shorted out a transformer. For some reason, this did not trigger a circuit breaker, and so the short fried the equipment at about five hundred degrees centigrade, igniting electrical cables. The ensuing fire caused 2.2 million marks of damage. It also caused the reactor's pumps to stop working, causing the temperature in the reactor to rise to danger-ous levels. A core meltdown was narrowly averted. Criminal charges were brought against the electrician.[60] Again, the secret police report does not mention issues other than personal responsibility. Why did the safety system fail? Did chaotic conditions in the plants, such as unsafe wiring, contribute to the accident? These questions do not appear to have been asked.

The secret police came up with two very different sorts of explanations for *Havarien.* The first was voluntaristic, moralistic, and surprisingly indi-vidualistic: Blame was assigned to particular individuals who needed to be punished. This sort of thinking went to the top: After a serious *Havarie* in 1982, Politburo members called for the responsible administrators at the Greifswald nuclear power station to be dismissed.[61]

The Stasi could not overlook the recurrence of certain kinds of acci-dents involving human error at the Greifswald plant, however. It therefore reached for a second explanatory framework that was political in nature. For example, a 1979 Stasi report found Greifswald workers lacking in a sense of the "honor" of their work,[62] a concept related to the ideal of the "socialist personality."[63] Management and plant foremen had failed in their responsibility to promote the "development of the socialist personality" and "political–ideological training." Thus, a lack of "political leadership" at the Greifswald plant was thought to play a major role in these *Havarien*[64] along with deficits in socialist consciousness and simple moral failings.

In its overall analysis of accidents at Greifswald, the Stasi also leaned heavily in the direction of human error. Of the 290 incidents that were investigated between January and October 1974 at the Greifswald nuclear power plant, 178 were attributed to in-house employee error,[65] although a not inconsiderable number—72—were blamed on design, construction, or manufacturing errors, which pointed more in the direction of errors on the part of the Soviets.[66]

The SAAS developed an alternate view of nuclear power plant accidents and safety in the 1980s. Basing its reports on its own inspections, particularly those conducted by the Permanent Oversight Committee for Plant Safety (Ständige Kontrollgruppe Anlagensicherheit), the SAAS formulated fundamental criticism of the East German nuclear power industry and its approach to safety. A 1983 SAAS report stated baldly that the GDR, not the Soviet Union, was responsible for East German nuclear plant and radiological safety. Therefore, the GDR had to conduct and apply its own nuclear safety research but also had to adopt and apply international (i.e., Western) nuclear safety standards.[67]

Over the years, SAAS reported on intractable problems that were not being taking care of and that the Soviets were unwilling or unable to provide assistance with. One of the most serious was the embrittlement of the Greifswald 1 reactor vessel, which could cause the vessel to crack or burst, leading to a core meltdown. This seems to have been an endemic problem with Soviet-made WWER-440/230 reactors. At Greifswald, radioactivity damaged the material out of which the reactor vessels was made, causing small cracks. This was the result of a Soviet design flaw: the fuel rods were too close to the wall of the reactor vessel, thus exposing it to too much bombardment by neutrons. By 1985, SAAS deemed "immediate countermeasures" necessary.[68]

Schmid points to a factor that may explain why the components of the reactor core were overcrowded within the reactor vessel: the Soviet railroad system could not transport loads over a certain size, which limited the size of reactor vessels that were manufactured in Leningrad and transported all across the Eastern Bloc by rail.[69] Soviet experts should have been familiar with the problem of reactor vessel embrittlement, yet they provided no advice.[70] Another problem: Soviet-made fuel rods caused a different safety crisis in Greifswald in 1984, when they unexpectedly overheated the coolant. Soviet industry professed ignorance.[71] Further problems in Greifswald included damaged pipes, corrosion in the steam generator, and seepage of salt water into the plant from the Baltic Sea.[72]

SAAS reports saw recruitment problems as a major explanation for Greifswald's poor safety record. "Nuclear power plants have to employ the best," the SAAS affirmed. But this was not the case, due to the "insufficient number of cadres with university and technical school training" available to the nuclear power industry. Graduates showed "little interest" in working in nuclear power plants, in part because there was an acute housing shortage in Greifswald but also because of limited opportunities to put scientific and technical training to use on the job and the poor state of R&D facilities in the Greifswald plant. Hence, Greifswald could not be highly selective in its recruitment practices.

Recent hires received on-the-job training, but this "can never replace a [good] theoretical foundation and understanding of the complexities."[73] The SAAS report presented a particularly dismal picture of conditions at the construction site of unit 75 at Greifswald: "Minimum requirements regarding order and cleanliness are not being consistently fulfilled, and fundamental breaches of basic rules of quality work, particularly with stainless steel, are occurring." No specialized qualifications were required of workers hired to work on this construction site, with the exception of welders. Most were young and inexperienced.[74]

SAAS attributed these and other failures in part to poor management and to the rush to meet deadlines. SAAS tried to put direct pressure on the general director of the Greifswald nuclear power station, Dr. Lehmann, to abide by national standards for highest allowable radiation levels, which Greifswald had apparently been ignoring.[75] At a 1985 meeting, SAAS put pressure on Lehmann to improve management, hire qualified cadres (engineers and managers), provide better housing for employees, and improve pay and the "political–moral recognition" of employees.[76]

What SAAS was asking for was a concerted effort to lift the nuclear industry out of the morass of inefficiency that had enveloped all of East German industry by the 1980s. The reports of the Permanent Oversight Committee for Plant Safety (of the SAAS) in particular are permeated with a deep sense of frustration, even desperation. How was it possible, one report asked, that after three years so little had been done to deal with the totally unacceptable conditions at Greifswald, despite directives coming from the highest level? Intractable problems were to blame: the paucity of R&D capacity in industry; the poor quality of coordination between the nuclear industry and the nuclear equipment industry, which were overseen by two different ministries; and the general lack of industrial capacity in the GDR. The solutions that were offered up—more research, improvements in management, Soviet help, better personnel—were reasonable. However, SAAS had little real authority, and indeed there was little that even those with real authority could do in the face of monumental systemic problems such as these.[77] Moreover, these reports were treated as state secrets, thus greatly limiting their scope of influence.

Conclusion

Intrinsically, the nuclear power plants used for electricity production in the two Germanys were similar in many ways. Their pressurized water reactors were similar although not identical in design. Typical problems of Soviet WWER-440 reactors can be traced to peculiarities in the Soviet production system. Soviet designs were not always worse than Western designs, however. Notably, the horizontal steam generators attached to WWERs had safety advantages over the vertical steam generators used in conjunction with Western pressurized water reactors.[78] If East Germany was plagued by more safety issues, this was not due to a lack of technical know-how.

The biggest differences lay in the technological and political systems in which nuclear power production was embedded. The Federal Republic had a much larger industrial and research capacity, allowing it to develop far better redundant safety features, termed "defense in depth." The GDR's small industrial base and R&D infrastructure would not, of course, have condemned it to backwardness had it like, say, Finland, been more thoroughly integrated into the world economy and global scientific and engineering communities.

Bloc thinking and fear of sabotage or Western espionage cut down on such exchanges, but the Western COCOM embargo also interfered with importation of much-needed equipment, as did the shortage of hard currencies. The GDR received some, but not enough technical help with issues affecting nuclear power plant safety and functionality from the Soviet Union. It is unclear whether this is the result of a general resentment and distrust of this German state or of some other factors such as putting Soviet needs first. International developments eventually made their way to the USSR and the GDR, but only quite late. Containment came too late for East German nuclear power plants. Nevertheless, there was a process of convergence and internationalization that brought some safety improvements to the GDR. Accidents occurred, but not the "big one."

Improvements were, however, limited by SED attempts to maintain its central control over the economy and society. The SED not only suppressed any outward criticism of nuclear power safety but also interfered on a profound level in the overseeing and development of technology. Until about 1975, loyalty to the Soviet Union stood in the way of an honest discussion of safety problems with Soviet technologies. The SED never entirely freed itself from this self-destructive attitude, but technical experts were essentially able to bring about a shift to the safety philosophy then dominant in the Western world, which was based on the use of redundant safety devices. In the GDR, these engineered safeguards were tailored to deal with typical problems of the Soviet-made nuclear power plants. East German engineers nonetheless faced severe limitations to their scope of activity, due to the nonnegotiability of Soviet technological domination and the dysfunctions of the East German industrial system.

In addition, the SED and secret police closely supervised the nuclear power industry, particularly when accidents or other mishaps occurred. Investigations focused on individual blame and on political failings. Systemic problems were largely ignored. Engineers and technical staff could not learn from nuclear power plant accidents because they were largely kept secret. The SAAS was not very successful in its attempts to remedy the poor flow of information and lack of mechanisms of self-correction in the East German industrial system. On the positive side, the SED leadership did come to accept an internationalization of its nuclear safety regime after Chernobyl.

The West German nuclear industry had numerous safety problems, some minor, others somewhat more worrisome. New technologies, such as the thorium and breeder reactors, posed high risks, as did attempts to dispose of nuclear waste in salt deposits. As far as can be determined, however, the Federal Republic had a better nuclear power safety record than did the GDR. The West German nuclear industry made better internal use of information on safety problems. Accidents were attributed more to technological problems than to human error, paving the way for technical solutions. Redundant safety systems appear to have cut down on the more serious categories of accidents and incidents. Additional factors that gave West Germany the edge over East Germany were its greater industrial and research capacity, the integration of its scientists and engineers into international scientific and engineering communities, and its easy access to Western technologies.

In West Germany, the relationship between the political and technological realms was problematic, but in a fundamentally different way than in the GDR. Facing stiff public criticism, West German state governments and the energy industry at times tried to cover up problems. However, information networks functioned better in the Federal Republic than in the GDR. A critical West German public demanded and got access to information on accidents and incidents. Public scrutiny put the industry under pressure to improve nuclear power safety. However, state governments and the nuclear industry also waged public relations campaigns and, from the mid-1970s onward, fiercely combated the protests of nuclear power opponents, as will be discussed in the following chapters.

Notes

1. Augustine, *Red Prometheus,* 22–25, 36 fn. 57; see literature cited therein.
2. Michael Schüring, "Advertising the Nuclear Venture: The Rhetorical and Visual Public Relation Strategies of the German Nuclear Industry in the 1970s and 1980s," *History and Technology* 29, no. 4 (2013): 374–75; article on 369–98.
3. Radkau, *Aufstieg und Krise,* 14.
4. Paul Laufs, *Reaktorsicherheit für Leistungskernkraftwerke: Die Entwicklung im politischen und technischen Umfeld der Bundesrepublik Deutschland* (Berlin: Springer Verlag, 2013); Wolfgang Müller, *Geschichte der Kernenergie in der Bundesrepublik Deutschland.* Vol. 1: *Anfänge und Weichenstellungen* (Stuttgart: Schäffer Verlag für Wirtschaft und Steuern, 1990). Major West German institutions discussed by Laufs include the German Atomic Commission (Deutsche Atomkommission, which existed from 1956 to 1971 and which laid the foundations for the creation of a nuclear energy research infrastructure and industry in West Germany); the German Radiation Protection Commission (Strahlenschutzkommission, which formulates and oversees regulations on human exposure to radiation); the Reactor Safety Commission (Reaktorsicherheitskommission, which provides expert advice on reactor safety); the Nuclear Technology Committee (Kerntechnischer Ausschuss, or KTA, which creates the specific rules for the nuclear industry); and independent expert organizations, including the Society for Plant and Reactor Safety (Gesellschaft für Anlagen- und Reaktorsicherheit), the Material Testing Institutes (Materialprüfungsanstalten, institutes connected with universities—notably the University of Stuttgart—that, among other things, test reactor and nuclear power plant components), and the Technical Inspection Association (Technischer Überwachungsverein, or TÜV, whose affiliates are independent consultants that inspect and monitor all sorts of equipment, including that used in nuclear power plants). To this list one could add the state-funded research institutes such as the nuclear research centers in Karlsruhe and Jülich.
5. Countess Marion Dönhoff, "Der Rechtsstaat in Gefahr?" *Die Zeit* (6 March 1981).
6. Weiss, "Nuclear Research"; Reichert, *Kernenergiewirtschaft.*
7. Peter Liewers, Johannes Abele, and Gerhard Barkleit, eds. *Zur Geschichte der Kernenergie in der DDR* (Frankfurt am Main: Peter Lang, 2000); Johannes Abele, *Kernkraft in der DDR: Zwischen nationaler Industriepolitik und sozialistischer Zusammenarbeit 1963–1990* (Dresden: Hannah-Arendt-Institut für Totalitarismusforschung, 2000).
8. Laufs, *Reaktorsicherheit,* 9.
9. Augustine, *Red Prometheus;* Nikolai Krementsov, *Stalinist Science* (Princeton, NJ: Princeton University Press, 1997); Asif Siddiqi, *The Soviet Space Race with Apollo* (Gainesville, FL: University Press of Florida, 2003); and Alexei Kojevnikov, *Stalin's Great Science: The Times and Adventures of Soviet Physicists* (London: Imperial College Press, 2004).
10. John Lancaster, *Engineering Catastrophes: Causes and Effects of Major Accidents,* 3rd ed. (Cambridge: Woodhead; Boca Raton, FL: CRC Press, 2005); Rebecca Solnit, *A*

Paradise Built in Hell: The Extraordinary Communities That Arise in Disaster (New York: Viking, 2009); Lee Clark, *Worst Cases: Terror and Catastrophe in the Popular Imagination* (Chicago, IL: University of Chicago Press, 2006).

11. Thomas Lindenberger, "'Havarie.' Reading East-German Society through the Violenc/se of Things," *Divinatio* 42–43 (2016): 301–369; "Havarie: Die sozialistische Betriebsgemeinschaft im Ausnahmezustand," in *German Zeitgeschichte. Konturen eines Forschungsfeldes*, ed. Thomas Lindenberger, Martin Sabrow (Wallstein: Göttingen 2016), 242–264.

12. Kate Brown, *Plutopia: Nuclear Families, Atomic Cities, and the Great Soviet and American Plutonium Disasters* (Oxford: Oxford University Press, 2012).

13. Radkau, *Aufstieg und Krise*, 71, 147–68, 186–93, 218–84, 364–71.

14. Charles Perrow, *Normal Accidents: Living with High-Risk Technologies* (New York: Basic Books, 1984), 4.

15. Thomas Wellock, "A Figure of Merit: Quantifying the Probability of a Nuclear Reactor Accident," *Technology and Culture* 58, no. 3 (2017): 678–721.

16. Radkau, *Aufstieg und Krise*, 259–64.

17. Laufs, *Reaktorsicherheit*, esp. 212–13, 343, 928–29.

18. Johannes Abele and Eckhard Hamper, "Kernenergiepolitik der DDR," in *Zur Geschichte der Kernenergie in der DDR*, ed. Peter Liewers, Johannes Abele, and Gerhard Barkleit (Frankfurt am Main: Peter Lang, 2000), 49; chapter on 29–89.

19. Bertram Köhler, "Schwerpunkte der Entwicklung im Kraftwerksanlagenbau der DDR," in *Zur Geschichte der Kernenergie in der DDR*, ed. Peter Liewers, Johannes Abele, and Gerhard Barkleit (Frankfurt am Main: Peter Lang, 2000), 149–50; chapter on 115–61.

20. Abele, *Kernkraft in der DDR*, 75.

21. Köhler, "Schwerpunkte," 150.

22. WWER is the German acronym for the *Wasser-Wasser-Energie-Reaktor*, that is, the water-cooled, water-moderated energy reactor. It is also known as the VVER reactor, standing for (in transliterated Russian) *Vodo-vodyanoi energetichesky reaktor*.

23. Bundesarchiv DC/20/12694, 128.

24. Bundesarchiv DC/20/12919, 125.

25. Sonja Schmid, *Producing Power: The Pre-Chernobyl History of the Soviet Nuclear Industry* (Cambridge, MA: MIT Press, 2015), 68, 108–10.

26. Dieter Hoffmann, "Physiker, Kommunist, Atomspion: die drei Leben des Klaus Fuchs (1911–1988)," *Physik-Journal* 11, no. 2 (2012): 43; article on 39–43; S. Collatz, D. Falkenberg, and P. Liewers, "Forschungs- und Entwicklungsarbeiten des Zentralinstituts für Kernforschung Rossendorf zur Kernenergienutzung," in *Zur Geschichte der Kernenergie in der DDR*, ed. Peter Liewers, Johannes Abele, and Gerhard Barkleit (Frankfurt am Main: Peter Lang, 2000), 447–50; article on 411–74. Also Bundesarchiv DC/20/12919, 126.

27. A.O. Adede, *The IAEA Notification and Assistance Conventions in Case of a Nuclear Accident: Landmarks in the Multi-Lateral Treaty-Making Process* (Boston: M. Nijhoff; London: Graham and Trotman, 1987), 139–41.

28. Abele and Hamper, *Kernenergiepolitik*, 64, 77; Alexander Schönherr, "Die ersten vier Blöcke des KKW Gleifswald von der Vorbereitung bis zur Abschaltung," in

Zur Geschichte der Kernenergie in der DDR, ed. Peter Liewers, Johannes Abele, and Gerhard Barkleit (Frankfurt am Main: Peter Lang, 2000), 229; article on 221–308.

29. Abele and Hamper, *Kernenergiepolitik,* 64–70. On the general decline of East German industry, Augustine, *Red Prometheus,* ch. 8.

30. Later known as the Society for Plant and Reactor Safety (Gesellschaft für Anlagen- und Reaktorsicherheit). This was an independent expert organization that prepared studies and collected statistics on behalf of federal German ministries.

31. "Schnell Erlamt," *Der Spiegel* (2 August 1982); Frank Bösch, "Taming Nuclear Power: The Accident Near Harrisburg and the Change in West German and International Nuclear Policy in the 1970s and early 1980s," *German History* 35, no. 1 (2017): 81; article on 71–95.

32. Bundesverband Bürgerinitiativen Umweltschutz, *Unfälle in deutschen Kernkraftwerken: Veröffentlichungen der vertraulichen Störfallberichte der Bundesregierung,* 2nd ed. (Ludwigshafen: Bundesverband Bürgerinitiativen Umweltschutz, 1979).

33. The most severe accidents, which necessitated immediate actions to ensure safety, were placed in category A (later S); Category B (later E) included less serious incidents involving safety issues that had to be addressed, but not immediately. Category C (later N) was for events that did not pose a threat to safety but that disrupted energy production. A fourth category, V, was introduced in 1985.

34. My calculations, based on annual reports found on the Bundesamt für Strahlenschutz website, retrieved 6 September 2016 from http://www.bfs.de/DE/themen/kt/ ereignisse/berichte/jahresberichte/jahresberichte.html.

35. "Übersicht über besondere Vorkommnisse in Kernkraftwerken der Bundesrepublik Deutschland für das Jahr 1986," retrieved 6 September 2016 from http://www.bfs. de/DE/themen/kt/ereignisse/berichte/jahresberichte/jahresberichte.html.

36. "Übersicht über besondere Vorkommnisse in Kernkraftwerken der Bundesrepublik Deutschland in den Jahren 1977 und 1978," 8, retrieved 6 September 2016 from http://www.bfs.de/DE/themen/kt/ereignisse/berichte/jahresberichte/jahresberichte. html.

37. Changes in the categories led to my decision to omit the 1979 report from this analysis.

38. "Auslegung."

39. "Bedienung, Wartung, Reparatur, Montage" in 1980 report.

40. "Betriebsweise" in the 1980 report, "Betriebsweise/Betriebsbedingung" in later reports.

41. The INES scale (International Nuclear and Radiological Event Scale), established by the IAEA (International Atomic Energy Agency) in 1990, rates events involving the civilian use of radioactive substances according to the level of human exposure, environmental contamination, and intactness of barriers and safety systems. See http://www-ns.iaea.org/tech-areas/emergency/ines.asp (retrieved 25 October 2017).

42. The part of the definition of an INES-2 event most relevant to the classification of the 1977 Gundremmingen incident is the following: "Significant failures in safety provisions but with no actual consequences." See http://www-ns.iaea.org/tech-areas/ emergency/ines.asp (accessed on 25 October 2017).

43. The US government was accused of promoting exports of US-manufactured reactors. Euratom collaboration with the United States caused bad blood with the French government. Radkau, *Aufstieg und Krise*, 178–79; 184–85; 199–200; 405.

44. This was a simple reactor type, akin to a large hot coil used to heat water. In most reactors, fission takes place in the nuclear fuel, usually uranium contained in fuel pins, which releases heat and radioactivity. This nuclear chain reaction has to be moderated to the point that no explosion (as in a nuclear bomb) or core meltdown can take place. This is done with fuel rods that are clad, in this case, in boron carbide, which is a moderator in boiling water reactors. When inserted into the nuclear fuel, these fuel rods absorb neutrons, thus lowering the rate at which atoms are split in the uranium. (The nuclear fuel and control rods together form the reactor core.) The reactor core sits in the reactor pressure vessel, which is filled with demineralized water. This water, flowing through what we may call loop A, transfers the heat. First it is heated and turns into steam. This steam powers a conventional turbine, which drives a conventional generator (both outside of the reactor), thus turning heat into electricity. This water is cooled down by water in a separate loop (which we can call loop C), fed by water from a river or the ocean. Thus, water is also a coolant in this kind of reactor. Once cooled and condensed, water from loop A is then pumped back into the reactor pressure vessel, and the cycle begins again. Water in loop A not only transfers heat but also serves as a second moderator whose rate of flow through the reactor can be adjusted. See the diagram in the Wikipedia article on boiling water reactors: https://commons.wikimedia.org/wiki/File:Boiling_water_reactor_no_text.svg. Author: Robert Steffens (alias RobbyBer 8 November 2004), SVG: Marlus_Gancher, Antonsusi (talk) using a file from Marlus_Gancher.

45. Laufs, *Reaktorsicherheit*, 326.

46. "Besondere Vorfälle in Kernkraftswerken in der Bundesrepublik Deutschland. Berichtszeitraum 1965–1976," dated July 1977. Retrieved 6 September 2016 from Bundesamt für Strahlenschutz, http://www.bfs.de/DE/themen/kt/ereignisse/berichte/jahresberichte/jahresberichte.html.

47. "Datum: 17. Januar 1977 Betr.: Serie, Essay," *Der Spiegel* (17 January 1977); "Störungen in Kernkraftwerken," *Die Zeit* (28 January 1977).

48. Laufs, *Reaktorsicherheit*, 18–24.

49. Robert Hargraves and Ralph Moir, "Liquid Fluoride Thorium Reactors," *American Scientist* 98 (2010), 304–13. Radkau also sees the failure to develop the thorium reactor as a lost opportunity. Radkau, *Aufstieg und Krise*, 258.

50. "Umweltfreundlich in Ballungszentren: Hoffnungen und Fehlschläge beim Hochtemperaturreaktor" and "Kernkraft: Funkelnde Augen," *Der Spiegel* (9 June 1986).

51. Radkau, *Aufstieg und Krise*, 278–81.

52. Anselm Tiggemann, *Die "Achilles-Ferse" der Kernenergie in der Bundesrepublik Deutschland: Zur Kernenergiekontroverse und Geschichte der nuklearen Entsorgung von den Anfängen bis Gorleben 1955 bis 1985* (Lauf an der Pegnitz: Europaforum-Verlag, 2004). On problems with Asse, "Umwelt: Expedition in ein Milliardengrab," *Der Spiegel* 8 (18 February 2013).

53. "Schleichweg zum Atomwaffenstaat?" *Der Spiegel* 46 (10 November 1986).

54. BStU, BV Rostock AKG, Nr. 178, T. 2, 52, 323.

55. Schmid, *Producing Power,* esp. 10–14, 21–22.
56. BStU, MfS-BCD, Nr. 2599, 360. The document refers to "criminal attacks and unauthorized action against a nuclear facility."
57. BStU, BV Rostock AS, Nr. 166, no. 78, 1–45 and BV Rostock AKG, Nr. 178, T. 2, 379–386 and 393–396.
58. BStU, BV Rostock AKG, Nr. 178, T. 2, 314–315.
59. IAEA, "INES: The International Nuclear and Radiological Scale" [brochure], Information Series 08-26941-E (Vienna: IAEA, no date). Retrieved 6 September 2016 from IAEA, https://www.iaea.org/sites/default/files/ines.pdf.
60. BStU MfS ZAIG 2462, 58–60.
61. Abele, *Kernkraft,* 75–77.
62. BStU Rostock AKG, Nr. 178, T. 1, 107.
63. See, for example, Fokko Ukena, "Sozialistische Persönlichkeit: Grundlagen, Ziele, Methoden und Resultate der sozialistischen Persönlichkeitskonzeption in der DDR" (Diss. University of Osnabrück, 1989); F.W. Willmann, "Moralische Gefühle und ihre Bedeutung für die sozialistische Persönlichkeit" (Diss. Karl-Marx-Stadt, Technische Hochschule, 1982).
64. BStU MfS ZAIG Nr. 2624, 67–73.
65. Wiring error/malfunction, operator error, maintenance error, repair error, installation error, or error while starting up.
66. BStU, BV Rostock AKG, Nr. 178, T. 2, 313–15, 323–36. In forty cases, the cause was unknown or the investigation was ongoing. In one case, a mishap was blamed on forces of nature.
67. Bundesarchiv DC/20/12694, 126–31.
68. Ibid., 58.
69. Schmid, *Producing Power,* 109–10.
70. Bundesarchiv DC/20/12629, 97.
71. Bundesarchiv DC/20/12694, 17–22.
72. Bundesarchiv DC/20/12629, 82, 98.
73. Ibid., 57, 122; quotations on 99 and 117.
74. Ibid., 120–22; quotation on 100.
75. Bundesarchiv DC/20/12694, 169.
76. Bundesarchiv DC/20/12629, 70.
77. Ibid., 106–7.
78. Schmid, *Producing Power,* 269, fn. 66.

Chapter 3

Dissenting Voices

The Emergence of Counterexperts in West Germany

A fundamental shift in popular views of nuclear power took place in West Germany in the 1970s, as issues of safety and risk took center stage. This transformation originated in the scientific and engineering communities— physicists and engineers concerned with the safety of nuclear reactors and power plants, on the one hand, and, on the other, medical researchers, radiologists, biophysicists, microbiologists, and chemists focusing on the impact of ionizing radiation[1] on human health. The scientists who uncovered and publicized the risks associated with atomic power often did so at considerable professional cost. They unleashed a transnational process of diffusion of criticism of nuclear power.[2] At the same time, they contributed to a decline in public trust in experts and growing confusion over what was and was not legitimate science.

In West Germany, critical voices in the scientific community and beyond, called "counterexperts" or "antiexperts" (*Gegenexperten*), had difficulties making themselves heard and being taken seriously in the public realm, although the nascent anti–nuclear power movement was very receptive to them. Political, scientific, and industrial elites resisted the popular claim to the right to question the safety of nuclear power, arguing that it was best to leave technology to the "experts." With the support of counterexperts, anti–nuclear power activists were increasingly able to challenge that argument. They became convinced of the importance of employing scientific and technical arguments to make a case against nuclear power. This infusion of science into the anti–nuclear power movement provided core principles and coherence and helped it break out of local and regional bounds and become a national and international movement.

The movement was a vehicle for popularization of science. Activists became convinced of the importance of understanding the underlying scientific and technical issues and attempted to educate their supporters and the public in these areas. This chapter of the movement's history provides important evidence that West Germany's turn away from nuclear power was not founded on a rejection of science. However, as will become evident, counterelites also mounted challenges to mainstream science, questioning the process whereby expert opinion was formulated. What became contentious was not so much the construction of abstract systems of knowledge as the political decisions regarding how knowledge was presented and applied to concrete problems.

In the GDR, the ruling SED long suppressed public discussion of nuclear power safety, and counterexperts only emerged there in the 1980s, as will be discussed in Chapter 7.

Experts and Counterexperts on Radiation and Human Health

One strand of criticism of nuclear power grew out of biomedical and radiological research. According to historian Alexander von Schwerin, the concept of the "endangered body" emerged in the nineteenth century with the rise of industrial society, science, public health, national insurance systems, and the interventionist state. Hiroshima and Nagasaki presented scientists and the public with a new level of danger to the human body. Increasing numbers of people were being regularly exposed to radiation as technologies of the Atomic Age proliferated in medicine, industry, and atomic energy production. Major resources were allocated for research on the health impact of radiation across the industrialized world.[3]

In the years after World War II, scientists understood the impact of radiation as primarily genetic in nature. This is not surprising, given the worldwide preoccupation with eugenics in the first half of the twentieth century. A US study in Hiroshima, begun in 1947 and overseen by the US Atomic Energy Commission (AEC), looked primarily for inheritable defects among the offspring of Hiroshima bomb survivors.[4] US geneticist Hermann Muller (1890–1967), who won the Nobel Prize in 1946 for his work on genetic mutations caused by x-rays, spoke during his acceptance speech and on later occasions about the perils of genetic degeneration of human populations due to radiation exposure. Muller was an unabashed proponent of negative eugenics (or weeding out of negative traits, including those resulting from radiation exposure), although only on a voluntary basis. The West German scientific establishment also worked within a eugenic paradigm, while at the same time breaking with the form that eugenics had taken in the Nazi era.[5]

Is there such a thing as a safe level of radiation exposure? Scientific thinking on this question was changing in this period. Geneticists adhered to a linear, nonthreshold model of genetic mutation resulting from radiation exposure. This meant that the risk of mutation increased proportionally with increasing levels of exposure and that there was no minimum level below which the risk of mutation was zero. By contrast, most experts believed, at least until the early 1950s, that there was a threshold for "somatic effects," that is, impacts on the health of exposed individuals, such as radiation sickness and cancer. In other words, they were convinced that low doses of radiation could not, for example, cause cancer. Partly because of this view, and partly due to pragmatic considerations and political expedience, nuclear standards had for decades been based on a threshold model that assumed that it was safe for humans to be exposed to radiation below a certain threshold level, called the "tolerance dose" (*Toleranzdosis*).[6]

These standards, developed since the 1930s, were initially based on observations of radiation burns, later on animal experiments, and even on human experiments. They were expressed in units of measure in an attempt to measure the biological impact of ionizing radiation exposure, based on the kinds of rays and energies involved and the distance from the source. The "rem" ("roentgen equivalent in man") and millirem (one-thousandth of a rem) were generally used in the United States, the "sievert" (named after physicist and medical research Rolf Sievert) and millisievert in Europe.[7] A distinction was made between whole-body exposure and exposure of individual organs. In the United States, the AEC established five rems per year as the upper limit for whole-body exposure to radiation as part of occupational duties and one-tenth this level (0.5 rems per year) for the general population in the mid-1950s.[8] Euratom also adopted this norm.[9]

The *Lucky Dragon* incident of March 1954, in which Japanese fishermen aboard a boat of this name were sickened or killed by the fallout from a US atomic test on the Bikini Atoll, awakened popular fears concerning aboveground nuclear weapons testing. These tests, which had been going on since 1946, drew scientists to the study of low-level radiation. Their attention was also shifting from inheritable genetic defects caused by radiation to its somatic effects. Research showing a causal relationship between low-level radiation exposure and leukemia was quite controversial because of its implications. When British medical researcher Alice Stewart (1906–2002) demonstrated in work conducted in 1953–1956 a correlation between exposure of fetuses to x-rays and later leukemia, the Medical Research Council, a British government agency, cut off her funding. The significance of her research was not widely acknowledged by the medical community until years later. Her work was instrumental in the discontinuation of the practice of x-raying pregnant women.[10]

Public concern over the health impacts of radiation exposure abated somewhat after the signing of the Nuclear Test Ban Treaty of 1963 but flared up again when another controversy broke out among US scientists in the late 1960s and early 1970s. US physicist and academic Ernest Sternglass widely publicized the link between radioactive fallout and childhood cancer, but the numbers he came up with were thought to be highly inaccurate, even among scientists who believed that radiation exposure carried grave health consequences with it. In their work at the AEC's Lawrence Livermore National Laboratory, US researchers John Gofman (1918–2007) and Arthur Tamplin came up with more conservative, but nonetheless alarming estimates of the impact of radiation exposure.

Scholar Ioanna Semendeferi has argued that they conducted highly valuable research, establishing that low-level radiation damaged chromosomes and linking such damage, not only to leukemia—as was already established—but to a wide variety of other sorts of cancer as well. Their findings indicated that there was no "threshold" of radiation exposure below which there was no risk of cancer; rather, even very low doses carried with them a small risk of cancer.

They argued that AEC standards of "safe" radiation exposure were responsible for fifteen thousand deaths a year in the United States that could be prevented if tighter standards were introduced. In retaliation, the AEC cut off funding to the Lawrence Livermore National Laboratory and waged a publicity campaign against them. Gofman and Tamplin were professionally marginalized and their findings were rejected by many in the scientific community at the time. In the long run, they proved to be quite influential, although the controversy over whether low levels of radiation exposure are harmful continues to the present day.[11] The popular impact of their work was immediate, however. They turned to a popular audience in their 1971 book, *Poisoned Power: The Case against Nuclear Power Plants*.[12] Their work was discussed in the popular press in Germany, even reaching the GDR, although their names were not mentioned there.[13]

In the postwar period, West German scientists conducted important research on radiation protection, genetics, cancer caused by radiation exposure, and other topics relevant to the impact of radiation on the human body but did not publicly express disapproval of radiation standards or nuclear power. Although well versed in the concept that even low levels of radiation exposure could damage DNA, West German scientists habitually deflected qualms about the use of nuclear technologies by referring to natural background radiation, an argument that ignores the fact that radiation exposure is cumulative. Von Schwerin believes they wanted to avoid conflict with the state, in part because of habits acquired during the Nazi period, but they also depended on funding from the state-financed German Research Foundation (Deutsche Forschungsgemeinschaft).[14] US researchers were somewhat more independent because of the US system of private universities, which offered a far larger number of job openings for academics.

Critics of Nuclear Power Plant Safety

Criticism of the prevailing system for nuclear reactor safety and risk calcula-tion helped kick off the anti–nuclear power movement. Here again, US sci-entists led the way. By the mid-1960s, there was considerable internal dissent within the AEC, which was in charge of both promoting nuclear power and ensuring safety. Staff scientists charged that oversight of the nuclear power industry was extremely lax. Of particular concern was the possibil-ity of nuclear reactor accidents, particularly core meltdowns. A 1967 AEC report highlighted problems with emergency reactor core cooling systems. Many scientists feared that a nuclear reactor accident resulting from a loss of coolant could lead to a core meltdown and breach of the containment building. The Union of Concerned Scientists (UCS), founded at MIT (the Massachusetts Institute of Technology) in 1969,[15] assailed the AEC on just this issue.

The AEC nonetheless gave preliminary approval to emergency reactor core cooling systems that researchers at two testing facilities believed were inadequate for a new generation of larger reactors. These scientists, along with members of the UCS and an association of some sixty environmental-ist organizations, testified during AEC hearings held in 1972–1973. Critics did not prevail, but the UCS started up an anti–nuclear power campaign. Increasingly, the AEC faced litigation brought by environmentalist organiza-tions. The then-powerful Joint Committee on Atomic Energy, which con-sisted of members of the US Senate and the US House of Representatives, also held high-profile hearings on nuclear power safety in 1973–1974. The AEC was split in 1974, leading to the creation of the Nuclear Regulatory Commission, which was solely responsible for nuclear power safety.[16]

Several prominent US scientists were scathing in their criticism of the US Nuclear Regulatory Commission's Rasmussen Report (WASH-1400) of 1975, which purported to demonstrate that the risks associated with nuclear power were negligible. One of these was Henry Kendall, a nuclear physicist, professor at MIT, founder of the UCS, then-advisor to Ralph Nader, and later Nobel laureate in physics. At his behest, a NASA scientist well versed in event tree methodology testified before the California legislature that this method could not be used to calculate overall risk, as the Rasmussen Report attempted to do. An entire "generation of activist scientists" sprung up in the United States in the 1960s.[17]

These critics of nuclear power safety attained considerable public visibility in the United States. For example, in a *New York Times* Sunday magazine section article, Kendall was quoted in 1974 as saying:

> The radioactive accumulation in a large power reactor is equivalent to the fallout from thousands of Hiroshima-size nuclear weapons ... Consider, for example, that 20 per cent of a reactor's radioactive material is gaseous in normal circumstances and, if released to the environment in one way or the other, could be swept along by the winds for many tens of miles to expose people outside of the reactor site boundaries to what could be lethal amounts of radioactivity. The lethal distance may approach 100 miles.[18]

The warnings of US experts such as Kendall were well known in West Germany, thanks to media coverage.[19] However unrealistic the scenario described by Kendall would appear to be given experiences in the ensuing decades, the important point here is that some of the more extreme statements made by nuclear power opponents in West Germany were not based on irrational imaginings but on public statements of scientists, generally from the United States.

West German scientists faced tremendous barriers to taking on public roles as opponents of nuclear power. West German physicists did become involved in high-profile activism against nuclear war in the 1950s and 1960s—for example in the Pugwash movement—but overwhelmingly supported "peaceful uses of the atom."[20] One such exceptional figure was Karl Bechert (1901–1981), a theoretical physicist who combined a top-flight academic career with a political career in a way unusual in Germany, serving as a member of the West German *Bundestag*, that is, the national parliament, from 1957 to 1972. In 1956, he brought one researcher's discovery of high levels of radioactive contamination on fields and in milk and other food products in Bavaria to the attention of the public. Atomic Minister Franz-Josef Strauß called the director of the institute where the researcher worked to complain.[21] Such incidents were exceptional, and they illustrate the political alarm that greeted criticism of nuclear power in that era.

Jens Scheer (1935–1994), a physicist and prominent anti–atomic power activist in the 1970s and 1980s, recalled that as a young professor in Bremen in the early 1970s, he was not acquainted with criticism of nuclear power safety: "The few scientific studies from the United States were not published for a long time because they were unwelcomed, and that's why [those findings] didn't make their way into textbooks." [22] Initially, he was a proponent of nuclear power. However, student activists in Bremen motivated him to read this critical literature. Also important was his direct contact with US opponents of nuclear power: "I visited people in America and [it] changed my mind." [23] A member of the KPD (Kommunistische Partei Deutschlands, or Communist Party of Germany), Scheer was suspended from his professorship from 1975 to 1980 under the West German Radicals' Decree of 1972 and 1976 prohibition of speech advocating violence. [24]

Scientists involved in the development of nuclear power faced particular scrutiny. In 1975, the Verfassungsschutz (the Office for the Protection of the Constitution, the West German equivalent of the FBI) conducted illegal surveillance of technical expert and nuclear manager Klaus Traube (born in 1928). An engineer, he worked for years in the development of nuclear power in West Germany and the United States. During his years as the director of Interatom, an important enterprise in the field of reactor construction, he was in charge of the development of the fast-breeder reactor in Kalkar. Minister of the Interior Werner Maihofer approved the surveillance operation because Traube had friendly contacts with a person connected to the Red Army Faction (also known as the Baader-Meinhof Gang). This operation, uncovered by *Der Spiegel* in 1977, generated much outcry because it violated constitutionally guaranteed rights. As it turned out, Traube was not involved in any terrorist plot, but he nonetheless left the nuclear industry and became a prominent anti–nuclear power activist. [25]

On the other hand, anti–nuclear power activists in West Germany initially faced high hurdles in terms of understanding the scientific issues and incorporating them into their movement. The movement grew out of local protests by the residents of areas where there were plans to build nuclear power plants. They represented a cross-section of the population. In the movement against the building of an atomic power plant in Wyhl, West Germany (to be discussed in greater depth in Chapter 4), people with something of a scientific education did play a certain role, as historian Stephen Milder has shown. Among those who started asking questions about the possible health consequences of an atomic power plant in Wyhl were Engelhard Bühler, a geneticist who had worked in the Kaiser Wilhelm Society during the Nazi period, and a group of young chemistry students at the University of Freiburg. However, most members of the citizens' initiative that provided the organizational backbone of the movement were not initially interested in the effects of low-dose radiation exposure or the risk of a nuclear power plant accident. They joined the movement out of concern for the impact of the proposed plant on agriculture. [26]

However, circumstances pushed them to seek out experts who could muster scientific arguments, speak with the media, and testify on their behalf at government hearings as well as in court.[27] This need became urgent when the Freiburg administrative court agreed to hear a suit to halt construction of the Wyhl nuclear power plant in 1977. A major anti–nuclear power umbrella organization, the Bundesverband Bürgerinitiativen Umweltschutz (BBU, or Federal Association of Citizens' Initiatives on Environmental Protection) very much bolstered the case of the opponents of the Wyhl plant by publishing results of a 1976 study of the Institute for Reactor Safety (Institut für Reaktorsicherheit, or IRS),[28] IRS Report 290, commissioned by the Federal German Ministry of the Interior.

The study's most important conclusion was that in the hypothetical case of the worst imaginable accident—involving total ejection of nuclear material from a breached reactor—anyone within a 100-kilometer radius of the plant would receive a lethal dose of radiation. The IRS found this to be an unrealistic scenario. The IRS conducted a second study that assumed an accident of far lesser proportions and that yielded far fewer potential fatalities. The ministry refused to release either of the reports, but the BBU got copies through backdoor channels. The BBU published IRS Report 290 in full and asserted, based on the calculations in that study, that 30.5 million West German residents might die in such an accident. Report 290 became the point of departure for questioning expert witnesses before the administrative court.[29]

Before the court, Wyhl activists felt outmanned and outgunned. "There are hardly any nuclear power opponents who have the necessary specialized knowledge," one activist lamented.[30] The citizens' initiatives actually brought over Ernest Sternglass and former AEC staff scientist Robert Pollard from the United States to testify, paying their way with contributions made by local citizens. Evidently, attempts to form alliances with professors from the nearby University of Freiburg had not been very successful. One professor backed out of testifying in court at the last minute. Of the fifty-three experts who testified, forty-two were considered proponents of nuclear power by the activists. Scientists not pegged as allies provided the decisive testimony, according to Laufs. While arguing that the scenario described in IRS Report 290 simply could not occur, they described possible situations in which a reactor *could* explode, killing people within fifteen kilometers. As good engineers, they argued in terms of probability, not absolute certainty.

Citing insufficient safety features to prevent the rupture of the reactor core in case of an accident, the court in fact stopped construction in Wyhl.[31] This decision was reversed on appeal in 1982. But what is interesting about the 1977 court case is that it makes clear that the scientific issues

were complex enough that judges could come down on the side of the anti–
nuclear power forces on the basis of testimony of major nuclear experts who
believed nuclear power to be safe, but who revealed details about the safety
of the nuclear power plants that made this assertion appear to be false. It
also marks an important step in the development of the West German anti–
atomic power movement, which was embracing science to an ever-greater
extent and trying to establish ties to experts who could help them in their
fight.

In 1977, activists from the Wyhl movement founded the Eco Institute
of Freiburg (Öko-Institut Freiburg), whose purpose was to provide the
movement with scientific expertise.[32] In 1980, this institute published
a translation of the UCS's rebuttal to the 1975 US Nuclear Regulatory
Commission's Rasmussen Report (WASH-1400).[33] This report analyzed
the various possible failures that could lead to an accident in a light water
reactor and concluded that the chances of a reactor core meltdown was one
in twenty thousand per reactor per year. Nevertheless, it concluded that the
risks of injury, death, and financial loss were far less in the case of a nuclear
power plant accident than in the case of more everyday disasters, such as
fire, earthquake, and airplane crash.[34] On the basis of documents obtained
through a Freedom of Information Act request, the UCS concluded that the
scientists involved in WASH-1400 had strongly tilted the results of their
research in a direction beneficial to industry—and it is this rebuttal that the
Eco Institute published.[35] Der Spiegel noted that US scientists such as Harold
Lewis, and even the US Nuclear Regulatory Commission itself, found that
WASH-1400 greatly understated the safety risks.[36]

Scientific ways of arguing fundamentally remolded the anti–nuclear
power movement in the 1970s. Student groups studied the scientific issues
in depth and self-published brochures on the scientific arguments against
nuclear power.[37] Previously, as shown in Chapter 1, arguments against
nuclear power had been characterized by "moralizing pathos" and cultural
arguments rooted in antimodern German intellectual traditions.[38] Now,
science became central to opposition to nuclear power.

The first truly scientifically argued anti–nuclear power tract that was
widely read in Germany was Holger Strohm's Friedlich in die Katastrophe
(Peaceably into the Catastrophe), which he published in 1973. It became
a very influential work in the anti–nuclear power movement and went
through numerous reprintings and new editions.[39] The book was based on
a wide variety of literature, not only studies written by US scientists who
were opposed to nuclear power but also official reports from West Germany,
Canada, and the United States. He pointed to pollution problems caused
by uranium mining as well as lung cancer cases among uranium miners.

He explained the way nuclear power plants of the various sorts functioned, pointing to major vulnerabilities: the possibility that an earthquake, sabotage, or human error could cause the cooling system to fail, leading to a core meltdown; embrittlement of the reactor pressure vessel; insufficiencies of emergency cooling systems; the possibility of nuclear leaks into the atmosphere and water; and defective fuel elements. All these problems had been encountered or at least discussed in the United States. In explaining the impact of radiation on the human body, Strohm embraced the no-threshold model, citing not only Gofman and Tamplin but also other participants in debates over low-level radiation and radiation standards.

Economic arguments also played a role in the book, which criticized the subsidies and other burdens taxpayers were forced to shoulder; these would be generated by high research and development costs, pollution, nuclear waste, the unwillingness of insurance companies to issue policies that would cover reactor catastrophes, and the general unprofitability of the nuclear industry. The book's principal political argument, based on the work of US public advocate Ralph Nader, was that governments that had made major investments in nuclear power could not be trusted to honestly face safety problems associated with it. Strohm's account was polemical, but it cited the arguments of government officials, the IAEA, and nuclear power advocates.

It also remained within the boundaries of the scientific debate of that era, and it provided anti–nuclear power activists with scientifically tenable arguments. At the same time, Strohm was not always able to distinguish between reasonable hypotheses and untenable theories that found no support among scientists. Sternglass, to name an example of a counterexpert cited by Strohm, was more of a media phenomenon than a serious scientist.

Graduating from college with a degree in production engineering, Strohm was not an expert in nuclear power. However, his education and a brief career in industry gave him enough of a grasp of the technological issues that he felt able to take on "the virtually united front of experts from industry, the scientific community and politics," which supported nuclear power. Along with physics professor Dieter von Ehrenstein—one of the few university professors who took a public stand against atomic power—Strohm testified before a committee on environmental protection of the Bundestag in 1972.[40] In the mid-1970s, other counterexperts also published numerous books that showcased scientific and technical arguments against nuclear power.[41]

Only slowly were opponents of nuclear power in the scientific community able to overcome the skepticism of the media and gain full recognition as experts in West Germany. Counterexperts enjoyed a triumph at the government's Gorleben hearings of 1979, which stopped the building of a nuclear reprocessing plant there.[42] The 1980 Bundestag Commission of Inquiry on Future Energy Policies paid considerable attention to dissenting views among scientific and technical experts.[43] By the early 1980s, journalists, critical scientists, and other counterexperts were arguing that the main force supporting nuclear power was not science but politics.[44] Counterexperts were fully embraced by the media and the public after the Chernobyl catastrophe in 1986. This was a democratizing development, but one that threatened to create uncertainty regarding who could or could not speak in the name of science.

Educating the Public

For the anti–nuclear power movement, science was not something for the elite but for all citizens, who should be empowered through education. Educating the public was important, one anti–nuclear power group asserted in 1977, because "every additional bit of information limits the possibility of manipulation and strengthens resistance against the atomic state."[45] In his 1977 book *Der Atomstaat* (The Atomic State), public intellectual and journalist Robert Jungk (1913–1994) posited that the buildup of nuclear weapons and nuclear power taking place across the Western world would necessitate a level of security that would destroy democracy and lead to the rise of a dictatorial security state.[46]

He saw comprehension of scientific issues as crucial in the fight to defend democracy, asking in 1975, "Should journalists study physics?" a question that he answered affirmatively. In fact, he was a regular contributor to *Bild der Wissenschaft*, a popular science magazine intended for a broad public.[47] Many activists also believed that science could be used to counteract what they saw as propaganda and manipulation on the part of the energy industry and the state. This preoccupation with mass manipulation, so typical of the 1960s and 1970s, can be traced back to the writings of public intellectuals and philosophers such as Theodor Adorno and Jürgen Habermas.[48]

A good example of science popularization by activists is a 1977 publication of the Hamburg chapter of the Bürgerinitiative Umweltschutz Unterelbe (Lower Elbe Citizens' Initiative on Environmental Protection, or BUU), the main organization behind the campaigns against the building of a nuclear power plant in Brokdorf (a town in the north of West Germany). This self-published booklet explained the science, economics, and politics of nuclear power in just under 150 pages. Written in clear, straightforward prose and illustrated with easy to understand hand-drawn diagrams and politically pointed cartoons, "Atomkraft Nein Danke: Informationen zur

Atomenergie" (Atomic Power No Thanks: Information on Atomic Energy) began its exploration of the physics behind nuclear power with simple explanations of the atom, matter, compounds, isotopes, nuclear fission, and radiation.

The booklet went into considerable depth on how the major kinds of modern nuclear power plants functioned and what could go wrong with them. It was particularly critical of breeder reactors, which in the 1970s were thought to be the reactors of the future because they recycled nuclear fuel, creating plutonium as an end product. However, the booklet pointed out, the sodium they used as a coolant could cause an explosion if it came into contact with air and water. It speculated that an uncontrolled critical mass in the reactor core of this kind of reactor could unleash a nuclear explosion that could be "twenty times as strong as the Hiroshima bomb."[49] This claim appears to come—directly or indirectly—from the US literature, in particular concerning a near disaster in the Detroit Fermi Reactor in 1966.[50]

The brochure also discussed the safety problems connected with the nuclear reprocessing and disposal of nuclear waste, referring to accidents in the United States and citing works by Jens Scheer and Holger Strohm. The booklet discussed the impact of radiation on human health in much the same vein as the body of scientific literature that asserted the validity of the linear nonthreshold model, which proposed that no amount of radioactivity was safe and that there was a direct correlation between the amount of radiation exposure and risk of chromosomal damage that could lead to cancer.

The nuclear power industry and state governments had been trying to counteract the effect of the anti–nuclear power movement and the publications of counterexperts with books, articles, and brochures of their own.[51] The BBU sought to systematically refute these pro–nuclear power publications and unmask what it considered to be dishonest propaganda. A 1975 BBU brochure in its series "Informationen zur Atomenergie" (Information on Atomic Energy) illustrates the lines of argumentation the organization used and made available to BBU members and those who visited BBU events.

One article in this brochure sought to unmask manipulative techniques developed by a Hamburg public relations firm working for the nuclear industry, analyzing the firm's four-point plan: it employed "education" to "conquer fears"; "dramatization" of the problems that would follow if nuclear power plants were not built to "cover up present fears with fears for the future"; "concealment" and "downplaying of problems" to minimize problems with nuclear power; and "improving the image" of nuclear power to "negate fears and build up a positive image" of nuclear power. The BBU newsletter also pointed out infantilizing use of language in pro–nuclear power literature, as well as coffee and cake events at which politicians, members of the clergy, physicians, and other opinion makers were presented with pro–nuclear power arguments.[52]

The Aktionsgemeinschaft für Umweltschutz Darmstadt (Action Committee for Environmental Protection of Darmstadt, West Germany) published a booklet that analyzed—mainly from a scientific and technical point of view—an exhibition about nuclear power at the information center of the Rhenish-Westphalian Power Corporation (Rheinisch-Westfälisches Elektrizitätswerk AG, or RWE) in Biblis.[53] It went through the exhibit, panel by panel, paragraph by paragraph, using simple, clear language and illustrations. For every point made in the energy company's presentation, this booklet had critical points to make, based on sound scientific and technical reasoning.

For example, in a section on problems with disposal of nuclear waste, it pointed to accidents in the transportation of nuclear waste, water damage to salt mines and salt stocks where waste was supposed to be stored, corrosion of spent nuclear fuel elements, and the instability of borosilicate glass containers for nuclear waste. Since no permanent waste depositories were available, much nuclear waste was stored at nuclear power plants, which greatly increased the risk of accidents at the plants. The brochure called nuclear waste the "skeleton in the closet of industrial society" that could never truly be disposed of.[54]

It became a point of honor among opponents of nuclear power to become acquainted with the scientific and technical arguments against nuclear power. Even publications as political in nature as the Communist League's *Arbeiterkampf* and *Graswurzelrevolution,* a periodical with a non-violent leftist orientation, felt it necessary to provide explanations of the scientific issues.[55]

Popularization was also a linguistic process. The anti–nuclear power movement appropriated, redefined, and popularized scientific terminology. In claiming scientific arguments as their own and appropriating scientific terminology, counterexperts, activists, the popular media, and the public remolded these concepts, turning them into arguments against nuclear power. This can be seen in the evolution of the term "GAU" (*größter anzunehmender Unfall,* or "maximum credible accident"). "Maximum credible accident" was a technical term developed by the Reactor Safeguards Committee of the US AEC, which was headed by Edward Teller in 1947–1953. It meant the worst accident that engineers thought possible—a kind of thought experiment to be used to develop technical means of preventing such an accident.[56] In West Germany, the concept became enmeshed in bureaucratic procedures.

West German authorities were willing to give energy companies permission to operate a nuclear power plant only if they provided assurance that this worst case scenario would not involve the release of high levels of radioactivity.

However, engineers complained that they could not, on the basis of their expertise, exclude the possibility of an out-of-control accident with absolute certainty. Such uncertainties, along with the subjective nature of defining a GAU, became fodder for opponents of nuclear power. In popular parlance, "Super-GAU" came to mean a core meltdown with catastrophic consequences for the environment. "GAU" and "Super-GAU" were widely used as metaphors for major catastrophes that had nothing to do with nuclear power, ensuring their place in the West German imagination. They proved to be far more long-lived and durable than the equivalent term "China syndrome"—taken from the name of a film—in the English-speaking world.[57]

According to scholar Matthias Jung, the term *Restrisiko* (residual risk) entered public life as part of political debate. In 1970, the West German science minister rejected a proposal to place a nuclear power plant in an urban area, arguing that the residual risk connected with human error, lack of experience, and statistical uncertainties were too great to allow the project to go forward. From the start, therefore, the question was how much of a residual risk was tolerable. This shifted the nuclear power debate "from the technical to the ethical and moral level," according to Jung.[58] Engineers adapted the concept to their own needs, defining it as an unforeseeable and unavoidable risk.

The courts also found it useful. The concept played a central role in the 1978 decision of the Federal Constitutional Court to allow the construction of a fast-breeder reactor in Kalkar, West Germany. The court argued that it was not humanly possible to guarantee absolute safety but that it was reasonable to expect the public to accept the residual risk connected with nuclear power.[59] According to legal expert Heike Mrasek-Robor, the state was required under the Atomic Law to intervene in the case of a danger to public safety. The German courts did not consider a technology to pose a danger if technical and scientific experts agreed that the risk of a major accident was "virtually" nonexistent. This is what the Federal Constitutional Court determined in the Kalkar case.[60] Nuclear power opponents vehemently disagreed. Judicial and professional attempts to turn around the meaning of "residual risk" were a failure and indeed engendered a sense of injustice and fear of collusion between the state and the nuclear industry that was ably deployed in numerous articles and speeches.

Conclusion

The nuclear power movement emerged, not from a rejection of science but from scientific controversies. Atmospheric nuclear weapons testing stimulated research and debate on low-level radiation. Many scientists came to accept the linear, no-threshold model with regard to genetic effects of radiation and somatic effects such as cancer. Generally, it was not the scientist working quietly in his or her laboratory or presenting research results at professional conferences who encountered professional difficulties as a result of such work. Rather, the scientists who had difficulties were the rare ones who addressed the public, criticized regulatory agencies, and extrapolated from research that nuclear power could pose public health problems.

Much of the early criticism of reactor safety came from the ranks of scientists working for the US AEC. This was above all a US phenomenon, made possible by the existence of private universities such as MIT and the tenure system, which protected scientists from dismissal. Conflicts over the Vietnam War doubtlessly contributed to the rebelliousness of US scientists. By contrast, the older generation of West German scientists involved in research on nuclear power and radiation had been conditioned by experiences during the Nazi regime to remain circumspect.

Since the experts largely supported nuclear power, it was left to the West German counterexperts, who had some scientific background but did not have top-flight careers, to spread knowledge and become public spokesmen of the nascent nuclear power movement. Activist leaders quickly realized the importance of making full use of this science-based critique of nuclear power and educating the members of the growing movement and the general public about these issues. They disseminated and popularized concepts such as the idea that there was no "safe" level of exposure to radiation, but rather a continuous risk curve.

Facing stiff public criticism, West German state governments and the energy industry waged major public relations campaigns, but they found they could not easily win science wars over nuclear power. Anti–nuclear power activists insisted that pro–nuclear power experts who laid claim to scientific objectivity were in fact beholden to political and economic elites. Popularization of science provided the anti–nuclear power movement with a set of arguments that could bridge ideological gaps and begin to influence mainstream society.

Notes

1. "Ionizing radiation" is what is commonly referred to simply as radiation, and it will be referred to as such in this chapter. Ionizing radiation is radiation that has enough energy that it can knock an electron out of its shell, leaving the atom charged. There are three kinds: alpha, beta, and gamma radiation. Examples are x-rays and the rays

emitted by radioactive elements such as uranium and plutonium. Examples of *non-ionizing* radiation are microwave radiation, visible light, and infrared.

2. For a discussion of the meaning of "transnational" in this context, see Astrid Mignon Kirchhof and Jan-Henrik Meyer, "Global Protest against Nuclear Power: Transfer and Transnational Exchange in the 1970s and 1980s," *Historical Social Research/ Historische Sozialforschung* 39, no. 1 (2014): 170–73, 175; article on 165–90.

3. Alexander von Schwerin, "Der gefährdete Organismus: Biologie und Regierung der Gefahren am Übergang vom 'Atomzeitalter' zur Umweltpolitik (1950–1970)," in *Wissensobjekt Mensch: Humanwissenschaftliche Praktiken im 20. Jahrhundert,* ed. Florence Vienne and Christina Brandt (Berlin: Kulturverlag Kadmos, 2008), 188–89, 195–96; article on 187–214.

4. The ABCC study was not able to detect significant heritable genetic mutations. M. Susan Lindee, *Suffering Made Real: American Science and the Survivors at Hiroshima* (Chicago, IL: University of Chicago Press, 1994), 60–61, 217–41.

5. Alexander von Schwerin, *Strahlenforschung: Bio- und Risikopolitik der DFG, 1920– 1970* (Stuttgart: Franz Steiner Verlag, 2015), 336–40.

6. Angela Creager, "Radiation, Cancer, and Mutation in the Atomic Age," *Historical Studies in the Natural Sciences* 45, no. 1 (2015): 14–48.

7. A rem was redefined in 1976 as a hundredth of a sievert. Thus, a millirem = 0.01 millisievert. (Previously, the rem was rather elastically defined.) Measures of radioactivity—becquerels and curies—and absorbed radiation—rads and grays—cannot be easily converted into millirems and millisieverts because the impact of radiation on the human body depends on many factors.

8. J. Samuel Walker, *Permissible Dose: A History of Radiation Protection in the Twentieth Century* (Berkeley, CA: University of California Press, 2000), 8–9, 16 (on human experiments conducted in the United States), 20–21, 27. The standards for rems were changed in 1976, making comparisons with older standards difficult.

9. Joachim Radkau, "Der Größte Anzunehmende Unfall," in *Ökologische Erinnerungsorte,* ed. Frank Uekötter (Göttingen: Vandenhoeck & Ruprecht, 2014), 54; article on 50–60.

10. Creager, "Radiation." On Stewart, see Lisa Rumiel, "'Random Murder by Technology': The Role of Scientific and Biomedical Experts in the Anti-Nuclear Movement, 1969–1992" (Ph.D. Dissertation, York University, Canada, 2009), 106, 155, 258; Gayle Greene, *The Woman Who Knew Too Much: Alice Stewart and the Secrets of Radiation* (Ann Arbor, MI: University of Michigan Press, 1999).

11. Rumiel, "Random Murder," 37–39, 105–10; Ioanna Semendeferi, "Legitimating a Nuclear Critic: John Gofman, Radiation Safety, and Cancer Risks," *Historical Studies in the Natural Sciences* 38, no. 2 (2008): 259–301; Radkau, *Aufstieg und Krise,* 436–37; Walker, *Permissible Dose,* 29–37.

12. John Gofman and Arthur Tamplin, *Poisoned Power: The Case against Nuclear Power Plants* (Emmaus, PA: Rodale, 1971).

13. For an example of popularization of Sternglass's work in the GDR, see "Ja zum Leben," NBI 43, 1958, 3–5. For an example of publicizing of Gofman and Tamplin's work in the Federal Republic, see Kai Krüger, "Strom aus der Bombe," *Die Zeit* (21 January 1972).

14. Von Schwerin, *Strahlenforschung,* 325–31, 336–59, 377–79.

15. The original purpose of the USC was to combat the spread of military research at institutions such as MIT. At the time, the Vietnam War was in full swing.
16. Gary L. Downey, "Reproducing Cultural Identity in Negotiating Nuclear Power: The Union of Concerned Scientists and Emergency Core Cooling," *Social Studies of Science* 18, no. 2 (1988): 242–56; article on 231–64; Richard Lyons, "Nuclear Experts Share Doubts on Power Plant Safety," *New York Times* (12 March 1972); Dean Abrahamson: Citizens vs. Atomic Power, *Bulletin of the Atomic Scientists* 29, no. 5 (1973): 43–45; Steven Del Sesto, "Uses of Knowledge and Values in Technical Controversies: The Case of Nuclear Reactor Safety in the US," *Social Studies of Science* 13, no. 3 (1983): 397–99; article on 395–416.
17. Wellock, "Figure of Merit," 705 and 702–7, 710–11.
18. Ralph Lapp, "Nuclear Salvation or Nuclear Folly?" *New York Times* (10 February 1974).
19. "Ein furchterregendes Unterfangen," *Der Spiegel* (21 July 1975).
20. Nehring, "Cold War," 160–66.
21. Radkau, *Aufstieg*, 434–36; Nehring, "Cold War"; "Atomstrahlen: Die Herbst-Wolke," *Der Spiegel* 44 (31 October 1956).
22. Ulrich Stock, "Jens Scheer: Der gefragte Gegner. Der Bremer Physiker und die Macht des Atoms," *Die Zeit* (23 May 1986).
23. Ibid.
24. Ibid. and Joachim Radkau, *Die Ära der Ökologie: Eine Weltgeschichte* (Munich: Beck, 2011), 364–65.
25. Hanshew, *Terror*, 166–67.
26. Milder, "Today the Fish," 34, 36, 44–46, 48, 49, 56, 67.
27. Radkau, *Aufstieg*, 439.
28. The Institute for Reactor Safety became part of the Society for Reactor Safety.
29. Laufs, *Reaktorsicherheit*, 107–8. Some might argue that a military or terrorist attack might precipitate such a catastrophe.
30. "Wyhl: Geballte Ladung," *Der Spiegel* 8 (14 February 1977).
31. Laufs, *Reaktorsicherheit*, 108–13; WDR Historisches Archiv, Archivnr. 0103355, Series: "Vor Ort"; Episode: "Wyhl-Prozess" (first shown on 31 March 1977).
32. Jens Ivo Engels, "'Inkorporierung' und 'Normalisierung' einer Protestbewegung am Beispiel der westdeutschen Umweltproteste in den 1980er Jahren," *Moving the Social: Journal of Social History and the History of Social Movements* 40 (2008): 81–100; retrieved 29 June 2016 from http://moving-the-social.ub.rub.de/index.php/Moving_the_social/issue/view/53. Also Joppke, *Mobilizing*, 126–28, 165–66, 239.
33. For the original report, see U.S. Atomic Energy Commission, *Reactor Safety Study: An Assessment of Accident Risks in U.S. Commercial Nuclear Power Plants: Summary Report* (Washington, DC: U.S. Government Printing Office, 1974). For the Union of Concerned Scientists' critique, see Henry Kendall, Richard Hubbard, and Gregory Minor, *The Risks of Nuclear Power Reactors: A Review of the NRC Reactor Safety Study, WASH-1400 (NUREG-75/014)* (Cambridge, MA: Union of Concerned Scientists, 1977). For the German translation, see Öko-Institut Freiburg, *Die Risiken der Atomkraftwerke: Der Anti-Rasmussen-Report der Union of Concerned Scientists*, trans. Richard Donderer (Fellbach: Adolf Bonz, 1980).
34. Laufs, *Reaktorsicherheit*, 662–69.

35. David Burnham, "Atom Safety Study by U.S. Is Questioned," *New York Times* (27 April 1977).
36. "Kronzeuge der Kernkraft," *Der Spiegel* 20 (14 May 1979). On the Lewis Committee, see Wellock, "Figure of Merit," 707.
37. For example, FU Berlin, UA, APO-Archiv: S, Sig. 037–038. "Schneller Brüter: Eine Informationsbroschüre über Funktionsweise Unfallgefahren Plutonium" (Freiburg: Arbeitskreis Umweltschutz an der Universität Freiburg, 1977).
38. Matthias Jung, *Öffentlichkeit und Sprachwandel: Zur Geschichte des Diskurses um die Atomenergie* (Wiesbaden: Springer Fachmedien, 1994), 84–85.
39. Holger Strohm, *Friedlich in die Katastrophe: Eine Dokumentation über Kernkraftwerke* (Hamburg: Verlag Association, 1973). By the 15th edition, the book had sold 132,000 copies. Holger Strohm, *Friedlich in die Katastrophe: Eine Dokumentation über Kernkraftwerke,* 15th rev. ed. (Frankfurt am Main: Zweitausendeins, 1988).
40. Frank Lübberding, "Kernkraftkritiker der ersten Stunde: Einer steht im Weg," *Frankfurter Allgemeine Zeitung,* 24 April 2011.
41. Arbeitsgruppe "Wiederaufarbeitung" (WAA) and der Universität Bremen, *Atommüll oder Der Abschied von einem teuren Traum* (Reinbek bei Hamburg: Rowohlt Verlag, 1977); Ralph Graeub, *Die sanften Mörder* (Rüschlikon-Zurich, Switzerland: Albert Müller Verlag, 1972); Peter Weish and Eduard Gruber, *Radioaktivität und Umwelt* (Stuttgart: Gustav Fischer Verlag, 1975).
42. Uekötter, *Deutschland in Grün,* 132.
43. Enquete-Kommission des Deutschen Bundestages, *Zukünftige Kernenergie-Politik: Kriterien—Möglichkeiten—Empfehlungen: Bericht der Enquete-Kommission des Deutschen Bundestages* (Bonn: Dt. Bundestag, Presse- und Informationszentrum, 1980).
44. Häusler, *Der Traum,* 149–55.
45. Jung, *Öffentlichkeit und Sprachwandel,* 175.
46. Robert Jungk, *Der Atom-Staat: Vom Fortschritt in die Unmenschlichkeit* (Munich: Kindler, 1977). English translation: Robert Jungk, *The New Tyranny: How Nuclear Power Enslaves Us,* trans. Christopher Trump (New York: F. Jordan Books/Grosset & Dunlap, 1979).
47. Reprinted in Robert Jungk, *Und Wasser bricht den Stein: Streitbare Beiträge zu drängenden Fragen der Zeit,* ed. Marianne Oesterreicher-Mollwo, paperback ed. (Munich: Deutscher Taschenbuchverlag, 1988), 165.
48. Jürgen Habermas, *Strukturwandel der Öffentlichkeit: Untersuchungen zu einer Kategorie der bürgerlichen Gesellschaft* (Neuwied: Luchterhand, 1962); English translation: Jürgen Habermas, *The Structural Transformation of the Public Sphere: An Inquiry into a Category of Bourgeois Society,* trans. Thomas Burger with Frederick Lawrence (Cambridge, MA: MIT Press, 1989), 175–77. Adorno's most important works in this context are: Theodor Adorno and Max Horkheimer, *Dialektik der Aufklärung,* hectograph manuscript, 1944; Theodor Adorno and Max Horkheimer, "The Culture Industry: Enlightenment as Mass Deception," in *Dialectic of Enlightenment: Philosophical Fragment,* trans. Edmund Jephcott (Stanford, CA: Stanford University Press, 2002).
49. FU Berlin, UA, APO-Archiv: S, Sig. 039–040, "Atomkraft Nein Danke. Informationen zur Atomenergie" (Hamburg: BUU Hamburg, 1977).

50. For an example of a book that the BUU team could have gotten its information from, see John Fuller, *We Almost Lost Detroit* (New York: Reader's Digest Press, 1975). The author was a popular nonfiction writer.

51. Der Bundesminister für Forschung und Technologie, Kernenergie. *Eine Bürgerinformation* (Bonn: Author, 1975); Hans Grupe and Winfried Koelzer, *Fragen und Antworten zur Kernenergie* (Bonn: Informationszentrale der Elektrizitätswirtschaft, 1975); Johannes Koppe, *Zum besseren Verständnis der Kernenergie: 66 Fragen: 66 Antworten* (Hamburg: HEW, 1971).

52. FU Berlin, UA, APO-Archiv: S, Sig. 029-032, BBU, "Informationen zur Kernenergie," Number K 8, "Die Atomenergie-Propaganda."

53. Biblis is a town in Germany (then West Germany), about twenty miles from Darmstadt.

54. FU Berlin, UA, APO-Archiv: S, Sig.039–040, Aktionsgemeinschaft für Umweltschutz Darmstadt, "Was nicht im 'Informations'-Zentrum in Biblis zu erfahren ist" (Darmstadt: Aktionsgemeinschaft für Umweltschutz Darmstadt, 1977), 30; also see 27–30.

55. Examples include "Warum werden Kernkraftwerke gebaut?" *Arbeiterkampf* (1 November 1976); "Schnelle Brüter: Die Kernkraftreaktoren die Päng können," *Graswurzelrevolution* 25–26 (1976): 12.

56. Laufs, *Reaktorsicherheit*, 258–259, 264.

57. Jung, *Öffentlichkeit und Sprachwandel*, 17, 70–72, 87–88. The 1979 film *The China Syndrome* describes the possibility of a reactor core meltdown that breaches the floor of the containment structure and pours into the earth, theoretically able to bore its way "all the way to China."

58. Jung, *Öffentlichkeit und Sprachwandel*, 74–75.

59. Laufs, *Reaktorsicherheit*, 116–18.

60. Heike Mrasek-Robor, "Technisches Risiko und Gewaltenteilung" (Dr. iur. Dissertation, University of Bielefeld, 1997), 60.

Chapter 4

From Local Roots to National Prominence

The Struggle over Wyhl

Wyhl was the birthplace of one of the most powerful and influential anti–nuclear power movements in the world. The movement that emerged there made a remarkable leap from the local stage to the national and international arenas; unleashed a fierce national debate about the nature of democracy and the role of experts in modern society; set in motion a wave of intense activism; and induced the media to cross boundaries that had constrained the public sphere since the founding of the Federal Republic of Germany.

Wyhl[1] is a village located in the Kaiserstuhl (Emperor's Chair), a hilly area of the Upper Rhine valley in southwest Baden-Württemberg, a West German state bordering on France. The Vosges mountains shelter the area from rain, giving it a relatively dry, Mediterranean climate atypical of Germany. This climate, along with its volcanic soil, has made it one of the prime wine growing regions of Germany since Roman times. However, to Hans Filbinger, minister-president of the West German state of Baden-Württemberg, it was part of an underdeveloped region whose economy needed to be jump-started with the building of a nuclear power plant.

Vintners were outraged, concerned that the plant's cooling towers could produce fog and rain, which could harm their grapes. Local grassroots environmentalist organizations with cross-border ties to France, long concerned with issues such as chemical pollution in the Rhine River, opposed plans to industrialize this rural area and to build an atomic power plant in Wyhl. They were concerned about the harm that could be done to the biotope of the Kaiserstuhl, a hiker's paradise and home to wild orchids, lizards, and other flora and fauna not found elsewhere in Germany.

What began as a local David and Goliath struggle was catapulted into national and international consciousness by a series of confrontations between authorities and demonstrators, broadcast on national television.

Illustration 4.1 Locals participating in the occupation of the nuclear power plant construction site in Wyhl, West Germany, 1975. HStAS, EA 1/924 Bü 1741. Courtesy of Hauptstaatsarchiv Stuttgart (Provincial Archives of Baden-Württemberg, Main State Archive, Stuttgart).

What viewers saw was police using high-powered water cannons and truncheons in an attempt to end the occupation of the Wyhl nuclear power plant construction site. Peaceful protesters were soaked to the skin on a brutally cold day, 20 February 1975. Police officers dragged many of them off. Among them were many locals, including an elderly woman and a fresh-faced farm woman who cried out in anger and despair before the TV camera. If the women's scarves and the men's porkpie hats and work jackets left any doubt as to their rural identity, their Allemanisch dialect did not. (Illustration 4.1.) These were salt of the earth, tradition-minded Germans whose love of their *Heimat* (home turf) was rooted in an often romanticized history of peasant ties to the land. (West Germans suffered from amnesia regarding the Nazi blood-and-soil variation of this narrative.) Local women who participated in protests did not challenge traditional gender roles.[2]

Minister-President Filbinger tried to blame the resonance of the protests in February 1975 on "communist" television journalists and executives. In the face of a seismic shift in public opinion, he proved no match for West German students and environmentalists, who flooded into Wyhl and participated in a long occupation of the site of the planned atomic power plant. Wyhl activists also mobilized science in a way that provided the movement with fundamental, universal arguments and gave nuclear power a leading role in the breakthrough of environmentalism as a major political cause. Antinuclear protesters in Germany and abroad tried to duplicate the occupation of the Wyhl construction site elsewhere—often without success. However, Wyhl set the stage for popular challenges to technocratic approaches to science and technology policymaking.

Wyhl and the Demise of the Nuclear Consensus: Issues and Actors

Powerful forces were arrayed against the anti–nuclear power movement. West German elites saw the acceleration of the nuclear power program as a way out of the crises caused by the oil embargos of 1973 and 1976. Social Democratic chancellors Willy Brandt (1969–1974) and Helmut Schmidt (1974–1982) very much wanted to develop nuclear power as an alternative to petroleum but also to coal, which was dirty and dangerous, particularly for coal miners, a major constituency of the Social Democratic Party. In a 1974 report for the West German government, experts projected a rise in the use of atomic power from 1 percent of energy sources in 1972 to 15 percent in 1985.[3]

Like most Christian Democratic leaders, Hans Filbinger, in office from 1966 to 1978, embraced this nuclear strategy with particular fervor. He saw the building of an atomic power plant in an "underdeveloped" corner of Baden as essential to the development of his state.[4] In addition, Filbinger was the chair of the board of directors of a utility company involved in building the nuclear power plant and as such was responsible to its shareholders. This conflict of interest was fully tolerated under German law as part of a system of partially state-owned utilities.

The Filbinger government laid claim to the right to decide where to build a nuclear power plant, based on its public mandate, its adherence to the law (in particular the Atomic Law of 1959), and its extensive reliance on experts. It presented itself as objective and rational. Expert reports, solicited by the government, purported to show that there were no major safety, seismological, hydrological, geological, meteorological, or agrometeorological issues that stood in the way of carrying out the project.[5] Filbinger claimed that conditions in Wyhl had been studied by experts "with a thoroughness unique in the world."[6]

This technocratic approach collided head on with a civil society that was rapidly expanding in the 1970s. First were the locals. Organized in

Bürgerinitiativen (citizens' initiatives), these farmers and small-town professionals were generally quite conservative. Although many were Christian
Democrats, they were willing to take on the Christian Democratic state government (Illustration 4.2.). Transborder ties with environmental activists in
France and Switzerland strengthened their resolve. On a December 1974 bus
ride to Stuttgart, where they wanted to meet with politicians, these middle-
aged small-town activists questioned the economic and scientific arguments
brought forth by the government, which they believed was acting in bad
faith: "They're greedy for money. Because it's thinly populated, they think
they can move industry here."

Moreover, they rejected the argument that the building of the nuclear
power plant would create jobs. Lacking in higher qualifications, locals could
at most work as janitors in the plant. Besides, they argued, there was little
joblessness among the rural population. In addition, they argued that due
diligence had not been done in the compiling of expert reports, underesti-
mating the danger of microclimate change that could endanger agriculture.
Claiming a "basic right to *Heimat,* to nature," they saw themselves as protec-
tors of the land: "We have to defend ourselves, for the sake of our children.
The land there is very fertile." After a meeting with unresponsive government
ministers in the state capital, Stuttgart, in December 1974, one Kaiserstuhl
resident exclaimed that the state's stance was a "mockery" and a "dictatorship." Another said, "I'm completely unsatisfied. What they're doing with us
is shameful." Yet another promised a "fight to the finish."[7]

A segment of the Christian Democratic Party of the Kaiserstuhl was
also against the Wyhl project. Alois Schätzle and Albert Burger, Christian
Democrats who represented the Emmendingen district (where Wyhl was
located) in the state assembly and the *Bundestag,* respectively, were profoundly critical of the way the government handled the Wyhl controversy.
In July 1974, they wrote state Economics Minister Rudolf Eberle, a major
proponent of the Wyhl nuclear power plant project, that the population was
more worried than ever. They and citizens who had approached the project
with an open mind were now dead set against it. "The insecurity of the
population and their fears, which cannot be allayed through discussion, have
totally taken over." Schätzle and Burger advocated placing "human beings at
the center of all decisions." "People's misgivings should be respected ... and
not just pushed aside for the sake of economic motives."[8]

Cultural differences may have played a role in the rebelliousness of
some members of the Kaiserstuhl CDU (Christlich Demokratische Union
Deutschlands, Christian Democratic Union of Germany, the more conservative of the major parties in West Germany). The population of the
Kaiserstuhl felt a particular affinity for the French Alsace and Switzerland

Illustration 4.2 Locals participating in a rally during the occupation of the nuclear power plant in Wyhl, West Germany. Annemarie Sacherer at the microphone. HStAS, EA 1/924 Bü 1741. Courtesy of Hauptstaatsarchiv Stuttgart (Provincial Archives of Baden-Württemberg, Main State Archive, Stuttgart).

due to their shared *Allemanisch* dialects and cultural heritage. By contrast, Stuttgart, the state capital of Baden-Württemberg, was culturally *Schwäbisch* (Suebian, a different German dialect and ethnicity). The CDU party organization in Wyhl was split, with the mayor and the majority of the town council in favor of the nuclear power plant, while other CDU members were very critical of the Filbinger government.[9] The Filbinger government angered the CDU party leadership greatly by not consulting with them before the partial authorization of the atomic power plant project in Wyhl on 5 November 1974.[10]

Both old media—notably magazines—and new media—television in particular—helped turn this local dispute into a national one. Here, the break with the Adenauer era and its repressive press politicies is quite evident.[11] At the same time, a tremendous political mobilization of the West German population was taking place. The student movement of the 1960s had drawn large numbers of young people into political activism. The New Left (ranging from orthodox Communists to "undogmatic" and antiauthoritarian socialists) was at first indifferent to environmentalism but discovered nuclear power as a major issue as a result of the Wyhl conflict. Even the Protestant Church was drawn into leftist politics and environmentalism, particularly through the activism of *Evangelische Studentengemeinden,* or Evangelical student congregations at universities.

Beyond these leftist circles, a much broader mobilization of the population was taking place, not least through *Bürgerinitiativen* (citizens' initiatives). These grassroots organizations were often involved in local issues—relating not only to environmental protection but also to traffic, playgrounds, nursery schools, and daycare centers. They drew in participants of all age groups and a broad range of political persuasions. Notable in the Wyhl movement was the balance between people from the region, many of them involved in citizens' initiatives, and outsiders—students, members of leftist organizations, and members of the Bundesverband Bürgerinitiativen Umweltschutz (BBU, or Federal Association of Environmentalist Citizens' Initiatives), a national umbrella organization founded in 1972.

By 1977, the Wyhl movement had moved beyond its local focus and started expressing some of the arguments and concerns that were to carry the West German anti–nuclear power movement to national and international prominence, such as the danger of radioactivity seeping into the area around the nuclear power plants, the risk of accidents, and the long half-lives of many radioactive substances found in nuclear power plants—a problem that would be handed down to future generations. Opponents of the Wyhl project also questioned whether nuclear power made economic sense. They pointed out that atomic power would not make West Germany less dependent on imports and the political pressure that went along with them because uranium had to be imported, often from problematic sources such as the Soviet Union. They also asked whether the Federal Republic's energy needs actually necessitated the expansion of nuclear power.[12]

Compelling although these arguments were, they would not have struck such a chord among West Germans had the nuclear power debate not become entangled with deep-seated preoccupations and controversies of that era. First, the nuclear power debate became intertwined with clashes relating to economic growth and consumption. Germany, but also Europe and the United States, have a long history of antimodernist, anticonsumerist, and antigrowth thought, both on the Left and on the Right.[13] Historian Alexander Sedlmaier has identified anticonsumerism as an important strand in leftist protest culture and revolutionary thought in postwar West Germany, inspired, among others, by Karl Marx, the Frankfurt School, and Herbert Marcuse.

This school's critique of consumerism was above all social, economic, and psychological in nature and rooted in a rejection of capitalism. They saw consumers' quest for material goods as a form of "voluntary servitude." The Red Army Faction, or RAF (popularly known as the Baader-Meinhof Gang) and lesser known groups and the squatters felt that this justified illegal and sometimes violent actions.[14] However, the proliferation of "alternative" lifestyles brought about a reorientation in leftist anticonsumerist thought, taking it in the direction of the search for authenticity, self-improvement, and community building. The personal became political, and the political started with the personal. The countercultural milieu was focused on creating a better society, starting with the individual.[15]

A powerful environmentalist critique of economic growth also came together in this period, spurred by the environmentalist turn, the end of the postwar boom with the oil crisis of 1973, and the 1972 Club of Rome report, *The Limits to Growth*. Using computer simulations, this report predicted that if high rates of economic and population growth were maintained, the result would be resource depletion, environmental disaster, and possible collapse of the world economy.[16] Some on the Left and Left–Center (including a few Social Democrats, such as Erhard Eppler) initially called for "zero growth" in West Germany. However, Social Democrats quickly retreated to a position advocating qualitative growth, or "humane growth," while industry, labor unions, and Christian Democrats rejected any attempts to slow or stop growth.

The far Left and the nascent Green movement also advocated "qualitative growth" and an improved standard of living for the lower echelons of society. Non-German authors, such as Barry Commoner, put forward the idea that growth and environmentalism were compatible as early as 1972. The concept of "sustainability" did not achieve prominence until the late 1980s, however.[17] Another strand in the debates about modernity and progress involved the concept of risk. Was material progress and consumerism worth the risks posed by nuclear power? This question did not attain a central place in debates until 1986, when Ulrich Beck's *Risk Society* appeared, but certainly began to be discussed in the era of Wyhl and Brokdorf.

The second major preoccupation that made the nuclear power controversy so acrimonious concerned the Nazi past and anxieties concerning the future of democracy in West Germany. The political establishment accused the student movement of embracing an antidemocratic tradition going back to the Nazis and—paradoxically—also allying itself with Communism. The mirror image of this view was prevalent among leftists: they viewed the West German elites and the state as the heirs of Germany's fascistic past. Karrin Hanshew has shown that controversies over the nature and viability of democracy in the Federal Republic became an explosive topic in connection with the rise of terrorism. A wave of spectacular attacks by the RAF spawned fears of subversion. While the Christian Democratic Party called for an expansion of state powers, the ruling Social Democratic Party turned to a modernization of policing but also oversaw the promulgation of the Radicals' Decree of 1972 and a ban on advocacy of acts of violence in 1976. Historians disagree over the extent to which the SPD in fact substantially restricted citizens' rights.[18]

Falco Werkentin and other historians have argued that the strategy of government and police in confronting demonstrations in the 1960s and 1970s was molded by an older, authoritarian model of policing, according to which police's prime duty was to protect the state, that is, quell domestic disorder.[19] Some studies have shown fairly clear continuities in policing between the Nazi era and the early Federal Republic, in terms of both personnel and thinking about policing.[20] Nevertheless, a wave of modernization took place from the 1960s to the 1980s involving computer-aided crime prevention strategies, which endangered citizens' rights in a new way.[21]

The police started to develop from the protector of the state to the protector of citizens from crime. However, as historian Klaus Weinhauer has argued, various forces conspired to return the police to their earlier mission to suppress domestic unrest. If anything, police brutality increased. He finds explanations in a conception of masculinity tied to strength and physical force, a sense of solidarity within the police force, anti-Communism, and the 1970s war on terror.[22] In some ways, however, police tactics were not as harsh as in the 1960s. The turning point was the shooting of student Benno Ohnesorg by police at a 1967 demonstration in West Berlin.

The courts had instated standards of reasonable use of force by the police, a principle that was understood to prohibit the use of deadly force, that is, firearms, in quelling even rowdy demonstrations.[23] In general, the police response to the anti–atomic power movement encompassed the use of massive force, intelligence gathering, and cooperative relationships with militarized services, as will be discussed in Chapter 5 in connection with the Brokdorf demonstrations. However, at Wyhl, mainly local and regional police were deployed, resulting in interactions between police and demonstrators that were more varied and subtle than has often been appreciated.

The February 1975 Clashes: Why Did the Police Not Stop the Occupation?

The way police broke up an attempted occupation of the Wyhl nuclear power plant construction site on 20 February 1975 shocked many West Germans. "We order you for the last time to leave the area," a police officer announced over a loudspeaker, just before police moved in to roughly drag off protesters. As far as one can tell from TV footage, occupiers did not use any force or actively try to resist but simply went limp. The police brought in huge, high-powered water cannons, which sprayed demonstrators with great force, completely soaking them and propelling them forward. (See Illustrations 4.3. and 4.4.)

Addressing the police, a young woman let out a plaintive wail in *Allemanisch*-inflected German: "This is shameful. If you had a heart in your chest, you wouldn't do such a thing!" Another young local woman cried out, "We aren't leftist radicals! We are peaceful citizens who are defending ..." Almost sobbing, an elderly woman burst in, "We are inhabitants of your region!" A middle-aged man said to the reporter, "Politicians exploit us, then get rid of us when it's convenient." Annemarie Sacherer, a young winegrower and activist, speaking before the camera in dialect, said she had seen how the police had treated a woman with a child: "So brutal!" she said. It had been "terrible." The police was only there for the powerful ("the big ones"), not for the "ordinary people" (the "little" people). "This is no longer a democracy," she asserted.[24] Police arrested fifty-four protesters and cleared the construction site.[25]

It was the political leadership that was responsible first and foremost for the handling of this and other demonstrations. From the start, the Filbinger government was utterly intransigent concerning the Wyhl project. In accordance with the Atomic Law of 1959, the government conducted public hearings in the region surrounding Wyhl. However, few locals had

Illustration 4.3 Police carry protesters off during an attempted occupation of the Wyhl nuclear power plant construction site, February 1975. Courtesy of Gerhard Auer, Emmendingen.

Illustration 4.4 Police use water cannon against protesters during an attempted occupation of the Wyhl nuclear power plant construction site, February 1975. Courtesy of Gerhard Auer, Emmendingen.

the opportunity to speak at hearings held in Wyhl on 9–10 July 1974. "Tumultuous scenes" ensued as angered citizens expressed their outrage.[26] By the fall of 1974, nine municipalities, 53 organizations, 330 individuals, and 89,400 petition signatories had registered their objections to the project.[27] At a 10 January 1975 hearing, government representatives again tried to greatly limit the audience's participation, taking turns holding long speeches from the podium.

The mayor of Wyhl, Wolfgang Zimmer, declared that this was a town hall meeting *for* the citizens and, by implication, not *by* them. "Don't make me turn off the microphone!" he admonished. A representative of the opposition stood at a microphone for a long time but was not allowed to speak. Discontented locals booed, whistled, and cried out, "This is a town hall meeting! Not a propaganda session!" The meeting ended in uproar.[28] Despite the protests and an impending court case against the project, the government authorized the beginning of construction on part of the plant on 22 January 1975. On 18 February 1975, some three hundred protesters, a large number of them women,[29] overcame police roadblocks and poured onto the Wyhl construction site, halting work on the power plant.

Police responses on 20 February do not so much reflect spontaneous reactions of the police as decisions on the political level. On 16 January 1975, representatives of the economics, state, interior, justice, and labor ministries of Baden-Württemberg had agreed that all legal means should be used to prevent protesters from occupying the construction site. The government tried to warn off protesters: "It should be made clear to the public that the occupations of sites are illegal actions that will be combated relentlessly."[30] The Filbinger cabinet specifically approved the use of roadblocks, barbed wire, water cannons, tear gas, dogs, and truncheons. State security offices, principally the Verfassungsschutz (the Office for the Protection of the Constitution, which conducted surveillance of radicals) were given the task of investigating the background of people considered to be ringleaders. Government officials assumed that most of the activists were fanatics and that some demonstrators would be armed, possibly with firearms—although this would have been extraordinarily unusual in West Germany.[31]

In contrast to government officials, the police was surprisingly cautious. In the weeks leading up to the 20 February clearing of the site, the local police were in close communication with the anti–atomic plant protesters. They arrested some in connection with actions such as blocking government officials' cars. Police repeatedly told occupiers to leave the site. But television cameras also caught protesters (many of them locals) conversing and joking around with police officers, who presumably were also from the Kaiserstuhl.[32] On 18 February, State Police Commissioner Günter Wöhrle recommended waiting to see if peaceful means—specifically, fines—might induce occupiers to go home. Ignoring this suggestion, the Filbinger government ordered the police to clear the site by force on 19 February 1975. After the fact, the police commissioner insisted that this had been a mistake and that it would have been better to watch and wait and to avoid an immediate confrontation.[33]

Police faced a difficult situation when demonstrators returned and occupied the Wyhl site on 23 February 1975. About four hundred police officers faced eight to ten thousand demonstrators, about a thousand of whom decided to storm and occupy the construction site.[34] The state police commissioner and the officer in charge of the deployment, Chief Inspector Schonhardt, had carefully planned and prepared this operation in coordination with the Minister of the Interior, with whom they stayed in contact all day via Telex and telephone.[35] Afterward, the police argued that they could not have foreseen the great numbers of demonstrators and "the massive resistance we would run into."[36] Festivities at the protest site degenerated into a "veritable drinking orgy" when activists brought in barrels of wine, loosening the inhibitions of those assembled.[37] Police unleashed "physical force, truncheons and water cannons" against protesters—to no avail. The police felt they had little choice but to withdraw.[38] Seventeen officers were injured. The number of injuries on the side of the demonstrators is unknown.

Neither the members of the police force sent to the demonstration nor the police officials in charge of the deployment showed any real inclination to take a hard line against the protesters. Perhaps the television programs about Wyhl that aired in January and February and the sympathy they engendered for the Wyhl protesters had something to do with this.[39] Ongoing contacts with the activists also played a role. The two sides engaged in talks before the 23 February demonstration in hopes that it would remain peaceful. [40] On 23 February the police wanted to deploy "a sufficient number, but not too many officers" to avoid provoking the demonstrators.[41] Overwhelmingly, these men were from local and district police forces. Only eighty-five were from the national riot police (*Bereitschaftspolizei*).[42]

Police Commissioner Wöhrle insisted that he had not been aware of the "tremendous importance that the state government attached to this project." He even claimed not to have known that Badenwerk (the corporation in charge of building and operating the nuclear power plant) was partially state owned.[43] Can he really have been so poorly informed, or was he being disingenuous? The latter interpretation seems more convincing, given that he was considered by the news magazine *Die Zeit* to be a "headstrong man" with considerable political skills.[44]

The police showed reluctance to use force on 23 February. Adhering to their understanding of what constituted reasonable force under the circumstances, police officials did not sanction the use of tear gas,[45] although the Filbinger cabinet had specifically given the police permission to do so. Police officers in the field were far less forceful than on 20 February. Many of them had close relationships with locals, who in some cases must have been friends, neighbors, or even relatives. It is true that this had not stopped them from employing harsh methods on 20 February. But it seems likely that the demonstrators' reactions of shock, pain, and anger on 20 February caused the police to hold back on 23 February. According to Wöhrle, "The decisive factor that doomed the police deployment was that not only unkempt young men showed up, but also winegrowers of the Kaiserstuhl."[46] Afterward, police hurled accusations at politicians and officials in Stuttgart "who dragged us into something and ran us into the ground."[47]

This debacle was of crucial importance. If the police had succeeded in ousting the occupiers on 23 February the government's chances of crushing the Wyhl anti–nuclear power plant movement would have been much greater. Instead, activists were able to continue the occupation for another, crucial seven months. The February 1975 demonstrations and clashes with police attracted many students from the nearby University of Freiburg. During the occupation of Wyhl, growing numbers of leftists from across West Germany were attracted to "the cause," which they initially saw as a struggle between a repressive state and citizens fighting for their rights. Gradually, however, the Left became more interested in environmentalist issues. Despite considerable tensions, outsiders and local activists found ways to work together at the Wyhl site in 1975. This led to a tremendous expansion of the anti–nuclear power movement, both in geographical terms and in terms of the issues.[48]

A Local Conflict Becomes a National Media War

The 20 and 23 February demonstrations and the ensuing occupation became a publicity disaster for the Filbinger government. Filbinger immediately went on the offensive, asserting on nationally broadcast evening news (*Tagesschau*) on 24 February 1975, "We want to separate the level-headed population from the extremist forces, which currently are engaged in terrible deeds in many localities, and who have succeeded in inciting the emotions." To convince the population of the necessity of nuclear power plants, he continued, it was necessary to sidestep the activities of the Communists, who were "cooking their own soup" (i.e., they were pursuing their own agenda).[49] Filbinger's repeated allegations that the Wyhl protests had been infiltrated by Communist subversives fit in with a widespread perception that West Germany was under siege by leftist protesters and terrorists.

Baader, Meinhof, and other members of the "first generation" of the Red Army Faction were vociferously protesting their prison conditions in this period. RAF member Holger Meins's hunger strike led to his death on 9 November 1974. On 27 February 1975, Peter Lorenz, the Christian Democratic candidate for mayor of Berlin, was kidnapped by a terrorist group allied with the RAF. The trial of Baader, Meinhof, and others began in Stuttgart, the capital of Baden-Württemberg, in May 1975. It was a wild spectacle, accompanied by the raucous behavior of the spectators, who largely sympathized with the RAF. This upsurge of radicalism reawakened fears of the overthrow of democracy going back to the Weimar era.[50] Thus, Filbinger's denunciation of the anti-Wyhl movement resonated ominously with wide swaths of the West German public.

Filbinger and his ministers also went on the offensive against what they saw as leftist media, expressing outrage at the way the Wyhl conflict was depicted on WDR (Westdeutscher Rundfunk, or West German Broadcasting), one of West Germany's regional television networks. WDR played a crucial role in the Wyhl controversy. Starting in January 1975, a series of documentaries on the conflict in Wyhl was shown as part of the series *Vor Ort* (At the Scene). The intention of this program was to give average citizens the opportunity to air their views on major issues. There was no commentary or voice-over, and even the journalists at the scene remained largely unseen and unheard.

These programs, featuring scenes of politicians' attempts to silence citizens and brutal police treatment of peaceful demonstrators, drew sharp criticism from Filbinger's government and its allies. They argued that every public television program on a political issue should be "balanced," that is, should allow both sides to explain their position. Instead, *At the Scene* highlighted the opinions of protesters and locals. The program's editor answered this critique by pointing out that most political programs on television gave far more air time to politicians and rarely let ordinary citizens express themselves. He felt—and the courts backed up this view—that individual programs did not have to be balanced but rather overall programming of public television stations needed to take viewpoints across the spectrum into account.[51]

By the 1970s, television dominated the media landscape as never before. By 1969, 83 percent of all West German households had a TV set.[52] There were no commercial TV channels, only public broadcasting stations—two national ones and one regional one in each region. Political influence was decisive in the naming of an *Intendant,* or general director, of a public broadcasting station, particularly on the state level. Süddeutscher Rundfunk (SDR, Southern German Broadcasting) and Südwestfunk (SWF, Southwest Broadcasting), both serving states governed by Christian Democrats, had Christian Democratic general directors. By contrast, WDR (Westdeutscher Rundfunk, or West German Broadcasting), the station for North Rhine-Westphalia (governed by the Social Democrats), was accused by conservatives such as Filbinger of being a "red broadcasting network."

However, WDR's director general, Klaus von Bismarck, was a liberal without party affiliation and a high official of the Protestant (Lutheran–Evangelical) Church. As head of WDR, he attempted to chart a politically neutral course. Werner Höfer, WDR programming director, also avoided clear-cut political ties. He saw WDR as a network that should provide a very high level of political and intellectual content.[53]

Filbinger and his ministers objected not only to what they saw as the leftist slant of WDR but also to the supposed emotionalism of its presentation of the Wyhl conflict. In a letter to Bismarck, Gerhard Mahler, state secretary in the state ministry of Baden-Württemberg, called the program "an almost classic case of manipulation." A former Christian Democratic campaign manager, Mahler was considered a heavy hitter in CDU circles. He found it "particularly worrisome and dangerous" that the program "called forth moods and emotions" rather than dealing rationally with the real issues.

He found the foregrounding of the reactions of female protesters particularly manipulative: "Crying women were supposed to make the viewer feel pity and sympathy for the 'oppressed.' Subliminally, [they] were used in a reprehensible manner to stir up public opinion against nuclear power. Crying women who demonstrate and fight against an atomic power plant on behalf of their children cannot be wrong in the eyes of the uninformed viewer." Somewhat menacingly, he interpreted what he called a failure on the part of WDR to promote "formation of objective opinion" a violation of a law that prohibited public broadcasting stations from serving a particular party, interest group, or Weltanschauung (world view).[54]

State Economics Minister Rudolf Eberle wrote of the 26 February program, "I have seldom encountered a case in which journalistic standards of objectivity and fairness were trampled upon so blatantly." He thought it biased that threats against the mayor of Wyhl, Zimmer, and his family were not mentioned. And he was particularly outraged that neither members of state and local government nor representatives of the utility company or the construction company had been asked to comment on the latest events. He complained that the clips had been taken out of context and did not represent what they really wanted to say. "Particularly serious," he wrote, was the broadcasting of this program at a time when "emotions and violence had to be avoided at all costs."[55]

The Filbinger government's attempts to depict the Wyhl movement as overly emotional was disingenuous. Filbinger and his colleagues bolstered their technocratic arguments in favor of atomic power with emotional appeals. In particular, they had developed arguments and imagery grounded in a sense of crisis and loss of security. Recalling the oil embargo of 1973, state Economics Minister Rudolf Eberle wrote on the day after the state government officially gave permission to build a nuclear power plant in Wyhl (5 November 1974): "Do you remember? ... the anxiety about whether you could keep your home warm, the worries about losing your job, the concerns for the future?"[56] Famously, Filbinger made a more spectacular emotional appeal to the population on 27 February 1975, when he claimed, "Without the atomic power plant in Wyhl, the lights will start going out in Baden-Württemberg by the end of the decade."[57] However, neither Filbinger nor his supporters admitted that these were emotional arguments.

A more conservative television network, SWF, also adopted this vilification of the emotional content of opposition to the Wyhl project. A report on Wyhl on the *Evening Journal* of SWF from 14 January 1975 described protesters at a hearing as "shouting down government representatives" and mentioned that the mayor of Wyhl had received death threats. A voice-over described a July 1974 meeting in terms that discounted the arguments of the opposition as pure emotion: "In the heated atmosphere created by the churches, objectivity was lost in a chaos of emotions."[58]

Some newspapers also went along with the Filbinger government's condemnation of WDR coverage of Wyhl as emotional and ideologically slanted. For example, Journalist Birgit Weidinger wrote in the relatively liberal *Süddeutsche Zeitung,* "Pictures of policemen in helmets, moving in on a crouching group of occupiers, elicit emotional reactions rather than the formation of opinion."[59] However, twelve of nineteen articles in prominent newspapers praised WDR for upholding freedom of opinion.[60]

At Eberle's behest, the program was brought up and severely criticized at meetings of the ARD, the consortium of public broadcasting stations, which exercised oversight over its members.[61] A public statement of the general directors asserted "that purely original sound features are limited in their ability to convey factual accounts. The danger of falsely presenting emotions as objective statements is particularly great in a feature of this kind."[62]

The insertion of emotion into the debates greatly complicated, but in some ways enriched, this conflict. For the opposition, emotion stood for ethical commitment and authenticity. "It is completely legitimate to show emotions," argued WDR television journalist Hans-Gerd Wiegand.[63] WDR defended itself well against its critics, and *At the Scene* eventually won a coveted prize.[64] This shift reflects a sea change in the perception of the role of emotions in public life. According to Frank Biess, the emotional repression of the 1950s gave way to a valorization of emotional expression in the 1960s. The result was a "clash of two very different emotional cultures"—that of the emotionally repressed older generation and that of the emotionally expressive younger generation.[65]

The ultimate showdown between the government and its opponents was a fight for public support. Thanks to the diligence of the WDR archive, we know a surprising amount about how the public reacted to this controversy. Of the one hundred surviving cards and letters sent to WDR in response to *At the Scene* programs on Wyhl, only a sixth were negative.[66] It is clear that both WDR and the Filbinger government took these viewer responses very seriously. *At the Scene* editor Wiegand answered a good many of the critical letters, trying to convince the authors that they were mistaken in their negative judgments of the program.

More surprisingly, a state ministry official wrote a letter to a viewer who had complained about the WDR program, expressing thanks in Filbinger's name.[67] Echoing assertions made by the Filbinger government, cards and letters from critics of the program on Wyhl accused WDR of bias and emotionalism. A viewer from Baden-Baden wrote, "Not once was even a minimal attempt made to present the standpoints of opponents and proponents [of the Wyhl nuclear power plant project]." He went on to say that he was not necessarily in favor of nuclear power.[68] Writing in the same vein, a man from Bonn called *At the Scene* a "blatantly red election locomotive."[69]

The anger springing from the generational conflict of the 1960s found its way into some letters. A viewer in Wuppertal wrote of WDR journalists, "These gentlemen remind me of my sons during puberty, when they were spouting phrases about industry bosses and managers. They should try doing some real work in industry … Or are you getting the German people ready for the gray misery of socialism?" He felt that the *At the Scene* program of 26 February misinterpreted the problems of West German society. In his mind, the kidnapping of politician Peter Lorenz by terrorists the next day illustrated this: "Do you still think it's a good idea to attack the police? Do you want to weaken the authority of the state even more?"[70]

Several letter writers critical of WDR defended the actions of the police as necessitated by the behavior of the demonstrators.[71] A woman from Weil am Rhein also emphasized the need for law and order, which was central to her understanding of democracy: "When anarchism gains the upper hand (as has been the case since 1969), democracy is doomed."[72] A man from Jülich found the methods of the Wyhl protesters reminiscent of those of the

Nazis—an argument that had often been leveled against activists from the 1968 student movement.[73]

Supporters of the Wyhl movement and WDR displayed very different understandings of objectivity, democracy, and security than did the critics. Many believed that *At the Scene* had redressed the imbalance in the public depiction of the conflict. Wyhl protesters themselves wrote: "No one who had been paying attention to Wyhl could overlook how biased the reporting of the state public broadcasting stations was."[74] One viewer wrote that, lacking the political elite's financial resources, the people could do little to make their views heard. Whatever evidence private citizens could gather on police response to the occupiers was ignored: "It was even more shocking that facts documented by bystanders' photos and films, such as the use of truncheons, were simply denied by our ministers in later debates in the state assembly." It took a television program to give the broad public a voice and to make the case in public that the Wyhl dispute was being mishandled.[75]

Viewers asserted that WDR's coverage of the Wyhl controversy promoted democracy. From Giessen, one viewer wrote that the program did a better job of "illustrating what democracy is than the other 90 percent of a month's worth of programs."[76] A man in Kiel wrote, "The fact that television reports such as this can be broadcast gives me hope that the democratic spirit can be put into practice."[77] A man in Nüsttal was not so optimistic. He wrote, "The pictures show more clearly than any commentary could, how people trying to exercise their constitutional rights were denounced and branded as heretics and criminals."[78] The scenes of demonstrators being dragged off by the police made a deep impression on many viewers, judging from this trove of correspondence. Referring to the way the police broke up the demonstration, a man from Hamburg asked, "Were these harmless elderly people the dangerous leftist extremists that Minister-President Filbinger was referring to? If so, we don't need to fear the 'leftists' any more. Or is in the eyes of the CDU/CSU every thinking person an extremist?"

Those who praised *At the Scene* were divided on its emotional content and objectivity. Thirteen of those who wrote positive letters explicitly termed the program "objective," often referring to Filbinger's assertions to the contrary. However, one member of a Protestant environmentalist group at the University of Freiburg praised an earlier *At the Scene* program about Wyhl but thought that emotional arguments had been overemphasized. In his opinion, it had not been made clear enough what rational arguments the Wyhl activists had.[79] A viewer in Berlin asserted that it was the government of Baden-Württemberg that had emotionalized the debate by "totally ignoring the will of the people."[80]

Although highly politicized, participants in the Wyhl controversy did not lose sight of the scientific issues.

Scientific Issues in the Conflict over Wyhl

Both opponents and proponents of atomic power attempted to ground their stance in science. In their press releases and public statements, the politicians made little use of scientific findings, relying instead on economic and legal arguments. They appear to have been poorly informed and little interested in the scientific and technical side of the issues. Experts served their purpose, however. Hoping to limit the population's questioning of atomic policy by referring the difficult questions to them, the Filbinger government solicited an array of scientific studies regarding Wyhl, all of whose results were very positive. Opponents argued that the government had sought out scientists friendly to the government and to nuclear power and that these experts had cherry-picked the data to ensure a positive evaluation.

In 1974, Christian Democratic representatives of the Kaiserstuhl, Alois Schätzle, and Albert Burger pointed to "contradictions and inadequacies" in the expert reports. "People's misgivings should be respected ... and not just pushed aside for the sake of economic motives." They asked the Filbinger government to commission another set of expert reports; to allow citizens of the area to testify; and to reduce the scope of the project (originally two blocks totaling 2,500 megawatts).[81] In November 1974, Schätzle requested that the Wyhl project be put on hold until a more thorough ecological report could be put together.[82]

What brought the movement against the building of the Wyhl reactor to life was the discovery that it could negatively affect winegrowing in the region. By contrast, activists were not initially very interested in the broader issues, such as the impact of radiation on the human body or the risk of a nuclear power plant accident.[83] Activists asserted that the danger of micro-climate change had been underestimated. They pointed to the tendency in the Kaiserstuhl toward temperature inversion, which could trap smog and smoke close to the ground and produce violent thunderstorms.[84] The citizens' initiatives struggled to find scientists who took this claim seriously or were skeptical about nuclear power and could serve as a counterweight to the government's experts.

To a certain extent they had loose ties with the professoriate at the nearby University of Freiburg. However, they did not have the financial resources to commission studies, although the government acquiesced to their demands for a second set of expert studies in 1976. It was also difficult for scientists who supported the anti–nuclear power activists to prevail against officially recognized experts and government institutions, for example the German Weather Service.[85] Moreover, counterexperts were not yet widely recognized as the peers of government-sponsored experts in the public realm. However, as discussed in Chapter 3, the Wyhl activists did bring over two US critics of nuclear power to testify before the Freiburg Administrative Court in 1977 and found unexpected support among West German expert witnesses.

The WDR letter collection provides interesting insights into the popular impact of scientific debates on nuclear power. Admittedly, the scientific and technical arguments found in the letters written to WDR in 1975 contained a liberal mixture of information and misinformation. Nonetheless, they reflect growing interest in general arguments against nuclear power that went beyond the particular circumstances in Wyhl. Two opponents of the power plant had extremely exaggerated notions of the possible dangers. One letter (from Berlin) spoke of four and a quarter million deaths in the case of a reactor accident in Wyhl.[86] An elderly woman from Frankfurt thought that the sinking of the ground water due to use in nuclear plants would turn West Germany into a Sahel zone.[87]

However, most letters from atomic power skeptics demonstrated a greater familiarity with the technical and scientific issues involved. A viewer from Worpswede asserted that radioactive substances were being released into the environment by nuclear power plants, nuclear reprocessing plants (which recycled spent nuclear fuel rods), and nuclear waste sites. He gave as an example Krypton 85, which, because it is heavier than air, tends to accumulate along the ground. He argued that every exposure to artificially produced radiation that goes beyond natural background radiation is harmful and that there are no safe levels of human-caused radiation. This was a controversial but arguable position. Pointing to the article of the German constitution that guaranteed the right to physical integrity, he argued that the involuntary exposure of German citizens to artificial radiation was unconstitutional and that force could be used to resist it. He also thought that nuclear energy did not make economic sense.[88]

A letter addressed to the Wyhl protesters and forwarded by them to WDR asserted that work on fifty-two nuclear power plant projects in the United States had been stopped in the preceding two years due to new findings concerning the impact of nuclear reactors on public health. It was claimed that in the vicinity of reactors in Schenectady and Idaho Falls, the incidence of all cancers was 300 percent higher than average, the incidence of leukemia was 1,500 percent higher, and the frequency of deaths due to deformities 500 percent higher. The source given was letter to the editor of the popular illustrated weekly, *Bunte*—certainly not a reliable source of scientific information.[89]

Another letter written in response to the WDR program—this one addressed to Filbinger and written by an ecology group in the town of Hohenhausen über Lemgo—repeated these statistics, adding other assertions. It was claimed that mutations caused by radiation exposure could become recessive traits that only in the third generation would cause deformities such as those caused in the 1960s by thalidomide—a particularly powerful image for West Germans.[90] Although these statements are plausible, it is clear that these two viewers got their information from a highly questionable source.

Two viewers asserted that solar power would soon make nuclear power unnecessary. Both greatly underestimated the time it would take to develop solar power into an economically viable alternative energy source.[91] However, the critics of WDR, if anything, had greater difficulties with the substantive scientific and economic issues. They also got their information from unreliable sources. A man from Miltenberg had asked an engineer, "who I know to be a conscientious man," for his opinion concerning nuclear power. This acquaintance assured him that nuclear power was not particularly dangerous, a view that the letter writer accepted as fact.[92]

A viewer from Linau bei Trittau (near Hamburg) pointed to several projects, including the building of a nuclear power plant in Stade (near Hamburg) that did not provoke the opposition of the inhabitants, an attitude that he felt showed "foresight." The essence of his argument seems to have been that nuclear power must be safe because it was familiar, because Northern Germans did not seem upset by it, and because it had been tested (for a short time) in the area where he lived.[93] Economic arguments also played a prominent role in the thinking of the critics of the WDR. The letter writer from Weil am Rhein saw a flaw in the activists' claim to be working "so that our children can live." She wrote, "Oh, my goodness, that's what we all want—and for us to live, we need to develop new sources of energy."[94] Several others saw the need for energy as a prime argument in favor of the building of atomic power plants.

Confronted with the overwhelming unpopularity of the Wyhl project, the government and utility companies launched a public relations campaign. A survey conducted by the Battelle Institute on behalf of the West German Ministry of Research had found that 75 percent of the population of the Kaiserstuhl opposed it.[95] How to combat negative views? The answer was thought to lie in advertisements in the national press, glossy brochures, a speaker service, newspaper inserts, reports, interviews, exhibitions, forums, information sessions for schools and the community at large, and tours of nuclear power plants. The government made a tremendous effort to win over Protestant and Catholic ministers, priests and church officials, teachers, politicians (including Social Democrats), members of the press, and other opinion makers.[96] If enough citizens could be won over, the government hoped, it could move in to clear the occupied site by force.[97]

The materials were not of high quality, particularly with regard to scientific content. One recipient of a brochure on atomic energy published by the government found numerous serious errors in it. A professor at the Institute for Medical Radiophysics and Radiobiology of the University of Essen, he noted that it was illogical to state, as the brochure did, that "everywhere, where radioactive material is used, exposure to radiation is far below the level in nature due to extensive safety measures." Exposure to radiation from manmade sources such as nuclear reactors had to be *added* to natural background radiation to calculate total exposure. He also pointed out that natural background radiation was not about 100 millirem per day, but 100 millirem per *year*. He was disappointed that these mistakes had found their way into this publication because he was a supporter of nuclear power.[98]

The US government also saw it as wise and appropriate to shore up support for nuclear power in Baden-Württemberg. This may be seen as an example of US "consensual hegemony" over science policies in Western Europe.[99] The American Consul General in Stuttgart met with Filbinger for a one-on-one conversation about nuclear power in April 1975. Afterward, the consulate sent Filbinger several pro–nuclear power US publications "that could be helpful for discussion and planning in Baden-Württemberg." In addition, the America House in Stuttgart brought in experts to talk to the public about atomic power. It is unclear whether Edward Teller had official US sponsorship for a public lecture that he gave in the Hotel Zeppelin (across from the main train station) in Stuttgart on 8 July 1975 in which he expressed his trust in the benefits of nuclear energy.[100]

Strong-Armed Methods of Persuasion

Filbinger, in turn, tried to enlist the press in combating the anti–nuclear power movement in Baden-Württemberg. He felt that the media had the duty to help uphold order: "The press has the job of making clear to the public that this obstinate resistance must end and that people have to be receptive to [rational] arguments. The press has to be induced to ease tensions." He and his government cultivated friendly journalists.[101] On the other hand, they attempted to intimidate journalists who would not go along with their program, for example Clemens Seiterich, editor in chief of the *Badische Bauern Zeitung*, who wrote of the cumulative effect of building several nuclear power plants on the Upper Rhine—in Fessenheim (France), Wyhl, Kaiseraugst (Switzerland), Gerstheim (France) and Freistett (West Germany): "We are shocked by the scale of the projects, which make it impossible to find analogies in past experience."[102]

In February 1977, Filbinger again scolded a newspaper editor—this time the editor in chief of the *Badische Zeitung*. He took exception to an article that asserted that not everything that is thought of as politically possible is actually doable in a democracy. Filbinger saw this as a "weakening of the rule of law." Expressing such an idea made the paper "responsible, to a not insignificant degree, for a dangerous escalation of things."[103]

Many of the methods used to get the government's point of view across to the public proved to be ineffectual. Officials sent to the Kaiserstuhl to speak to locals encountered tremendous hostility. In mid-June 1975, a delegation that included Eberle and Erwin Teufel, state secretary in the Ministry of the Environment of Baden-Württemberg, was surrounded by "rioters" who shouted, "Kill them!" They became involved in a fierce back-and-forth that lasted two hours.[104] Even in the more controlled medium of the press, officials were not able to lead the discussion in the direction they would have liked. In particular, some officials' statements enflamed oppositional sentiments rather than calming the waters.

In response, the Filbinger government redoubled efforts to maintain control over the message being disseminated to the public. A memo warned members of the government to be careful to avoid public gaffes and be cautious when talking to journalists.[105] Filbinger scolded Minister Eberle about appearing to apologize for mistakes made by the government.[106] The leadership realized that it was disadvantageous that Filbinger had become the very negative symbol of Baden-Württemberg's pursuit of atomic power. Others began stepping forward as spokespersons for official policies.[107] However, try as it might, the government seemed to be making no impression on the Wyhl protesters, who were still ensconced in their makeshift quarters on the construction site, or their many supporters.

Ousting the Occupiers and Denouement

Persuasion having failed, the Filbinger government and the utility companies turned to the use of power. Facing threats of criminal prosecution, the occupiers left the Wyhl construction site on 7 November 1975. Difficult negotiations led to the conclusion of the Offenburg Agreement on 31 January 1976. It promised that construction work would not be restarted before a new set of experts' reports were presented. The citizens' initiatives had to renounce the use of force and illegal actions.[108] The citizens' initiatives had some input into the choice of experts, who compiled numerous additional reports. At least one of the new reports contradicted the original report. These concessions meant little to the citizens' initiatives, which were biding their time, still hoping to block the building of the Wyhl nuclear power plant, by force if necessary.[109]

The continuing bitter opposition of the citizens' initiatives and the population becomes abundantly clear in a letter to Filbinger from a confidant, Albert Burger, a Christian Democratic member of the West German Bundestag from the electoral district of Emmendingen. He asserted that the crisis was greatly weakening the CDU and that many CDU members had left the party. The Wyhl controversy was ripping families apart, and "the pros and cons shape all aspects of everyday life in this area." Burger argued that it would be unthinkable to try to restart construction of the atomic power plant in Wyhl by force, particularly since the population was filled with "a sense of anguish and existential fear." He wrote:

> It would be a big mistake for the government to think that this is a matter of the usual emotions. The farmers and winegrowers in these districts are disappointed to the core; they feel misunderstood, and cannot fathom that a state government headed by the CDU wants to build a nuclear power plant that endangers their existence.[110]

The conundrums of the Wyhl impasse were rendered irrelevant by the Freiburg Administrative Court, which, in a decision of 14 March 1977 revoked the construction permit for Wyhl. This decision came as a great surprise to the government, the utilities companies, and the antinuclear activists. The ruling was based on the court's finding, based on expert testimony, that the danger of a rupture of the reactor vessel (part of the reactor core) was too great. (Chapter 3.)

Heinz Steuber of the *Stuttgarter Nachrichten* asserted that the decision of the Freiburg court on Wyhl "not only puts all other similar projects on ice, but also calls into question the nuclear power plants that are already in operation." If other courts were to follow suit, "the lights could really go out" in West Germany, he warned. By contrast, Michael Doelfs of the *Badische Zeitung* saw the decision in the context of heightened awareness of risk brought about by the Seveso disaster, an industrial accident in 1976 in which tens of thousands of Italians were exposed to dioxin. He thought that the decision opened the way for a rethinking of risk, which up until then had been overly dominated by technical standards.[111]

The Freiburg court's decision on Wyhl did not prove to be the total game changer many expected because it was vacated on appeal in 1982 by a higher administrative court (in Mannheim). This 1982 decision limited the power of the courts, leaving decisions concerning risk to the executive branch. It was up to government agencies to establish standards and call on experts to determine the facts. Courts could not step in to establish their own standards. Their job was to determine whether proper procedures had been followed. Although not widely recognized as such at the time, this became an important legal precedent.[112]

It was, however, the 1977 decision that ultimately determined the fate of the Wyhl nuclear power project because it came at a time of rapidly changing perceptions of nuclear power, as noted in a letter to Filbinger from the editor in chief of the *Badische Zeitung*. The national party congress of the SPD in November 1977 passed a resolution making coal the centerpiece of West German energy policy. By early 1978, there was a surplus of energy supplies in the Federal Republic, caused by the economic downturn of 1976–1978 and three mild winters in a row.[113] But was this really the reason? Such shifts did not derail plans to build a nuclear power plant in Brokdorf (to be discussed in the next chapter).

Decisive, in my opinion, was the massive political resistance across the political spectrum. The Christian Democratic leadership of Baden-Württemberg was highly vulnerable to criticism from within the ranks of Christian Democratic voters and politicians. In 1975, Eberle had said before television cameras that the government was ready to use force if necessary to begin construction. In a television interview after the conclusion of the Offenburg Agreement, he said, however, "For us, nothing is more important than avoiding violence." Nor could the CDU rely on outside support. West German Secretary of Research Hans Matthöfer wrote in 1977 that "nuclear power cannot be implemented against the will of the majority." A representative of *Badenwerke* (one of the utility companies), Eberhard Benz, said in 1985 that industry had undergone a learning process and now realized that it was necessary to convince the populace before starting such a project.[114]

Filbinger resigned in disgrace in 1978, after it became widely known that he had, as a Navy prosecutor and judge, participated in the passing of seven death sentences (five in absentia) during the Nazi era or shortly thereafter.[115] His successor, Lothar Späth, initially wanted to allow the Wyhl project to go forward but evidently could not face the massive resistance that was expected.[116] He declared in 1983 that the atomic power plant would not be needed until 1993. Later, he again postponed the project. Economics Minister Rudolf Eberle died in office of a heart attack on November 17,

1984. The Wyhl nuclear power plant was never built. And the Wyhl movement became a model for antinuclear protests across the globe.[117]

Conclusion

The Filbinger government attempted to push through its technocratically conceived nuclear power program on two levels—through exercise of state power and persuasion. Strong local roots helped prevent the Wyhl movement from being steamrollered by the government. Filbinger and his allies could not brush off the adamant resistance of a core constituency of their own party. Police officers with local ties followed orders to forcibly remove occupiers from the nuclear power plant construction site on 20 February. But their behavior three days later shows that they did not have the stomach for a repeat performance, especially given the festive mood among the locals, rowdy and drunken although they were. Faced with the populist anger of the rural population—in many people's minds the backbone of society—even the police chief wavered in his sympathies.

The tremendous political ferment of that era and mobilization of the citizenry was also crucial to the success of the Wyhl movement. The Left "discovered" environmentalism as a result of Wyhl and in doing so gave it a more central role in West German political culture. But environmental activism has very diverse roots. Hans-Helmut Wüstenhagen, for example, was originally a member of the Freie Demokratische Partei (FDP, or Free Democratic Party), a classical liberal party. His organization, the BBU, was to play a major role in the political consolidation of environmentalism in Germany. Cross-border collaboration with French and Swiss environmental organizations, but also trans-Atlantic connections, enabled the West German anti–atomic power movement to gain influence internationally as well.

Filbinger attempted to gain control over the discursive realm by restricting expression of oppositional views on television, as well as through public relations. One public broadcasting station, WDR, managed to resist state control and made it possible for Wyhl activists to speak directly to television audiences across West Germany. Not only did the space for public debate on nuclear power expand, but the nature of the debate changed as a result of greater popular access to the media. Filbinger tried to depict these protesters as subversives. Access to television enabled them to claim a more diverse identity. They could assert, as did one member of the citizens' initiative for Wyhl, "We are the upright housewives and mothers ... We are the element that preserves and protects."[118]

Nonetheless, public opinion concerning the Wyhl plant was polarized, as viewer letters indicate. Segments of the public also disagreed fundamentally over what was more important about democracy: the rule of the majority through an elected government or the ability of those whose fundamental needs were ignored by the government to make their voices heard, a quarrel that went back to the 1960s. The Wyhl controversy opened up channels of communication and established a model of protest. The basic issues were not settled and in fact were only becoming truly contentious in this period. A deeply divided media landscape fostered the development of a polarized debate.

Many began to reject the idea that only experts could make technical decisions. A wave of popularization of science brought with it information and misinformation, based on sources of varying quality. With this came a refinement and expansion of public thinking concerning nuclear power that went far beyond the original, local concerns of Kaiserstuhl winegrowers. The West German public was beginning to dare to grapple with complex scientific issues. Social Democratic leader of Baden-Württemberg Erhard Eppler, in a speech given in Wyhl on 26 June 1975, said:

> Anyone who is somewhat acquainted with the topic of nuclear power knows how thin the ice we've been skating on is, how imperfect our knowledge is … Anyone who approaches the subject with the attitude that he has a monopoly on truth and can throw a few crumbs of it to the people, should not participate in the debate.

He questioned the wisdom of allowing experts to make the big decisions about atomic power, something that happened because the politicians did not have the technical and scientific knowledge to understand the issues.[119]

This was the beginning of a total rethinking of the meaning of risk. No longer was risk a purely scientific category to be calculated by scientists, but an ethical, political, and cultural category. It was here that the people could and should participate in decisions concerning technologies that would affect their lives, West Germans increasingly argued.[120] The Wyhl movement drew strength from several sources: an anti-technocracy, pro-science narrative that resonated nationally and internationally; access to the media; cooperation among a great variety of political organizations; clever negotiating strategies; and disruption through a long and successful occupation of the site of the planned atomic power plant.

Notes

1. "Wyhl" is pronounced like the English word "veal."
2. Jens Ivo Engels, "Gender Roles and German Anti-Nuclear Protest: The Women of Wyhl," in *Le demon moderne: la pollution dans les sociétés urbaines et industrielles d'Europe*, ed. Christoph Bernhardt (Clermont-Ferrand: Presses univ. Blaise Pascal, 2002), 407–424.
3. Hauptstaatsarchiv Stuttgart (Provincial Archives of Baden-Württemberg, Main State Archive, Stuttgart, abbreviated throughout as HstAS), EA 1/924 Bü 1739, "Deutscher Bundestag—7. Wahlperiode. Drucksache 7/2802," 148.
4. Notably under a plan of 22 June 1971. HStAS, EA 1/924, Nr. 1741, letter of Ministry of Interior of Baden-Württemberg to president of the State Assembly, 8 April 1975, 2.
5. HstAS, EA 1/924 Bü 1738, "Auszug aus dem Protokoll ... vom 4.10.1974," 22–23. Under the Atomic Law, the individual German states approved the building of atomic power plants. The approval of the Federal (i.e., national) Ministry of the Interior was also needed. The approval process was divided into partial approvals. Before the project in its entirety was approved, the state government had to commission a series of expert reports. The Federal Ministry of the Interior also had to seek the advice of experts, who sat on various commissions, notably the Reactor Safety Commission.
6. HStAS, EA 1/924 Bü 1739, Filbinger to Hauff, 9 January 1974, 1.
7. WDR Historisches Archiv (West German Broadcasting Historical Archive), Archive number 0165708.
8. HStAS, EA 1/924 Bü 1738, Burger and Schätzle to Eberle, 11 July 1974.
9. HStAS, EA 1/924 Bü 1738, State Ministry to Filbinger, 28 November 1974.
10. HStAS, EA 1/924 Bü 1738, draft of letter, Filbinger to Gerstner, 16 December 1974.
11. Hodenberg, *Konsens und Krise*, 9–12.
12. WDR Historisches Archiv, Archivnr. 0103355, Series: *Vor Ort*; Episode: Wyhl-Prozess; first shown 31 March 1977.
13. Reinhard Steurer, *Der Wachstumsdiskurs in Wissenschaft und Politik: Von der Wachstumseuphorie über 'Grenzen des Wachstums' zur Nachhaltigkeit* (Berlin: VWF Verlag für Wissenschaft und Forschung, 2002). Also "Der Wachstumsdiskurs in Wissenschaft und Politik: Von der Wachstumseuphorie über 'Grenzen des Wachstums' zur Nachhaltigkeit" (Dr.phil. diss., University of Salzburg, 2001), retrieved 8 January 2018 from https://www.wiso.boku.ac.at/fileadmin/data/H03000/H73000/H73200/_TEMP_/Steurer_Wachstumsdiskurs_in_Wissenschaft_und_Politik_Diss_01.pdf.
14. Alexander Sedlmaier, *Consumption and Violence: Radical Protest in Cold-War West Germany* (Ann Arbor, MI: University of Michigan Press, 2014).
15. Reichardt, *Authentizität und Gemeinschaft*, esp. chs. 2, 3, and 4, and sections 7.2, 9.1.
16. Donella Meadows, Dennis Meadows, Jørgen Randers, and William Behrens, *The Limits to Growth: A Report for the Club of Rome's Project on the Predicament*

of Mankind (New York: Universe Books, 1972). German edition: Donella Meadows, Dennis Meadows, Jørgen Randers, and William Behrens, *Die Grenzen des Wachstums: Bericht des Club of Rome zur Lage der Menschheit,* trans. Hans-Dieter Heck (Stuttgart: Deutsche Verlags-Anstalt, 1972). For a thorough critique of *The Limits to Growth* and analysis of its impact, especially in West Germany, see Steurer, "Wachstumsdiskurs" 2001, 153–243. Also Kai Hünemörder, *Die Frühgeschichte der globalen Umweltkrise und die Formierung der deutschen Umweltpolitik (1950– 1973)* (Wiesbaden: Franz Steiner Verlag, 2004), 222–27.

17. Georg Stötzel and Martin Wengeler, *Kontroverse Begriffe: Geschichte des öffentlichen Sprachgebrauchs in der Bundesrepublik Deutschland* (Berlin and New York: de Gruyter, 1995), 78–81; Steurer, *Wachstumsdiskurs,* 2001, 244–49, 477–89. On the position of the Left and the Greens, which was a decentralized movement until 1980, see Silke Mende, *"Nicht rechts, nicht links, sondern vorn." Eine Geschichte der Grünen* (Munich: Oldenbourg Verlag, 2011), 447–52.

18. For a sympathetic interpretation, see Karrin Hanshew, *Terror and Democracy in West Germany* (Cambridge, UK and New York: Cambridge University Press, 2012). For a more critical approach, see Larry Frohman, *"Datenschutz,* the Defense of Law, and the Debate over Precautionary Surveillance: The Reform of Police Law and the Changing Parameters of State Action in West Germany," *German Studies Review* 38, no. 2 (2015): 307–27. On the student movement, see Andrea Ludwig, *Neue oder deutsche Linke?: Nation und Nationalismus im Denken von Linken und Grünen* (Opladen: Westdeutscher Verlag, 1995), 28–30. On the 1976 law, see Renate Wiggershaus and Rolf Wiggershaus, "Beim 'Gewaltschutzparagraphen' geht es nicht nur um Gewalt," *Gewerkschaftliche Monatshefte* 27, no. 10 (1976): 597–602, retrieved 10 January 2018 from http://library.fes.de/gmh/main/pdf-files/gmh/1976/1976-10-a-597.pdf.

19. Falco Werkentin, *Die Restauration der deutschen Polizei: Innere Rüstung von 1945 bis zur Notstandsgesetzgebung* (Frankfurt and New York: Campus Verlag, 1984).

20. Gerhard Fürmetz, Herbert Reinke, and Klaus Weinhauer, eds., *Nachkriegspolizei: Sicherheit und Ordnung in Ost- und Westdeutschland 1945–1969* (Hamburg: Ergebnisse, 2001).

21. Frohman, "Datenschutz."

22. Klaus Weinhauer, "Staatsgewalt, Massen, Männlichkeit: Polizeieinsätze gegen Jugend- und Studentenproteste in der Bundesrepublik der 1960er Jahre," in *Polizei, Gewalt und Staat im 20. Jahrhundert,* ed. Alf Lüdtke, Herbert Reinke, and Michael Sturm, 301–24 (Wiesbaden: VS Verlag für Sozialwissenschaften, 2011).

23. West German courts derived this legal standard from the Basic Law (the West German constitution) and the *Polizeigesetz,* or laws governing policing in the various states. Mathias Graßmann, "Die Polizei in der Bundesrepublik Deutschland," in *Beiträge zu einer vergleichenden Soziologie der Polizei,* ed. Jonas Grutzpalk Franziska Harnisch, Christiane Mochan, Björn Schülzke, and Tanja Zischk, 89–107 (Potsdam: Universitätsverlag Potsdam, 2009).

24. WDR Historisches Archiv, Archive number 0165239. On Sacherer, see Eith, Nai hämmer gsait, 52.

25. HStAS, EA 1/924 Bü 1744, Entwurf-Bericht über die Polizeieinsätze in Wyhl, S. 15.

26. HStAS, EA 1/924 Bü 1738, Memorandum for the minister-president (Filbinger), 10 July 1974; Memorandum for the minister-president (Filbinger), 11 July 1974.

27. HStAS, EA 1/924 Bü 1738, "Auszug aus dem Protokoll ... vom 4.10.1974," 22–23.

28. WDR Historisches Archiv, Archive number 0161681, series *Vor Ort*, episode "Wyhl Baden," first shown on 22 February 1975. Also HStAS, EA 1/924 Bü 1739, police telex, 20 January 1975.

29. Stephen Milder, "'Today the Fish, Tomorrow Us': Anti-Nuclear Activism in the Rhine Valley and Beyond, 1970–1979 (Ph.D. diss., University of North Carolina at Chapel Hill, 2012), 195–98.

30. HStAS, EA 1/924 Bü 1744, Eberle to Filbinger, 5 March 1975, 2. Statement made at meeting on 16 January 1975.

31. HStAS, EA 1/924 Bü 1739, "Tagesordnung für die Sitzung des Ministerrats," 21 January 1975, 1–4.

32. WDR Historisches Archiv, Archive number 0161681, series *Vor Ort*, episode "Wyhl Baden," first shown on 22 February 1975.

33. HStAS, EA 1/924 Bü 1744, Nr. III 8112/34, "Betr.: Polizeieinsatz in Wyhl," 18 July 1975, 7.

34. HStAS, EA 1/924 Bü 1744, "Vermerk über die Sitzung des MD-Arbeitskreises 'Polizeiliches Vorgehen in Wyhl' am 4. Juni 1975," dated 6 June 1975, 2; EA 1/924 Bü 1744, Nr. III 8112/34 dated 19 June 1975. The number of demonstrators has been estimated as much higher, e.g., in Milder, "Today the Fish," 215–16.

35. HStAS, EA 1/924 Bü 1744, memorandum from the State Police Commissioner in Freiburg to the Minister of the Interior of Baden-Württemberg, 22 April 1975, 4.

36. HStAS, EA 1/924 Bü 1744, Nr. III 8112/34, 18 July 1975, 3.

37. HStAS, EA 1/924 Bü 1744, "Vermerk über die Sitzung des MD-Arbeitskreises 'Polizeiliches Vorgehen in Wyhl am 4. Juni 1975," dated 6 June 1975, 2. Also HStAS, EA 1/924 Bü 1744, Nr. III 8112/34 dated 19 June 1975, 6.

38. HStAS, EA 1/924 Bü 1744, "Entwurf—Bericht über die Polizeieinsätze in Wyhl," 17.

39. Deutsche Kinemathek, WDR series *Vor Ort*, program titled "Bürger gegen Atomkraftwerk in Wyhl," first broadcast on 11 January 1975. WDR Historisches Archiv, Archive number 0161681, series *Vor Ort*, program titled "Wyhl Baden," first shown on 22 February 1975.

40. HStAS, EA 1/924 Bü 1744, memorandum from the State Police Commissioner in Freiburg to the Ministry of the Interior of Baden-Württemberg, 22 April 1975, 1–2.

41. HStAS, EA 1/924 Bü 1744, Nr. III 8112/34 dated 19 June 1975, 5, 6.

42. HStAS, EA 1/924 Bü 1744, "Entwurf-Bericht über die Polizeieinsätze in Wyhl," 19. Riot police consisted both of federal police and officers from various states.

43. HStAS, EA 1/924 Bü 1744, Nr. III 8112/34 dated 18 July 1975, 18 and 1, 2.

44. Jörg Bischoff, "Gestolpert über Wyhl? Minister Schiess opfert seine Polizisten," *Die Zeit*, 26 November 1976.

45. HStAS, EA 1/924 Bü 1744, memorandum from the State Police Commissioner in Freiburg to the Minister of the Interior of Baden-Württemberg, 22 April 1975, 2.

46. HStAS, EA 1/924 Bü 1744, Nr. III 8112/34 dated 18 July 1975, 7.

47. HStAS, EA 1/924 Bü 1744, "Vermerk über die Sitzung des MD-Arbeitskreises 'Polizeiliches Vorgehen in Wyhl' am 4. Juni 1975," 6 June 1975, 2.
48. Milder, "First the Fish," 207–70.
49. WDR Historisches Archiv, Archive number 0014837.
50. Milford, *Greening Democracy,* 99.
51. WDR Historisches Archiv 12523, Höfer to Mahler, 10 March 1975, 2.
52. Meyen and Hillman, "Communication Needs," 464.
53. Interview with Ludwig Metzger, 11 June 2012.
54. WDR Historisches Archiv 12523, Mahler to Bismarck, 28 February 1975, 4, 1, 2.
55. WDR Historisches Archiv, 12523, Eberle to Bausch, 27 February 1975.
56. HStAS, EA 1/924, Nr. 1738, Eberle, "Warum das Kernkraftwerk Wyhl genehmigt wird?" 6 November 1974.
57. WDR Historisches Archiv, Signatur 12523, Deutsche Presse-Agentur telex, 27 February 1975.
58. WDR Historisches Archiv, Archive number 0014837.
59. Birgit Weidinger, *Süddeutsche Zeitung,* 28 February 1975.
60. WDR Historisches Archiv, Signatur 12523.
61. WDR Historisches Archiv, 12523, "Gemeinsame Beratungen von Ständiger Fernsehprogrammkonferenz und Fernsehprogrammbeirat"; Hammerschmidt to Bausch, 1 March 1975.
62. "'Bürger gegen Atomkraftwerk' auch vor ARD-Hauptversammlung," *FUNK-Korrespondenz,* 19 March 1975.
63. WDR Historisches Archiv, Signatur 12523, transcript of "Glashaus—TV Intern," dated 1 April 1975.
64. The Adolf-Grimme Prize. Günter Schloz, "Adolf-Grimme-Preis 1976," in *Das Fernsehen und sein Preis,* ed. Lutz Hachmeister, 60 (Bad Heilbrunn: Klinkhardt, 1994).
65. Biess, "Feelings in the Aftermath," 43.
66. Exact data: seventy positive and supportive of the protesters; seventeen negative and critical of the protesters; two negative and supportive of the protesters; eleven neutral (requested copies, which could be interpreted as a positive evaluation of the film). WDR Historisches Archiv, 12523, 7210, 11853.
67. WDR Historisches Archiv, 7210, Schurig to WDR, dated 12 March 1975.
68. WDR Historisches Archiv, 7210, letter to WDR dated 27 February 1975. Names of the writers of this and the following letters deleted to protect their privacy.
69. WDR Historisches Archiv, 7210, postcard to WDR dated 26 February 1975
70. WDR Historisches Archiv, 7210, letter to WDR dated 5 March 1975.
71. WDR Historisches Archiv, 7210, letter to WDR dated 27 February 1975.
72. Ibid.
73. WDR Historisches Archiv, 7210, letter to WDR dated 26 February 1975.
74. WDR Historisches Archiv 12523, Carola Burg et al. to Höfer, 13 March 1975.
75. WDR Historisches Archiv, 7210, letter to WDR dated 25 March 1975.
76. WDR Historisches Archiv, 7210, letter dated 26 February 1975.
77. WDR Historisches Archiv, 7210, letter dated 27 February 1975.
78. WDR Historisches Archiv, 7210, letter dated 26 February 1975.
79. WDR Historisches Archiv, 11853, letter dated 14 November 1974.

80. WDR Historisches Archiv, 7210, letter dated 4 April 1975.
81. HStAS, EA 1/924 Bü 1738, Burger and Schätzle to Eberle, 11 July 1974.
82. HStAS, EA 1/924 Bü 1738, State Ministry to Filbinger, 28 November 1974.
83. Milder, "Today the Fish," 34, 36, 44–46, 48, 49, 56, 67.
84. WDR Historisches Archiv, Archive number 0165708.
85. Example in EA 1/924, letter of the German Weather Service [Deutscher Wetterdienst], Stuttgart Office, to the Economics Ministry, dated 22 October 1974.
86. WDR Historisches Archiv, 7210, letter dated 26 February 1975.
87. WDR Historisches Archiv, 7210, letter dated 27 February 1975.
88. WDR Historisches Archiv, 7210, letter dated 7 April 1975.
89. WDR Historisches Archiv, 7209, letter dated 27 February 1975. Source given: Letter to the editor concerning "Die Angst vor der Bombe," *Bunte,* Nr. 34, 1974.
90. WDR Historisches Archiv, 7209, letter dated 27 February 1975.
91. WDR Historisches Archiv 7209, letter to Filbinger dated 2 March 1975, sent from Spandau [a section of West Berlin] and letter to citizens' initiative in Wyhl dated 24 February 1975, sent from Stuttgart.
92. WDR Historisches Archiv, 7210, letter to WDR dated 26 February 1975.
93. WDR Historisches Archiv, 7210, letter to WDR dated 26 February 1975.
94. WDR Historisches Archiv, 7210, letter to WDR dated 27 February 1975.
95. Battelle Institute, *Einstellungen und Verhalten der Bevölkerung gegenüber verschiedenen Energiegewinnungsarten: Bericht für das Bundesministerium für Forschung und Technologie, Abteilung Sozialwissenschaften* Teil A: *Die empirischen Untersuchungen* (Frankfurt am Main: Battelle Institute, 1977).
96. HStAS, EA 1/924 Bü 1740, Abteilung VI, "Vermerk: Betr.: Kernkraftwerk Wyhl. Hier: Weiter Massnahmen der Öffentlichkeitsarbeit," quotations on 1 and 5; HStAS, EA 1/924 Bü 1744, KWS, "Übersicht Öffentlichkeitsarbeit," 8 October 1975 and Abteilung VI, "Vermerk. Betr.: Öffentlichkeitsmaßnahmen zum Thema Kernenergie und Wyhl," 13 October 1975.
97. HStAS, EA 1/924 Bü 1741, Abteilung IV, "Vermerk über das weitere Vorgehen beim Bau des Kernkraftwerks in Wyhl," 1 April 1975; quotations on 1, 5, 7.
98. HStAS, EA 1/924 Bü 1741, letter to Filbinger dated 28 May 1975 [author's name omitted in conformity with privacy laws]; quotation on 1. Natural background radiation is now calculated to average about 2.4 millisievert or 240 millirem per year.
99. Krige, *American Hegemony.*
100. HStAS, EA1/924 Bü 1743, letter dated 2 July 1975 from Walter E. Jenkins, Jr., general consul of USA in Stuttgart to Filbinger.
101. HStAS, EA 1/924 Bü 1742, transcript of Filbinger's dictated memorandum, 16 June 1975.
102. HStAS, EA 1/924 Bü 1742, Seiterich to Hildebrand, 5 June 1975; quotation on 3. Planned nuclear power plants in Freistett, Kaiseraugst, and Gerstheim were never built.
103. HStAS, EA 1/924 Bü 1745, Filbinger to Fürst (draft), 28 February 1977. Editor's reply: HStAS, EA 1/924 Bü 1745, Fürst to Filbinger, 4 March 1977.

104. HStAS, EA 1/924 Bü 1742, Letter to Mayer-Vorfelder [name of author illegible], 19 June 1975.

105. HStAS, EA 1/924 Bü 1743, State Ministry, memorandum of 7 July 1975.

106. HStAS, EA 1/924 Bü 1742, Filbinger to Eberle, no date. Letter was not sent, but Eberle was informed in person (according to an attached note).

107. HStAS, EA 1/924 Bü 1741, Abteilung IV, "Vermerk über das weitere Vorgehen beim Bau des Kernkraftwerks in Wyhl," 1 April 1975.

108. HStAS, EA 1/924 Bü 1762, Abteilung III, "Betr.: Kernkraft Wyhl; hier: Zustimmung zur Offenburger Vereinbarung," 20 February 1976, appendix ("Vereinbarung").

109. HStAS, EA 1/924 Bü 1762, Abteilung III, Badenwerk to Filbinger, 22 January 1976; Abteilung III, "Vermerk für die Sitzung des Ministerrats am 1. Juni 1976"; Eberle to Richter, 19 May 1976; Telex Ministry of the Interior to State Ministry, 18 June 1976; "Betreff: Verhandlungen ... hier: Gespräch am 23. August 1976 im Landratsamt Emmendingen. Ergebnisprotokoll."

110. HStAS, EA 1/924 Bü 1763, Burger to Filbinger, 11 November 1976; quotation on 3.

111. Heinz Steuber, "Der Schock von Freiburg," *Stuttgarter Nachrichten,* 15 March 1977; Michael Doelfs, "Plausibler Spruch," *Badische Zeitung,* 15 March 1977.

112. Heike Mrasek-Robor, "Technisches Risiko und Gewaltenteilung," (Dr.phil. diss., University of Bielefeld, 1997).

113. HStAS, EA 1/924 Bü 1745, Fürst to Filbinger, 4 March 1977; Economics Ministry to State Ministry, 6 March 1978; Stavenhagen to Filbinger, 11 April 1978.

114. On Eberle and Benz: WDR Historisches Archiv, Archive number 0133760; program "Aktuelle Stunde," 17 March 1985. Matthöfer quotation in Hans Matthöfer, *Interviews und Gespräche zur Kernenergie* (Karlsruhe: Mueller, 1977), 11. Quoted according to Joppke, *Mobilizing,* 100.

115. Wolfram Wette, ed., *Filbinger—Eine deutsche Karriere* (Springe, Germany: zu Klampen Verlag, 2006). Controversy has swirled around the meaning of his wartime career.

116. "Grüne Hölle," *Der Spiegel* 18 (2 May 1983).

117. An example is the Clamshell Alliance, which fought the building of a nuclear power plant in Seabrook, New Hampshire.

118. WDR Historisches Archiv, Archive number 0014837.

119. HStAS, EA 1/924 Bü 1742, "SPD-Presse-Information, 26. June 1975/Nr. 117," 1.

120. The most important book on this subject did not appear until 1986: Ulrich Beck, *Risikogesellschaft* (Frankfurt am Main: Suhrkamp, 1986).

Chapter 5

Environmentalism as Civil War

Brokdorf—and the Consequences

Demonstrators desperately throwing volley after volley of Molotov cocktails, rocks, and iron bars against the police. Police water cannons spraying high-volume geysers that knock protesters off their feet. Clouds and clouds of tear gas, some thrown from low-flying helicopters, just above the activists' heads. In back of the wild melee, a high fence, almost an apparition of the Berlin Wall—tall, topped with razor wire, with moats in front. Wave after wave of demonstrators charge it, but police in riot gear defend it to the hilt, at times standing shoulder to shoulder, forming a human shield. Such were the pitched battles fought on bitterly cold fall and winter days and nights in 1976–1977. Tens of thousands of demonstrators faced off against the largest police deployment that West Germany had seen up until then.[1] This near civil war was fought over the building of a nuclear power plant in Brokdorf, in the far north of the Federal Republic.

Opinion in West Germany was sharply divided. The conservative *Rheinischer Merkur* referred to the demonstrators as "a hard core of rowdies" and "well-known professionals who are not interested in nuclear power, but in destabilizing the state."[2] By contrast, an article in the Berlin daily *Tagesspiegel* asserted, "The sight of armed state power provoked the anger of protesters who had wanted to demonstrate peacefully." Criticizing the police (and forgetting Schleswig-Holstein's brutal Nazi past[3]), the mayor of a neighboring village commented, "We have never before seen such brutality here."[4] What is remarkable is not this polarization but the fact that these violent demonstrations did not permanently discredit the movement. This chapter attempts to explain how this opposition to nuclear power gradually began to make inroads among the broad public following developments up to the eve of the Chernobyl disaster of 1986.

While the Wyhl conflict had been notable for its relatively peaceful nature, armed conflict and accompanying media wars reached a fever pitch at Brokdorf, in the state of Schleswig-Holstein, where the Christian Democratic government of Gerhard Stoltenberg wanted to construct a nuclear power plant that would serve both the states of Schleswig-Holstein and Hamburg. Many of the same elements observable in the Wyhl case were also present in the Brokdorf conflict: determination on the part of the government to defend this project both as an important component of a modernization program and as a profit-making public utility; government attempts to intimidate the media and discredit the protesters as dangerous radicals; the development of opposition into a major social movement; and polarization of the public.

However, what created near civil war conditions at Brokdorf was Stoltenberg's resolute commitment to avoid a "second Wyhl" and the protest movement's utter determination to achieve the ultimate form of disruption: the occupation of the construction site on which the Brokdorf atomic power plant was to be built. It was to be a long fight. Brokdorf went into operation on 8 October 1986, less than six months after the Chernobyl disaster, but the protests continued for years. Despite this ostensible failure, Brokdorf gave the anti–nuclear power movement coherence and direction, maintaining the momentum achieved by activism in Wyhl.

Politicians, Activists, the Media, and the Public in the Brokdorf Conflict

Gerhard Stoltenberg was dedicated to a vision of modernity that involved nuclear power as well as to resolute resistance against radicals whom he thought posed an existential threat to democracy. As the Federal Minister for Scientific Research from 1965 to 1969, he had promoted research on nuclear power.[5] Unlike Hans Filbinger, however, Stoltenberg did not place himself at the head of the atomic power effort once he became the minister-president of Schleswig-Holstein, nor did he produce the kind of strident rallying cries so typical of Filbinger; yet in the end he proved to be both cannier and more ruthless. Bland competence, combined with uncompromising rigidity, eventually won Stoltenberg a place in Chancellor Helmut Kohl's 1982 cabinet, although not the chancellorship he hoped for.[6]

Stoltenberg, who became the minister-president of Schleswig-Holstein in 1971, was an enthusiastic supporter of plans to transform his largely rural state into a major industrial region whose energy supply would come from the largest concentration of nuclear power plants in capitalist

Western Europe. Eleven atomic power plants and several gigantic industrial plants were planned for the seventy-mile stretch of the River Elbe between Hamburg and the North Sea, an area that had been an idyllic, sparsely populated farming and tourist area up until then. Another five were to be built upriver from Hamburg.[7] The Stoltenberg government chose Brokdorf as the site for one of these plants in October 1973.

Neither enthusiasm for nuclear power nor links between the state and the nuclear power industry can fully explain the state's extreme response to the Brokdorf protests. Also of considerable importance were the CDU's resentment over the decline of Christian Democracy and ascendancy of the Left, its desire to reclaim control over the national government, Stoltenberg's careerism, and fears concerning terrorism and the decline of law and order in West Germany. A wave of spectacular attacks by the RAF were linked in the minds of many mainstream politicians with the danger of subversion.

By contrast, Austrian journalist Robert Jungk attributed the state's harsh crackdown on the Brokdorf protests to the emergence of the "Atomic State," a term that he coined at a protest rally at Brokdorf in February 1977 and that carried with it echoes of the Nazi SS State.[8] He argued that the use of atomic energy made security measures necessary that would rob citizens of their rights. A Jewish refugee from Nazi Germany, he argued that the "atomic state" was on its way to erecting concentration camps. He saw reliance on atomic energy as the result of "overtechnification," i.e., the uncontrolled development of technology, excessive consumption, and the alliance between industry and the state.[9] Jungk's thesis that an all-powerful atomic industry enjoyed the unwavering support of the state, although influential at the time, has been at least partially discredited by scholars. However, the concept of the Atomic State enjoyed widespread popularity in the late 1970s and early 1980s.[10] This compelling antitechnocratic narrative provided focus to organizations with disparate agendas.

The anti-Brokdorf movement was not radical from the start. Locals, many quite conservative, were initially the dominant force. They enjoyed widespread legitimacy and were depicted sympathetically, even, for example, by the stridently right-wing *Bild Zeitung*, which portrayed them as colorful Plattdeutsch (Low German dialect)-speaking "marshland people."[11] Seventy-five percent of locals opposed the Brokdorf project, according to a 1973 survey. Opponents of the nuclear power station formed an umbrella organization of citizens' initiatives, the *Bürgerinitiative Umweltschutz Unterelbe* (Lower Elbe Citizens' Initiative on Environmental Protection, or BUU) in 1973. It was organized in four regional associations, to which local citizens' initiatives belonged. As historian Roger Karapin points out, they failed to win over the kind of support of local elites that had been so important in the case of Wyhl. They also found it difficult to deal with the massive show of force and politicians' stonewalling at public hearings in November 1974 and March 1976. Because of the weakness of the local movement, outside activists became the dominant force in the movement.[12]

By this time, the West German anti–atomic power movement was rooted in six milieus: the local, often rural and often conservative citizenry; Protestant activism[13]; New Left elements who emerged out of the APO (*Außerparlamentarische Opposition,* or extraparliamentary opposition) movement of the 1960s; the *Autonomen,* or "autonomous," antiauthoritarian Marxists and post-Marxist anarchists inspired by Antonio Negri (among others), who at times embraced violence and who continued to play a major role in political activism up through the 1980s; pacifistic, "libertarian socialists" or nonviolent anarchists, such as the Grassroots (*Graswurzel*) movement; and ecological activists that were part of the new social movements of the postboom era. These groups could not agree on whether it was legitimate or even necessary for activists to resort to violence. The outcome of this debate played a role in the eventual widespread acceptance of the antinuclear position.

Another major party to the Brokdorf conflict was the media. Angered by coverage of the Brokdorf conflict, the Stoltenberg government waged a very aggressive campaign to halt what it saw as one-sided, leftist TV journalism. Journalists at the NDR (Norddeutscher Rundfunk, or Northern German Broadcasting) network were motivated by the same conception of critical journalism as their colleagues at WDR who reported on the Wyhl protests. In the short term, NDR succumbed to political pressure, although careful analysis is necessary to determine whether it capitulated or whether a more subtle form of accommodation and negotiation was at work.

A "countercultural public sphere" provided an alternative to the predominantly commercial media world. Alternative media included not only leftist newspapers but also posters, leaflets, and wall newspapers widely available at universities as well as pirate radio stations. Pirate radio stations were actually illegal until the ending of the monopoly of state-run radio. The authorities sent out detector vans to find where they were being transmitted from and shut them down. Therefore, pirate radio stations such as Radio Zebra in Bremen, which existed from 1980 to 1983, could only transmit for fifteen minutes at a time. The research of media historian Inge Marszolek shows that this was enough to win a core of young listeners and to create an experimental alternative to official broadcast culture, of which anti–nuclear power songs and declarations were a part. These stations also played an important role in getting out information about upcoming demonstrations and meetings.[14]

Another experimental form of communication was the wall newspaper. One example, preserved in the State Archives of Hamburg, spoke of the "total militarization" of the police response to the demonstration on 13 November 1976. The purpose of this wall newspaper was, according to its publishers, to counteract what were seen as the lies and propaganda of the "bourgeois" press.[15] Alternative media such as these not only promoted the internal cohesion of the anti–nuclear power movement but also expanded the boundaries of civil society. Particularly successful were the leftist newspapers, which promoted the self-reflection and political development of the anti–nuclear power movement.

The German public was drawn into the Brokdorf conflict primarily by the commercial media. On many nights, antinuclear protests were the first or most prominent item on the evening news.[16] Competition to influence public opinion was fierce. West Germans were very much concerned about the terrorist threat, but they were also becoming more receptive to environmentalist concerns, which were awakened by Rachel Carson's *Silent Spring* and the 1972 Club of Rome report. On 10 July 1976, a grave environmental catastrophe took place in Seveso, Italy. An explosion at a chemical plant owned by an affiliate of the Hoffmann-Laroche corporation spread dioxin over a wide area. The company tried to cover up the release of this highly toxic substance, much to the detriment of the local population. This incident became an iconic symbol of the existential threat posed by industrial accidents and irresponsible corporate behavior.[17]

By the 1970s, the German public came to believe that the risks posed by nuclear power were of a fundamentally different nature than those posed by other technologies. Historian Christoph Wehner has shown that the public was concerned by the refusal of private insurance companies—first in the United States, then in West Germany—to insure nuclear facilities. "If it can't be insured, it can't be safe" was the popular reaction.[18] The Three Mile Island nuclear power plant accident in the United States rallied public opposition to atomic energy as never before.

Scientific issues continued to play a role in this phase of the history of the nuclear power movement. Antinuclear groups from Hamburg and Bremen sought to gain a deeper understanding of the scientific issues through study groups, collaboration with science majors at the universities, and contacts with scientists. For example, a group of physics students at the University of Hamburg put together a brochure on nuclear power that sold for one D-mark. It contained explanations of nuclear chain reactions, how reactors worked, reactor types, radiation leakage during normal operation, risk, nuclear accidents, and nuclear waste. It cited sources discussed in Chapter 3 such as Strohm, as well as Gofman and Tamplin, but also unpublished conference papers and information directly given to the students by professors.[19] Such study of scientific issues was flanked by publications of the BBU (Bundesverband Bürgerinitiativen Umweltschutz, or Federal Association of Citizens' Initiatives on Environmental Protection—a West German organization), particularly its series "Informationen zur Kernenergie" (Information on Nuclear Energy).

"We have never before seen such brutality here": The 30 October 1976 Demonstration

The Stoltenberg government allowed construction work on the Brokdorf nuclear power plant to begin on 26 October 1976, before a legal challenge was decided by the courts. It ordered the police to erect a massive fence around the construction site and to prevent demonstrators from occupying it. Widely decried as a "cloak-and-dagger operation," the start of construction under massive police protection was, according to *Der Spiegel,* West Germany's most influential news magazine, a "rash maneuver" intended to trick the opponents of the nuclear power plant.[20] The liberal minister of the interior of Nordrhein-Westfalen, Burkhard Hirsch, was quoted as saying that construction of the Brokdorf atomic power station should not have been started until the problem of atomic waste disposal was solved and until legal challenges had been decided.[21]

Opponents quickly called for a demonstration and an attempt to occupy the site, which took place on 30 October. The five thousand demonstrators faced a police force of four hundred, twelve on horseback and twelve with police dogs. *Spiegel* found this deployment "unprecedented."[22] Chancellor Helmut Schmidt, too, termed the massive police deployment in Brokdorf "idiocy."[23] However, it was later eclipsed by far more massive state reaction to protests.

Police initially stood back while peaceful demonstrators cut open the fence and streamed into a corner of the construction site.[24] By nightfall, they had built fires, put up nine tents, and gotten food and other supplies from protesters on the outside. Police waited until many demonstrators had left the area to force the occupiers to leave, using water cannons.[25] The widely read weekly illustrated magazine *Stern* printed huge, spectacular photos of these events.[26] In a sense these picture spanned the gap between television and print media. One photo shows police huddled together as if poised to charge against the demonstrators. The reader sees their white helmets, but not their faces.

A water cannon sprays demonstrators. In the foreground are rolls of razor wire. According to the text, "While the demonstrators sat in front

Illustration 5.1 "The cool, clear one out of the North." Political cartoon featuring Gerhard Stoltenberg, *Stern.* "In this unpleasant fight ... about nuclear energy, much too complicated an issue ... we simply have ... the stronger arguments!" by Fritz Wolf. "Der kühle Klare aus dem Norden," *Stern,* 49/1976. Courtesy of the Fritz-Wolf-Gesellschaft e.V., Osnabrück.

of their tents and sang, police officers cut open the razor wire, mounted police charged through the camp, officers with mace and truncheons following them. They beat and sprayed demonstrators indiscriminately, and several demonstrators—completely unarmed—got mace in their faces." A follow-up article in *Stern* compared the fortifications around the Brokdorf construction site to the Berlin Wall, an image that was to be evoked often in media references to Brokdorf.[27] Forty-five demonstrators and one police officer were injured in the melee.[28] A Stern cartoon showed Stoltenberg holding a police truncheon and claiming to have the "stronger arguments" in the fight over Brokdorf. (See Illustration 5.1.)

An article in the middle-of-the-road West Berlin newspaper *Tagesspiegel* presented police behavior on 30 October 1976 as outrageous. "At the End, the Tents Were Burned," the headline read, referring to the accusation that the police burned the demonstrators' tents after forcing them from the construction site on 30 October—an accusation that the police denied in internal reports. One woman told the reporter, "Some of us were hit on the head with rubber truncheons; one woman cried the whole time. A traffic police officer, whom we encountered on the way out told us that for the first time in his life, he was ashamed of his profession." The article went on to state that the authorities had gone into the confrontation prepared for violence and in fact had provoked it with an uncalled-for show of police force: "The sight of armed state power provoked the anger of protesters who had wanted to demonstrate peacefully." Out of anger they joined the crowd that tried to occupy the site.[29]

A massive police presence continued at the site and in the surrounding area. *Der Spiegel* asserted, "The nighttime action of the electrical company and the deployment of state disciplinary measures gave the inhabitants of the Lower Elbe region insight into omnipotence and impotence such as they had never gotten before." Many "marshland people" now felt they were living in a "police state."[30]

The conservative press depicted anti–nuclear power activists in a highly negative light, however. Down-market but high-circulation newspapers such as *Bild* and *Die Welt* published overwhelmingly negative articles about Brokdorf protesters, many of them highly derogatory, according to a content analysis by the Battelle Institute.[31] The *Frankfurter Allgemeine Zeitung*, a respected, widely read conservative newspaper with a reputation for a high intellectual level, generally took a more differentiated approach. For example, journalist Klaus Wiborg wrote that he recognized that not all anti-Brokdorf protesters were leftist radicals but that "provocateurs" had played a disproportionate role there.

He pointed out the hypocrisy of Moscow-oriented groups involved in the Brokdorf conflict that ignored the proliferation of atomic power plants in the Eastern Bloc. Here, even non–Moscow-oriented "K-groups" agreed with him. Wiborg argued that the Brokdorf power station was needed to cover energy needs.[32] Chancellor Helmut Schmidt had spoken out in the days after the demonstration, saying that Brokdorf had to be built to ensure sufficient electricity but, more importantly, to protect jobs: "I'm telling you, your jobs depend on whether or not we build enough power plants."[33]

A conservative newspaper, the *Bayern Kurier*, asserted that anti–nuclear power activism was a "classic example of Communist agitation."[34] This argument contradicted the facts: among the fifty-two demonstrators arrested, only seven were members of Communist organizations.[35] "The continual defamation of the demonstrators bodes ill for the future of political rights in this country," proclaimed a public letter signed by Günther Grass and other well-known authors.[36]

The Stoltenberg government was particularly concerned about the depiction on television of the events of 30 October 1976. The NDR program "Brokdorf—Ein Zweites Wyhl? Protokoll eines Konflikts" (Brokdorf—A Second Wyhl? Report on a Conflict) in fact placed police actions in a negative light.[37] "It all began peacefully," the voice-over stated at the beginning. The program very much downplayed the role of Communist groups, while emphasizing the role of locals concerned with radioactive contamination of fields and livestock. A Protestant minister in his clerical vestments, acting as a spokesman for the demonstrators, added to the sense of legitimacy of the protests.

In this context, police use of force—scenes of police using truncheons, tear gas, and water cannons—seemed brutal and unjust. The conservative

Rheinischer Merkur ridiculed NDR coverage of the 30 October demonstration: "[It was] Brokdorf as they wanted to see it: martial-looking police officers, water cannons, German shepherds, then the poor demonstrators who were clubbed when the demonstration was disbanded, a minister—naturally opposed to atomic power—in his black robes."[38] The Stoltenberg government demanded the right to broadcast a rebuttal to this program on NDR, threatening to take the case to court.[39] At this point, NDR resisted but could not in the long run withstand the pressure exerted by Schleswig-Holstein.

The political fallout of the 30 October crackdown was, however, damaging to the Stoltenberg government. Members of the committee for the interior of the Schleswig-Holstein state parliament sharply questioned Rudolf Titzck, Minister of the Interior, regarding the use of force at the 30 October demonstration. One member pointed out that if allegations that mace had been used on demonstrators who were peacefully sitting on the ground—to force them to get up—this was in violation of standing orders.[40]

The Social Democratic Party in Schleswig-Holstein turned against the Brokdorf project, bringing a proposal before the state parliament on 23 November 1976 that would have permanently terminated it.[41] Violence on 30 October 1976 also caused members of Helmut Schmidt's cabinet to begin to doubt the wisdom of forcing nuclear power on an unwilling citizenry. State secretary Volker Hauff of the Federal Ministry for Research and Technology commented in the wake of the 30 October demonstration, "If we again have something like in Brokdorf, we'll have to write off nuclear power."[42] Criticism of actions of the Stoltenberg government and of the police, coming not only from the political Left but also from more moderate Social Democratic and liberal circles and media outlets, intensified in the wake of the 13 November demonstration.

"A form of violent demonstration unknown up until now": The Role of the Police in the 13 November 1976 Occupation Attempt

A tremendous escalation of violence took place on 13 November when twenty to thirty thousand demonstrators converged on Brokdorf. This demonstration is particularly well documented, allowing a close analysis of policing methods and media responses. More police were deployed than ever before in the history of the Federal Republic: 1,300. Police units from neighboring states were called in, as were rapid response battalions of the federal police (*Bundespolizeiabteilungen,* or BPA) and Federal Border Guard units. The BPA were national riot police, while the Federal Border Guard, as described by historian Falco Werkentin, was a cross between a military force and a police force, not unlike the National Guard in the United States.[43]

The Federal Border Guard brought military thinking, tactics, and equipment to the task. They were the ones who commanded and operated the helicopters and water cannons at Brokdorf. Testimony in the media as well as police reports indicate that this militarization of police response, rather than encouraging participants in the protest to demonstrate peacefully, made a peaceful demonstration impossible.

The BUU had called for a peaceful demonstration, and in fact the rally began with an open-air church service conducted by Protestant ministers, who also begged protesters not to resort to violence. A Social Democrat, Freimuth Duwe, gave a speech in which he accused Stoltenberg of "destroying the rule of law." Environmentalist and Christian Democratic member of the *Bundestag*, Herbert Gruhl, also spoke at the rally, calling for the "defense of the last intact vestiges of our homeland." An atomic physicist of the University of Bremen, Dieter von Ehrenstein, spoke out against leaving it to technocrats to solve the major problems faced by society.[44] The majority of demonstrators simply marched peacefully—most as part of a citizens' initiative. One witness recalls that there were about as many women as men among the nonviolent demonstrators.[45]

The BUU lost control over the demonstration, however, as a group of some three to four thousand demonstrators began a siege on the construction site, which they hoped to occupy. Some members of the BUU apparently spontaneously decided to join in this assault. Between 2:00 and 9:00 p.m., wave after wave of demonstrators advanced to the fence, using planks and sticks to cross moats, ladders to climb the fence, and heavy-duty cutting tools to break through the fence. Overwhelmingly men, these militants assaulted the police with rocks, iron bars, and Molotov cocktails. During this and later demonstrations, the reinforced fence around the construction site was the focus of pitched battles. The police defended it with great tenacity and were successful in keeping the protesters out. The police used tear gas and mace to push them back. The active ingredient in tear gas, phenacyl chloride, was also added to the water sprayed under high pressure from water cannons. Where protesters had actually broken through the fence, police formed human chains to hold them back and rolled out more razor wire.

At one point, when charging demonstrators seemed about to break through the fence, police helicopters dropped tear gas canisters. Some retreating protesters were also gassed. The crowd destroyed a water cannon and tipped over and burned a police van. Miraculously, no one was killed. Estimates of the number of injuries vary widely. The police reported that 81 police officers and about 100 demonstrators were injured.[46] By contrast, the

BUU asserted that at least 500 demonstrators had been injured by the use of tear gas and mace, some quite seriously; 16 had suffered trauma to the head or spine after being hit by police officers; and 150 experienced hypothermia after being soaked by water cannons. Further, the BUU claimed that police had not allowed ambulances through. At least one journalist was injured in the melee. Protesters came forward to testify that police had pursued and attacked peaceful demonstrators on foot, using both truncheons and chemical sprays, sometimes spraying mace into the face of a demonstrators at close range. Police officers appeared to be angry and "upset."[47]

The SPD was quite critical of the Stoltenberg government's handling of the demonstration. Head of the Schleswig-Holstein SPD, Günther Jansen, threatened to press charges against the Schleswig-Holstein Minister of the Interior, Titzck, several police officers, and a helicopter crew on the grounds of criminal assault, as was reported on the evening news.[48] A more detailed and sober critique of the government's policies took place during a Schleswig-Holstein parliamentary committee meeting. SPD official Klaus Klingner criticized the police for not allowing a BUU car equipped with loudspeakers to enter the area where the rally took place, thus preventing the BUU from communicating with the demonstrators and trying to keep things under control.

Gert Börnsen, a Social Democratic member of the Schleswig-Holstein parliament, pointed to other ways in which the police had undermined peaceful demonstrators and contributed to the escalation of violence: police blockades that forced demonstrators to walk over two miles—thus throwing the schedule for the rally into confusion—and the disruption of the rally and the open-air church service by helicopters circling overhead. In committee, Social Democrats grilled government representatives, who admitted that tear gas had been dropped on demonstrators who were trying to leave the area.[49]

Leftist, liberal, and politically moderate sectors of the press excoriated police and the Stoltenberg government for their handling of the demonstration. The *Hamburger Abendblatt* depicted families participating in the demonstration "with children and the dog" and recounted the experiences of an older woman who had come with her husband and adult children. Around 7:00 p.m., they joined crowds leaving the area. This is when helicopters dropped tear gas bombs on them. Taking off running, the woman found herself facing a police cordon. A police officer hit her over the head with a truncheon, causing a four-inch gash on her head. She fell into a water ditch, where she lost her shoes.[50]

Stern evoked the image of the Berlin Wall in explaining the desperate anger of demonstrators trying to breech it. The magazine depicted the 13

November demonstration as a "civil war" and an "atomic war with conventional weapons," waged by the state against citizens. When the state ordered "critical citizens to be bludgeoned," it acted like a "police state" instead of a "constitutional system." This would be "poison" for the politicians, *Stern* predicted.[51] Meanwhile, the conservative, right-wing and Christian Democratic press clung to the image of the demonstrators as Communists, "professional revolutionaries," and "thugs."[52] A *Frankfurter Allgemeine Zeitung* op-ed titled "Please, No Excuses" sought to blame the Protestant pastors and two members of the West German parliament who participated in the demonstration for the violence. They, along with other demonstrators, should have known better, the piece argued, because authorities, expecting violence, had warned against participating.[53]

The police use of force at Brokdorf was felt by many to violate generally accepted norms of police behavior. One of the basic problems, which sociologist Martin Winter has noted, was their fixation on protesters who were militant, violent, and seen as a threat to democracy.[54] Government and police also failed to recognize that the use of massive force could alienate and radicalize protesters. Police reports distinguished between "dedicated" environmentalists, organized in citizens' initiatives, and Communist groups bent on the overthrow of the capitalist system. But in the field, the police at times treated peaceful demonstrators as if they were militants, based on the idea that K-groups were sometimes *disguised* as citizens' initiatives. Moreover, police argued that the BUU—nominally in charge of the 30 October and 13 November demonstrations—lost control over events, but they declared solidarity with those attempting to storm the construction site, thus allowing the Communist groups to become the "driving force" behind the 13 November demonstration.[55]

Police maintained that the Brokdorf demonstrations of late 1976 and early 1977 inaugurated "a form of violent demonstration unknown until now." The police believed a "breakdown in the sense of right and wrong among participants" had taken place, promoted by certain political movements and segments of the media. This attitude, the police claimed, was encapsulated in the slogan "When justice becomes injustice, resistance becomes a duty." The "criminal energy" of the Communist K-groups surpassed "previous experiences in the Federal Republic." Although professing to only use violence against things, not people, the demonstrators had brought weapons with them—almost all of which were confiscated by the police—that police felt were clearly meant to be used against the police.[56] This idea was called into question in a Schleswig-Holstein parliamentary committee meeting by state representative Gert Börnsen, who argued that clubs could also be used to try to knock the fence down.[57]

There were specific circumstances on the ground that also contributed to police overreaction. It came out in a parliamentary hearing that—astonishingly enough—the Federal Border Guard units sent to Brokdorf consisted overwhelmingly of seventeen- to eighteen-year-olds.[58] In addition, police officers were allowed to drink alcohol on duty during the 13 November demonstration. Due to the "shameful consequences" of this loosening of regulations, officials decided to ban the consumption of alcohol by police officers during future demonstrations.[59] The police, Federal Border Guard, and federal riot police (BPA) also found themselves in a very stressful situation in which they were subjected not only to injury but also to insult: peaceful demonstrators whistled and hooted at the police in a way that was considered highly disrespectful.[60] It is also important to note that many of the forces deployed in Brokdorf were far from home and thus did not feel the kind of sympathy for locals that we have seen in the case of Wyhl. The distinction between "good" and "bad" demonstrators—problematic in the best of times—was evidently lost in the confusion of the moment. And it is not unreasonable to conjecture that the age and inexperience of some personnel, as well as inebriation, contributed to police brutality and overall poor judgment.

Other aspects of police handling of the Brokdorf demonstrations were, however, carefully planned and prepared, particularly the gathering of intelligence. They deployed undercover police officers and informants but also had the help of the *Verfassungsschutz*, the West German domestic security agency, the equivalent of the FBI in the United States.[61] It is not clear how the security service got their information, whether they simply gathered up mimeographed flyers at the Universities of Hamburg and Bremen or actual spies or informants sat in on meetings of Communist and environmentalist groups. In any case, they were well informed about plans to storm the Brokdorf construction site.[62]

However, activists were adept at fighting back against police surveillance and intelligence gathering. They infiltrated construction crews and security personnel, posing as journalists. Demonstrators used disguises and makeup to make identification after arrest nearly impossible. They often used much of the same gear as police officers: helmets, padded and waterproof clothing, truncheons, and shields. Undercover police trying to infiltrate the ranks of the demonstrators had a hard time because the groups had come up with clever ways of revealing who they were. For example, when a watchword was called out during a demonstration, all legitimate members of a group ran to the side, leaving the hapless police infiltrator standing alone.[63]

Police strategies were molded by their fixation on violent demonstrators. The police were aware of the large nonviolent segment of the movement but did not incorporate that knowledge into their strategies in any meaningful way in 1976–1977. They saw it as their role to prepare for the onslaught and to rebuff attempts to occupy the construction site, which they did very effectively. However, all too often peaceful demonstrators found themselves on the receiving end of police use of force. Police argued that they had remained within the bounds of West German law in their use of force on 13 November. However, they, along with the Stoltenberg government, faced a barrage of criticism.[64]

In the short run, a segment of the SPD turned against the Brokdorf nuclear power project, and a court injunction forced a hiatus in construction, which lasted from 1977 to 1981.[65] In the long run, protest policing at Brokdorf contributed to distrust, fear, and an escalation of violence on both sides of the conflict. This was amplified by similarly harsh policing measures used in Grohnde, where at a 19 March 1977 protest violent clashes between fifteen thousand demonstrators and five thousand police officers led to many physical injuries.[66] Some in the anti–nuclear power movement felt that activists themselves needed to play a major role in overcoming this pattern of violence and counterviolence.

Violence and Its Discontents: Debates among Anti–Nuclear Power Activists

The Brokdorf demonstrations of 1976–1977 unleashed intense debates concerning violence among activists. The strongest advocates of nonviolent methods were anarchopacifists, feminists, Protestant activists, and—outside of Hamburg—the citizens' initiatives and their umbrella organizations, the BUU (whose focus was Brokdorf) and the BBU, a national organization. Older people and demonstrators who came as individuals rather than as part of a group were also little inclined to participate in violence. A BBU booklet from 1977 laid out a game plan for citizens' initiatives, involving carefully planned actions and campaigns aimed at convincing the citizenry rather than seeking out confrontations with the police. It argued that by perpetuating animosity between the police and demonstrators, activists were playing into the hands of the state and industry and ensuring that the police remained pliant tools of the powers that be. They should focus on changing society from the inside: "Every social change begins with change in individuals or groups." To win over the population, activists should employ positive methods: "Constructive, creative actions are better than destructive ones; humorous ones better than bitterly earnest ones; enlightening and reasoning ones better than threatening ones; and those that seek understanding and sympathy better than hurtful or insulting ones."67 Admittedly, as Andrew Tompkins has pointed out, supposedly pacifistic anti–nuclear power organizations in fact at times behaved quite opportunistically, tacitly supporting forms of behavior that would normally be considered violent or disingenuously defining violence very loosely.68

A segment of the Left also embraced nonviolence. These decentralized anarchopacifist groups (often linked to citizens' initiatives) pursued strategies different from those of hard-line militants of the hierarchically organized Marxist K-groups (many of them Maoist or Trotskyist) and did not cooperate with them during demonstrations. "It has become clear that, given massive state power, an occupation in Brokdorf is no longer possible," wrote nonviolent protester Gerd Knapienski in December 1976.69 Such activists argued that violence only served to turn the West German public against the movement. Furthermore, police officers were human beings, they argued, and it was intrinsically wrong to commit violence against them or any other person. "We want to fight for our rights in imaginative ways, not by [indulging in] the mindlessness of violence." They would be untrue to their utopian goals if they employed violence, they argued.70 "Grassroots revolution" activists such as Wolfgang Hertle worked to foster cooperation between the pacifistically inclined peace movement and the anti–nuclear power movement.71

By contrast, *Arbeiterkampf* (Workers' Struggle), the newspaper of the Communist League (Kommunistischer Bund, or KB), declared ambiguously, "Police terror in Brokdorf demonstrates that this state represents the interests of finance capital, not the people. This state is based on brutal violence. The people cannot uphold their interests through 'peaceful resistance,' but also not through putschist kamikaze actions of isolated minorities."72 A radical Maoist K-group with roots in Hamburg, the Communist League reached a peak of about 2,500 members in this period, making the fight against nuclear power an important part of its program.73 It tried to dominate the citizens' initiatives that formed the foundation of the BUU in Hamburg but found itself embroiled in fierce infighting with rival K-groups. Wrangling between the Communist League and the Communist Party of Germany (Kommunistische Partei Deutschland, or KPD) threatened to paralyze attempts of local citizens' initiatives to mobilize the citizenry against Brokdorf.74

The split in the anti-Brokdorf movement deepened over demonstrations planned for 19 February 1977. A court order had just temporarily suspended construction, but most organizations wanted to go ahead with a major demonstration. While Schleswig-Holstein chapters of the BUU, leftist Social Democrats, and the German Communist Party (Deutsche Kommunistische

Partei, or DKP) advocated a peaceful demonstration in Itzehoe, a town kilometers away from Brokdorf, the Hamburg BUU and a great variety of other groups insisted on another protest in Brokdorf. According to *Graswurzelrevolution,* 50 percent of the delegates to the first West German conference on Brokdorf in 1977 favored a nonviolent demonstration.[75]

In the lead-up to the 19 February demonstration, the Communist League sharply rejected calls of the KPD and some "Spontis," or antiauthoritarian leftists,[76] to again try to storm the construction site and initiate an occupation, deeming such an attempt highly unlikely to succeed given the momentary circumstances.[77] The Communist League's approach was opportunistic, fueled by rivalries with other K-groups, but it also represented a slow backing away from violent methods and a growing awareness spreading among many activists that battles with the police were counterproductive and futile. The coverage in *Arbeiterkampf* also makes clear that many demonstrators were afraid and did not want to be physically harmed. Only a small minority seems to have found the prospect of confrontations with the police exhilarating.[78] In light of the massive police response to anti–nuclear power demonstrations in the West German town of Kalkar and Malville, France in 1977,[79] Sponti and later Green Party leader and Foreign Minister Joschka Fischer wrote that direct action had reached a "military dead end." Disillusioned by the futility of armed conflict, he and other leftists gradually became involved in politics.[80] One journalist argued that the anti–nuclear power movement needed to break out of its political ghetto and build bridges to average citizens who opposed atomic power but were not of the Left. This journalist saw nonviolence as absolutely essential to this broadening of the movement. [81]

However, the debates of the late 1970s concerning violence were very fraught. As historian Andrew Tompkins points out, the extreme police response to demonstrators in Kalkar and Malville, resulting in at least one death and several debilitating injuries, produced a deep sense of frustration and anger among activists.[82] In addition, the German Left felt under siege. The West German state reacted very harshly to an obituary for a victim of RAF terrorism, the West German attorney general, written by the "Göttingen Mescalero,"[83] an anonymous student, and published in a student government newspaper on 25 April 1977. The West German press and the authorities understood Mescalero to be advocating political assassination, although in fact he was arguing against it. The state pressed charges against over a hundred people involved in the publication or reprinting of the article. This muzzling of free speech fomented a deep sense of alienation.[84]

In addition, the West German anti–nuclear power movement was largely shut out of mainstream politics in this period.

Fighting a Two-Front War: Social Democratic Supporters and Opponents of Nuclear Power

Opponents of the Brokdorf project were constantly being told that they should not take direct action but should work within the political system to promote their objectives. However, massive political forces stood arrayed against any attempts to halt Brokdorf. The leadership of both major parties—SPD and CDU—viewed the atomic power project there as the decisive element that would either make or break the Federal Republic's nuclear energy program.[85] The energy crisis loomed large in the minds of members of the political establishment. They were also much influenced by the nuclear power industry and its lobbyists, who pushed the idea that West German prosperity was tied to nuclear power.

West German Chancellor Helmut Schmidt (in office 1974–1982) was a staunch supporter of atomic power, which he viewed as essential to economic growth.[86] Growth was crucial to maintaining employment levels, balancing government budgets, financing the West German social state, overcoming balance of payments problems, reducing social inequality in West Germany, helping developing countries, and paying for environmentalist measures—all top SPD goals.[87] In this, Schmidt had the full support of the labor unions, particularly the Confederation of German Trade Unions (Deutscher Gewerkschaftsbund, or DGB). Schmidt discounted nuclear risk and did not tire in pointing out that far more human lives were lost each year to coal mining accidents than to nuclear power.

Had it not been for Schmidt's steady support of nuclear power, the Brokdorf project might have been shelved. The Schmidt government realized that a profound shift in public opinion was taking place but hoped to overcome the "confusion and emotionalism" of the population by showing leadership.[88] Both publicly and privately, Schmidt and his Minister for Research and Technology, Hans Matthöfer, encouraged Stoltenberg not to give up when, in 1977, the courts forced a stop to construction, pending the resolution of the problem of what to do with nuclear waste from the Brokdorf plant.[89] Stoltenberg waited out this period of legal limbo, ordering work on the project to restart in 1981.

There were dissenters in the ranks of the Social Democratic Party, however, particularly in Northern Germany. The Schleswig-Holstein SPD brought a proposal to abandon the Brokdorf project before the state parliament on 23 November 1976[90] and called on citizens to participate in a peaceful demonstration against Brokdorf in Itzehoe in February 1977.[91] In a 1977 interview in *Der Spiegel*, Günther Jansen, head of the Schleswig-Holstein

SPD, expressed the opinion that Helmut Schmidt had "gone on a few too many foreign trips with industrialists and bankers."[92] He later had to apologize for this statement but did not give up his opposition to Brokdorf. In the wake of the Three Mile Island nuclear power plant accident on 28 March 1979 near Harrisburg, Pennsylvania, the SPD tried to make the state elections in Schleswig-Holstein on April 29, 1979 into a referendum on Brokdorf. Stoltenberg barely won, perhaps because he signaled that he intended to approach nuclear power with greater caution in the future, but in fact he soon resumed his quest to build the Brokdorf plant.[93]

By 1980, Hamburg Mayor Hans-Ulrich Klose, a Social Democrat, had turned against the project, at least in part as a reaction to Three Mile Island. "We know—despite Harrisburg and other accidents—too little about the residual risks of atomic technologies and about possible dangers resulting from technical failures," he wrote.[94] Already served by three nuclear power plants, Hamburg was already overreliant on nuclear power, he argued.[95] This exposed Hamburg to too great a risk. The times demanded new thinking about energy, growth, and conservation: "Anyone who wants Brokdorf is choosing the old energy policies, based on waste."[96] He was also concerned that frustration over Brokdorf would increase popular support for the Greens, then a small but growing party.[97]

Nonetheless, in February 1981, he was able to convince the Hamburg Senate—the city-state's governing body—to vote to pull Hamburg Electrical Works (HEW) out of the Brokdorf project.[98] This effort failed in the face of massive CDU opposition and claims of the ARD television network and the newspaper *Bild* that Hamburg would be liable for hundreds of millions of Deutschmarks in compensation for damages if HEW were forced to abandon Brokdorf.[99] Klose had difficulties maintaining the support of his own party, which feared that his politics were too leftist, particularly for voters of the older generation.[100] He felt compelled to step down as mayor on 25 May 1981 in light of his failure to pull Hamburg out of the Brokdorf project.[101]

Die Zeit termed the fight over Brokdorf an "ideological war concerning nuclear power."[102] Many viewed this as a clash between two different visions of West Germany's future, one centered on growth, the other on environmentalism. With a touch of drama, *Der Spiegel* asserted, "That divide—the barbed wire, concrete wall and moats that delineate the Brokdorf construction site—goes right through the entire West German people."[103] The Northern German television network, NDR, was pulled into this ideological struggle.

Brokdorf Media Wars, 1976–1986

NDR was nearly destroyed as a result of the fights over its reporting. NDR's public status as the regional broadcasting system of three states—Schleswig-Holstein, Hamburg, and Lower Saxon—made it subject to government pressure, unlike private print media, which enjoyed freedom of the press and looked back on a long tradition of assertively partisan reporting. The Stoltenberg government demanded that government spokespersons be given more airtime and be allowed to present rebuttals to NDR programs.[104] The governments of Hamburg and Lower Saxony did not support his demands for programming changes.[105] He nonetheless pressed ahead with full force. Top Schleswig-Holstein officials met with NDR's director, Martin Neuffer, on several occasions, threatening to pull out of the broadcasting consortium, a step that would have destroyed NDR's financial foundations.[106] Stoltenberg repeatedly spoke out in the press against NDR, saying that Neuffer would not get away with shrugging off "grave journalistic errors." He asserted that several NDR journalists had violated the constitutional foundations of the public broadcasting system and that their views were out of line with those of the "overwhelming majority of the population."[107]

Top NDR officials defended themselves. They noted that although it was illegal for public broadcasting stations to advocate the violent overthrow of the democratic constitutional order, they could certainly report on unconstitutional activities. NDR did not have the responsibility to defend government policies. An NDR analysis of program content revealed that twenty-nine programs presented nuclear power in a positive light, twenty-nine negatively, and eighty-seven in a neutral manner. NDR provided a long list of politicians who had been interviewed on the air about Brokdorf. In a vote on whether or not to censure NDR, the board split along party lines.[108]

However, the NDR board replaced Neuffer with Friedrich-Wilhelm Räuker, a Christian Democrat in 1980. According to *Der Spiegel,* Räuker was notorious for suppressing politically critical television programs, limiting the influence of overly independent-minded journalists, and promoting employees who exhibited the "right" attitudes.[109] The tenor of NDR programs on Brokdorf—as well as on a good many other topics—changed. Anti–nuclear power activists were now frequently termed *Chaoten* (roughly, "anarchists"[110]) or *Polit-Rocker* (political bikers or motorcycle gang members). In one 1981 program, Hamburg activists who printed pamphlets and flyers and distributed video tapes calling on the public to participate in anti-Brokdorf demonstrations were termed *Schreibtischtäter der Gewalt,* or "desk perpetrators of violence," a term that usually referred to Nazi era bureaucrats who helped organize the Holocaust from their desks.[111] Some NDR programs presented the police perspective. When fifty thousand demonstrated against the Brokdorf plant in June 1986, Räuker allowed only very limited NDR coverage.[112]

However, there still were quite a few programs that portrayed government policies and police tactics in a negative light and that questioned the wisdom of pursuing atomic power. This is a reflection of the courage and professionalism of NDR journalists, but it is also a consequence of left–liberal Hamburg's participation in the NDR consortium and its government's growing opposition to the Brokdorf project. An episode of the series *Panorama*, first broadcast on 20 January 1981, asserted that "times had changed": energy needs were sinking; the costs of nuclear power were rising; the safety of nuclear power seemed ever more questionable; and older atomic power plants were showing "signs of age" that necessitated repairs.[113] When the right-wing tabloid *Bild am Sonntag* vehemently attacked this program,[114] Director Räuker went to bat for NDR, forcing *Bild am Sonntag* to print NDR's rebuttal to this article.[115] Critical reporting continued under renowned journalists such as Stefan Aust and Karsten Biehl.[116] Christian Democrats eventually circumvented NDR by introducing private, commercial television stations. This development was to totally change the media landscape of the Federal Republic, depoliticizing television in the late 1980s and early 1990s.[117] By then, however, other strongholds of critical opinion had established themselves, notably high-circulation leftist newspapers and magazines.

An Upsurge in Anti–Nuclear Power Protest, 1979–1981

A skull superimposed on a dark, stormy sky looms over an atomic power plant—Three Mile Island, where on 28 March 1979 the worst civilian nuclear power plant accident up until then took place. The big, bold letters across this cover of *Der Spiegel* read "Alptraum Atomkraft"—"the nightmare of atomic power." One article in this issue, which discussed the political fallout of the accident in the United States, was titled "The Future of Atomic Power is Over." Another article expressed skepticism over the reassurances of the West German nuclear industry that such an accident could never occur in West Germany and explained what impact a super-GAU would have on human life.[118] Other West German media outlets also reported extensively on the accident, leading to widespread questioning of the safety of nuclear power.[119]

In the wake of Three Mile Island, Gorleben, a town in Lower Saxony, became a symbol of the West German nuclear power project and of resistance against it. Gorleben was slated to become the final resting place of nuclear waste produced by nuclear power plants. This involved two separate projects: one to reprocess uranium, another to store radioactive waste. In 1979, in protest, five hundred tractors made their way from various towns to Hannover, the state capital. One hundred thousand demonstrators met the participants in this "Gorleben Trek" in Hannover. State minister-president Ernst Albrecht promptly gave up plans for a reprocessing plant but insisted on moving forward with plans to turn naturally occurring salt domes in Gorleben into storage facilities for radioactive waste.

The Gorleben protests remained largely nonviolent, at least on the side of the activists, who included large numbers of women.[120] They brought a maternalist and ecofeminist orientation to the movement, focusing on women's concern for children.[121] Women dominated the leadership of the Environmentalist Citizens' Initiative of Lüchow-Dannenberg (Bürgerinitiative Umweltschutz Lüchow-Dannenberg), the organization that coordinated protests in Gorleben. Marianne Fritzen (1924–2016), a cofounder of the organization and its head around 1980, presented herself as the mother of five children and a woman of the people. "A short woman wearing a wool cap in front of a row of police officers. That was me," she said of a picture taken during a Gorleben protest that became an iconic image of the movement.[122] Starting out in a conventional role, she and others grew beyond it as activists: "Just having to speak in public and to deal with the male world, which we feared, has taught us something."[123] The Gorleben movement drew in local women. Activist Eva Quistorp remembers: "These so-called 'just housewives' turned out to be a good deal more clever than I as a big-city feminist had thought ... They dressed simply, were married; we would have tea together, and they would [suddenly] come up with the most amazing ideas."[124] By contrast, Undine von Blottnitz, a member of the Silesian aristocracy, was a very polished spokesperson of the Gorleben movement.[125]

In 1980 over the Easter weekend, five thousand women converged on Gorleben to take part in an anti–atomic power event. According to one woman, the exclusion of men allowed participants to break out of the traditional mold of female reticence and to freely talk and debate the issues. Over coffee and cake, they were able to get to know each other on a personal level but also to raise their awareness regarding nuclear power and to talk through fears regarding potentially violent demonstrations.[126] One of the local participants in the Gorleben Women's Group wrote in the feminist magazine *Courage* that the women needed to be by themselves so that men could not talk over them. Some husbands complained that they were being deprived of homecooked suppers when their wives went off canvassing. These female activists "saved their marriages" by getting their husbands involved in the movement, so that they would not feel neglected. This was even true of leading figures in the movement such as Undine von Blottnitz, the author recounted.[127]

Protesters of both sexes occupied one of the drilling sites for the nuclear waste disposal facility in May. Activists proclaimed the site the "Free Republic of Wendland" and, during the month-long occupation (3 May to 4 June 1980), created a village, replete with imaginative wooden structures from which banners of all sorts flew, cultivated communal life centering on building projects, communal meals, street theater and songs, and in general maintained good cheer and high spirits. Women built a "Women's House," while theology students erected a church. Occupiers manned a pirate radio station and published a newspaper. Antinuclear farmers expressed their support for the village, and one donated a slaughtered pig. Even the mail carrier delivered letters. Here was a happy, harmonious counterexample to the terrors of Brokdorf. Legions of protest tourists, including the very young and the old, trooped through the village. This photogenic experiment and its sad conclusion—police clearing of the site—were featured in a positive light on numerous TV programs,[128] although average citizens were not necessarily enthusiastic about the hippyish goings-on in "Wendland."

Police ended the occupation by force in June 1980. The demise of the "Free Republic of Wendland" led to a round of soul-searching on the Left. Some saw this experience as a repudiation of pacifism: "The moralists are asking no less from us than to surrender without a struggle to our own executioners."[129] An article in *Graswurzelrevolution* asked whether a woman who was raped should simply remain passive and not try to defend herself. The obvious answer was no. Similarly, the movement had to defend itself.[130] However, a different article in the same issue called for patience: "We need to let go of the idea that we could get rid of the nuclear program with two or three campaigns and a couple of explosive charges. To achieve a nuclear-free Germany/Europe, fundamental social changes will be necessary. A hundred thousand, a million people will need to become active in this process of liberation."[131]

Brokdorf again became the focus of anti–nuclear power activism. On February 28, 1981, somewhere between 50,000 and 100,000 demonstrators braved the bitter cold to march in the largest Brokdorf demonstration to date. Work on the Brokdorf nuclear power plant, which had been suspended for four years, had resumed earlier in the month. Police, activists, and the West German public held their breath in anticipation of what was feared would be the most violent demonstration yet. Some locals expressed fear of violence, although others expressed support of the demonstration.[132] A police report claimed that at an organizational meeting on 14 February, those who advocated a peaceful demonstration were silenced by the shouts of "militants."[133] Organizers decided to go forward with a march to the Brokdorf construction site, in defiance of a court-ordered ban on protest within a five-kilometer radius of the site.

On the evening before the big February 1981 demonstration, University of Hamburg students sitting in a pub drinking beer told a TV reporter that they were determined to demonstrate the next day, despite the court order forbidding the demonstration. One asserted, "You can't always respect government orders." In the Weimar Republic, he argued, citizens had to ignore a government order to get the eight-hour workday. Another student declared that he was going for the sake of his three-year-old daughter. The journalist asked why the prospect of violence did not deter this group of students. One answered, "I have only experienced violence committed by institutions," and she named schools, the workplace, and the police. Would she throw stones at the police?, asked the reporter. She would try to avoid it, she replied, but would defend herself if threatened.

But the program also gave police officers the chance to articulate their point of view. One police officer thought that citizens had the right to express their opinions by demonstrating but that violence could not be construed as a form of speech. He was unhappy that it was sometimes so difficult to discern which demonstrators were the troublemakers because that meant that the police sometimes went after peaceful demonstrators by mistake.[134]

As it happened, the 28 February demonstration turned out to be far more peaceful than previous ones, despite scattered violence.[135] Representatives of the Jusos and Young Democrats (youth organizations of the Social Democrats and the FDP), the Falcons (a socialist youth group), the Green Party and the BBU called on demonstrators to avoid violence at all costs. They provided the demonstrators with flowers, which they handed out to police officers at the demonstration.[136]

Historian Jan-Hendrik Schulz argues that the 1981 demonstration revealed deep fissures in the anti-Brokdorf movement. While demonstration leader Jo Leinen negotiated with the police and worked to prevent violence, "autonomous," antiauthoritarian Marxists and members of various K-groups sought confrontation with the police. Placed in the larger historical context presented in this chapter, Schulz's findings point to something more than just discord within the movement: In 1981, the "militants" played a much less central role than in 1976–1977. They were greatly outnumbered by the nonviolent demonstrators and outmaneuvered by the police. The police presence was massive: Some ten thousand police officers were deployed, along with Federal Border Guard units and special response units (*Spezialeinsatzkommandos*, or SEK), the equivalent of American SWAT teams. Police made effective use of nonviolent methods such as roadblocks, control posts, and searches. However, Federal Border Guard units are reported to have engaged in unprovoked attacks on groups of stragglers leaving the demonstration, and their helicopters flew so close over the heads of departing demonstrators that they knocked some off their feet—another example of the pernicious effect of the Federal Border Guard's militarized approach. According to one interview conducted by Schulz, the SEK units played a more constructive role, picking out and removing the most aggressive demonstrators with almost surgical precision.[137]

Liberal politicians and the media felt that both police and demonstrators had undergone a learning process: that both had ratcheted down the level of physical force and confrontation. Social Democrats and FDP leaders praised both. Even the BBU lauded the "astonishingly flexible" behavior of the police. In turn, the head of the police union (Gewerkschaft der Polizei, or GdP), Helmut Schirrmacher, criticized the politicians who refused to yield on the Brokdorf project and implied that he sympathized with the protesters, declaring, "Over 50,000 people marching to Brokdorf in this cold—that shows that something has been missing from the political discussion. The worst part is that the police always have to pay for it."[138]

Marion Dönhoff, publisher of the highly respected weekly commentary broadsheet *Die Zeit*, wrote, "The prognosis is hopeful in light of the events of last weekend" because both sides had come to their senses and realized what they needed to do to keep West German society from descending into a police state.[139] Television coverage also argued that activists had turned to peaceful methods. On Northern German Broadcasting, the head of the environmentalist BBU asserted, "With the exception of a few marginal groups," the Brokdorf activists were "normal, respectable citizens," adding, "We're not against the state, just against the nuclear industry."[140]

A new style of anti–nuclear power activism was emerging, characterized by the decline of the K-groups, the expansion of citizens' initiatives, the rise of the nascent Green Party, and the growing significance of nonviolent and feminist points of view in the movement. Links between the anti–nuclear power movement and the peace movement intensified. Largely nonviolent in nature, the peace movement took off in reaction to the threat of the deployment of middle-range nuclear weapons under the NATO Double-Track Decision of 1979, which was called that because it involved NATO rearmament in response to the Warsaw Pact's buildup of SS-20 middle-range missiles, combined with NATO offers of disarmament talks. The BBU brought together opponents of nuclear power and the peace movement at a 1981 peace conference. Protestant activists, such as Pastor Jörg Zink, also contributed to the coming together of the ecology and peace movements.

Their cooperation grew out of the twin threats of the destruction of nature and destruction through war. The two movements emphasized the links between civilian and military nuclear technologies, for example the dangers of plutonium production. They had a uniting vision of *Ökopax*—a world in which people and nature could coexist in peace. The peace movement had a pacifying impact on the anti–nuclear power movement.[141] International ties with European and US peace movements, often through intellectuals, scientists, physicians, or churches, also reinforced the nonviolent ethos of the West German peace movement, however different the specific aims in the various countries were.[142] The basically nonviolent peace demonstrations of the early 1980s were the largest demonstrations that West Germany had ever seen. More than a million West Germans took to the streets in the fall of 1983.

This was not by any means the end of militancy in the West German anti–nuclear power movement. But militancy and political violence narrowed in scope. According to government statistics, only 4.3 percent of all demonstrations in 1982 were marred by violence.[143] Environmentalism was achieving a broader appeal than ever before. A successful campaign against forest dieback due to acid rain (*Waldsterben*) was thought to prove the value of peaceful campaigns. This and other environmentalist issues turned the Green Party (founded in 1980) into a major political force, helping them to garner over 5 percent of the vote and thus seats in parliament in 1983.[144]

Conclusion: Protest and Public Opinion

The Brokdorf protests of the 1970s were characterized by violence and by their size. Initially, militants eclipsed the huge crowds of peaceful demonstrators. The local, rural population, whose protests in Wyhl were so appealing to older, more conservative segments of the population, played a far smaller role in Brokdorf. Instead, masked militants wearing helmets and throwing stones initially captured the attention of the police, politicians, and the media. This allowed Stoltenberg and others to portray the movement as related to the terrorism that was supposedly sweeping the land. Stoltenberg tried to stymie the movement in the four areas that sociologists see as crucial to the success of social movements: by preventing disruption through occupation of the construction site; by avoiding negotiations with activists; by sowing divisions between militants and "peaceful" organizations; and by undercutting the main ideas of the movement with accusations of subversion.

Stoltenberg overplayed his hand with a massive militarization of police deployments, however. He brought in large contingents of the Federal Border Guard, federal riot police, and SWAT teams, which contributed to an escalation of the violence. Local police—who might have approached the demonstrators in a more differentiated way—were lost among the endless ranks of militarily trained and equipped outsiders. The Stoltenberg government antagonized the movement by building a "Berlin Wall" around the Brokdorf site before the courts could decide the case, as well as with a huge, militarized deployment on 13 November 1976. Police were given strict orders not to retreat. Their aggressive tactics at that demonstration were apparently amplified by alcohol as well as the deployment of underaged recruits.

The policing forces' evident anger elicited anger among the activists but also acted as a magnet, drawing militant political groups that had been indifferent to environmentalism to the Brokdorf demonstrations. The fence around the Brokdorf nuclear power plant construction site became the focal point for attacks. However, the great majority of the demonstrators—women and older demonstrators in particular—avoided physical altercations. In later recollections, many demonstrators distanced themselves from "militants," a term that they themselves used. Police failure to distinguish between peaceful and violent demonstrators called forth media criticism and the beginnings of defections of Social Democrats from their party's pronuclear policies.

The anti–nuclear power movement would be anathema to the majority of West Germans as long as it was associated with violence. In the wake of the 1976–1977 demonstrations, the Left itself began to question the efficacy and morality of violence, particularly against people. Nonviolent anarchists were particularly clear in their stance, although the more violence-prone Communist groups also began to rethink their tactics. Brokdorf represented the high point of violent confrontation between anti–nuclear power activists and the state but also a turning point that eventually led to greater acceptance of anti–atomic power activism. The constant drumbeat of anti–nuclear power protests also grabbed the public's attention.

Even before Three Mile Island and Chernobyl, a persuasive negative narrative regarding nuclear power had made its way into mainstream popular culture—a reflection of the impact of the Wyhl and Brokdorf protests. Broad-ranging analysis of the press shows that by 1974–1977, newspaper coverage on nuclear power was overwhelmingly negative.[145] Concerns about nuclear power were surfacing in highly visible places. On a 1978 Christmas special, the very popular TV comic Loriot appeared in a skit called "We're Building an Atomic Power Plant." For Christmas, a grandfather buys his grandson a kit to build a model of a nuclear power plant. In the glow of the Christmas tree, the child's father encourages his son to put together the new toy with him:

> Look, Dicky—We're going to build an atomic power plant. There are the trees and the houses and the cows. They would like to have a nice nuclear power plant. And we're going to build one ... And now we carefully put the tiny uranium bar into the combustion chamber, and put the safety dome on top ... If we've done something wrong, it will go "poof" ... then all the houses and all the cows fall down.[146]

And because they have done something wrong, the little model atomic power plant explodes. "It's gone 'poof,'" the father exclaims.

When news of the Three Mile Island accident reached Germany, West Germans were coincidentally confronted at the newsstands with the latest issue of *Der Spiegel*, published just before the disaster, featuring on its cover a large warning sign with the radioactivity symbol and the word "Radioactive!" Published only days before, this issue featured a cover story about the conflict over Gorleben, titled "March to the Atomic State?" "Gorleben brings a new, explosive quality to the fight over nuclear energy," according to this article. "This is the unavoidable consequence of an energy policy that has relied too much on the nuclear power plant." The risks involved were "uncontrollable."[147]

Around 1980, antinuclear protest was becoming more palatable to moderate, middle-class, and older West Germans. Movies such as *The China Syndrome* stoked not only fears regarding the technology but also distrust against the elites that promoted it. New approaches to activism emerged from the debate about political violence on the Left. The Gorleben "Republic of Wendland" presented an idealistic, peaceful form of protest. Feminism had a moderating influence on the movement. The ever-expanding environmentalist movement was pioneering new forms of political mobilization and protest that did not rely on violence. On the other hand, support for the far-left Communist political scene was dwindling. This decrease in militancy was much in evidence in the February 1981 anti-Brokdorf demonstration.

When pollsters asked West Germans whether they favored the expansion of nuclear power, 53 percent said yes in 1976–1977, 36 percent in 1981, and 19 percent in 1984. A shift in the political culture of the Federal Republic had also taken place, with 54 percent of the 1984 survey participants asserting that nuclear power was a *political* issue. This refusal to yield to "expert" opinion was strongest among Social Democrats and Greens. While the overwhelming majority of scientists and representatives of the nuclear industry believed that the chances of a serious nuclear accident in West Germany were virtually nil, only 54 percent of the general population shared their optimism.[148] Such optimism was then crushed by Chernobyl.

Notes

1. LASH, Abt. 605, Nr. 5510, memorandum "Nächste Besprechung der Regierungschefs von Bund und Ländern."
2. Birgit Laprell, "Aktion Brokdorf. Bürgerinitiative wurde umfunktioniert," *Rheinischer Merkur* (5 November 1976). Similar argument in (no author), "Polizei: Brokdorfer Bauplatz-Besetzer waren keine friendlichen Demonstranten," *Flensburger Tageblatt* (5 November 1976).
3. Uwe Danker and Astrid Schwabe, *Schleswig-Holstein und der Nationalsozialismus* (Neumünster: Wachholtz, 2005).
4. Dieter Stäcker, "Zum Schluß wurden die Zelte verbrannt," *Tagesspiegel* (2 November 1976).
5. Radkau, *Aufstieg,* 210.
6. On Stoltenberg's ambitions, see "Die wären froh, wenn ich fortginge," *Der Spiegel* (19 May 1986).
7. "Wehren, versteken, weglopen," *Der Spiegel* (5 April 1976), 89–92.
8. Jung, *Öffentlichkeit und Sprachwandel,* 96.
9. Jungk, *Der Atom-Staat* and English translation: *The New Tyranny.*
10. Patrick Kupper, *Atomenergie und gespaltene Gesellschaft: Die Geschichte des gescheiterten Projektes Kernkraftwerk Kaiseraugst* (Zurich: Chronos, 2003). See also comments in Kai Hünemörder, "Zwischen Bewegungsforschung und Historisierungsversuch: Anmerkungen zum Anti-Atomkraft-Protest aus umwelthistorischer Perspektive," in Robert Kretzschmar, Clemens Rehm, and Andreas Pilger, *1968 und die Anti-Atomkraft-Bewegung der 1970er Jahre: Überlieferungsbildung und Forschung im Dialog* (Stuttgart: Verlag W. Kohlhammer, 2008), 162; article on 151–67.
11. "Warum Elbfischer Bertram Angst vor dem Atom-Kraftwerk hat," *Bild Zeitung* (9 March 1976). Brokdorf is located in Wilstermarsch, one of the Elbe marshes. Residents of the area are sometimes called "Marschmenschen," or marshland people.
12. Roger Karapin, *Protest Politics in Germany: Movements on the Left and Right since the 1960s* (University Park, PA: Pennsylvania State University Press, 2007), 117–59.
13. Schüring, *Bekennen.*
14. Inge Marszolek, "Radio Zebra in Bremen. Amateur-Radios und soziale Bewegungen in den frühen 80er Jahren," *Rundfunk und Geschichte* 39, nos. 1–2 (2013): 41; article on 41–55. Retrieved 26 October 2017 from http://rundfunkundgeschichte.de/zeitschrift/archiv/.
15. Staatsarchiv Hamburg, 224_3_P1977_28, "Wandzeitung: KKW Nein," 1 (January 1977).
16. NDR Archive, *Tagesschau*, 30 October 1976; 11, 13, 14, 16, 23 November 1976; 18 December 1976 (copies on VHS tape).

17. Frank Uekötter and Claas Kirchhelle, "Wie Seveso nach Deutschland kam: Umweltskandale und ökologische Debatte von 1976 bis 1986," in *Wandel des Politischen: Die Bundesrepublik Deutschland während der 1980er Jahre*, ed. Meik Woyke (Bonn: Verlag J.H.W. Dietz Nachf., 2013), 321–38.

18. Wehner, "Grenzen der Versicherbarkeit," 595.

19. Arbeitsgruppe Kernkraftwerke an der Universität Hamburg-Physik, Kernkraftwerke gefährden unsere Umwelt (Hamburg: photocopy, 1975). Mao-Projekt, Free University of Berlin, retrieved 6 December 2016 from http://www.mao-projekt. de/BRD/NOR/HBG/Hamburg_AKW_Uni_Physik_1975.shtml. According to footnotes on 20, 21.

20. "Mit Hauruck," *Der Spiegel* 45 (1 November 1976), 123–25.

21. "Brokdorf: Eis ohne Energie," *Der Spiegel* 46 (8 November 1976), 115–20; quotation on 115.

22. "Mit Hauruck," *Der Spiegel* 45 (1 November 1976), 123–25.

23. Peter Assmus, "Bundeskanzler zum Atomkraftwerk: Brokdorf muß gebaut werden!" *Hamburger Abendblatt* (November 9, 1976).

24. On the role of nonviolent groups in the occupation, see Gerd Knapienski, "Brokdorf—Übungsplatz für Bürgerinitiativen," *Graswurzelrevolution* 25–26 (December 1976), 4–5.

25. LASH, Abt. 621, Nr. 573, "Zusammenfassender Bericht über den Ablauf der Demonstrationen gegen den beabsichtigten Bau eines Kernkraftwerkes in Brokdorf," 1–12.

26. Wulf Beleites and Claus Lutterbeck, "Diese Woche," *Stern* (4 November 1976).

27. "Diese Woche: Atomenergie: Bürger, Prügel und Bonzen," *Stern* (11 November 1976).

28. LASH, Abt. 621, Nr. 573, "Zusammenfassender Bericht über den Ablauf der Demonstrationen gegen den beabsichtigten Bau eines Kernkraftwerkes in Brokdorf," 1–12.

29. Dieter Stäcker, "Zum Schluß wurden die Zelte verbrannt," *Tagesspiegel* (2 November 1976).

30. "Brokdorf: Eis ohne Energie," *Der Spiegel* 46 (8 November 1976), 115–20; quotation on 115.

31. Shirley van Buiren, *Die Kernenergie-Kontroverse im Spiegel der Tageszeitungen: Inhaltsanalytische Auswertung eines exemplarischen Teils der Informationsmedien: Empirisch Untersuchung im Auftrag des Bundesministers des Inneren* (Munich and Vienna: R. Oldenbourg, 1980), 94, 117–18.

32. Klaus Wiborg, "Brokdorf soll zu einem zweiten Wyhl gemacht werden," *Frankfurter Allgemeine Zeitung* (12 November 1976).

33. Peter Assmus, "Bundeskanzler zum Atomkraftwerk: Brokdorf muß gebaut werden!" *Hamburger Abendblatt* (November 9, 1976). On this occasion, Schmidt alluded to the political problems connected with relying on petroleum and asserted that coal was more expensive than nuclear power.

34. "Mißbrauchte Angst der Bürger: Ein Musterbeispiel kommunistischer Agitation," *Bayern Kurier* (13 November 1976).

35. LASH, Abt. 621, Nr. 573, "Zusammenfassender Bericht über den Ablauf der Demonstrationen gegen den beabsichtigten Bau eines Kernkraftwerkes in Brokdorf," 1–12.

36. "Documentation: Bauplatz für das Kernkraftwerk wurde zur Festung," *Frankfurter Rundschau* (9 November 1976).
37. NDR Archive, digitized film on computer hard drive, 1976.
38. Birgit Laprell, "Brokdorf: Bürgerinitiative wurde umfunktioniert," *Rheinischer Merkur* (November 5, 1976).
39. "Landesregierung fordert Gegendarstellung," *Flensburger Tageblatt* (10 November 1976).
40. LASH, Abt. 605, Nr. 5510, "Niederschrift über die 19. Sitzung des Innenausschusses des schleswig-holsteinischen Landtages am 4. November 1976."
41. LASH, Abt. 605, Nr. 5509, internal document titled "Zeittafel über die Entwicklung des Genehmigungsverfahrens beim Kernkraftwerk Brokdorf."
42. "Diese Woche: Atomenergie: Bürger, Prügel und Bonzen," *Stern* (11 November 1976).
43. Falco Werkentin, *Die Restauration der deutschen Polizei: Innere Rüstung von 1945 bis zur Notstandsgesetzgebung* (Frankfurt am Main, New York: Campus Verlag, 1984), 86.
44. "Sturmangriff von 3000 Radikalen auf Bauplatz Brokdorf gescheitert," *Flensburger Tageblatt* (15 November 1976).
45. Interview with Inge Marszolek, 7 January 2016.
46. LASH, Abt. 621, Nr. 573, "Zusammenfassender Bericht über den Ablauf der Demonstrationen gegen den beabsichtigten Bau eines Kernkraftwerkes in Brokdorf." Also Gert Kistenmacher, "Schwere Zusammenstöße in Brokdorf zwischen Polizei und Atomkraftgegnern," *Süddeutsche Zeitung* (15 November 1976). The number of wounded demonstrators accord to "Sturm auf Brokdorf abgewehrt," *Frankfurter Rundschau* (15 November 1976). On the journalist, see Claus Lutterbeck, "Der Bürgerkrieg in Brokdorf," *Stern* (18 November 1976).
47. FU Berlin, UA: Ermittlungsausschuß der BUU-1976, "Augenzeugenberichte aus Brokdorf," 3rd ed. (Hamburg: self-published, 1977), unnumbered pages.
48. NDR Archive, *Tagesschau*, 14 November 1976 (copy on VHS tape). "Die Schlacht von Brokdorf," *Hamburger Abendblatt* (15 November 1976).
49. LASH, Abt. 605, Nr. 5510, "Niederschrift über die 23. Sitzung des Innenausschusses des schleswig-holsteinischen Landtages am 6. Dezember 1976."
50. "Brokdorf—Eine Idylle ist zerstört," *Hamburger Abendblatt* (15 November 1976).
51. Claus Lutterbeck, "Der Bürgerkrieg in Brokdorf," *Stern* (18 November 1976).
52. For example, "Eine neue APO?" *Bayern Kurier* (20 November 1976); Elimar Schubbe, "Bürgerinitiativen vor dem Scheideweg," *Rheinischer Merkur* (19 November 1976).
53. "Keine Ausreden, bitte," *Frankfurter Allgemeine Zeigung* (15 November 1976).
54. Martin Winter, "Police Philosophy and Protest Policing in the Federal Republic of Germany, 1960–1990," in *Policing Protest: The Control of Mass Demonstrations in Western Democracies*, ed. Donatella Della Porta and Herbert Retter (Minneapolis, MN: University of Minnesota Press, 1998), 197, 199–200; article on 188–212.
55. LASH, Abt. 621, Nr. 576, Erfahrungsbericht, 9–11.
56. Ibid., 13, 10, 11.
57. LASH, Abt. 605, Nr. 5510, "Niederschrift über die 19. Sitzung des Innenausschusses des schleswig-holsteinischen Landtages am 4. November1976."

58. LASH, Abt. 605, Nr. 5510, "Niederschrift über die 23. Sitzung des Innenausschusses des schleswig-holsteinischen Landtages am 6. Dezember 1976."

59. LASH, Abt. 621, Nr. 576, Erfahrungsbericht, 35.

60. "Sturmangriff von 3000 Radikalen auf Bauplatz Brokdorf gescheitert," *Flensburger Tageblatt* (15 November 1976).

61. LASH, Abt. 621, Nr. 576, Erfahrungsbericht, 17.

62. Examples: LASH, Abt. 621, Nr. 573, "Zusammenfassender Bericht über den Ablauf der Demonstrationen gegen den beabsichtigten Bau eines Kernkraftwerkes in Brokdorf"; LASH, Abt. 605, Nr. 5510, untitled flier ("An allen Orten . . ."); LASH, Abt. 605, Nr. 5510, report of Norderstedter Initiativgruppe der Regionalgruppe Hamburg der BUU, 17 January 1977.

63. LASH, Abt. 621, Nr. 574, 57–58; Abt. 621, Nr. 576, Erfahrungsbericht, 12–13, 17–18.

64. The police report refers specifically to *das Prinzip der Verhältnismäßigkeit*, or the proportionality principle. LASH, Abt. 621, Nr. 573, "Zusammenfassender Bericht über den Ablauf der Demonstrationen gegen den beabsichtigten Bau eines Kernkraftwerkes in Brokdorf."

65. Court decisions of February and October 1977 forced construction to stop until the problem of nuclear waste disposal was resolved.

66. Protesters against the building of a nuclear power plant in Grohnde (Lower Saxony, Germany) tried to occupy the building site and were repelled by police.

67. FU Berlin, UA, APO-Archiv: S, Sig. 029–032, Umweltwissenschaftliches Institut des BBU e.V., "Aktionskatalog des Bundesverbandes Bürgerinitiativen Umweltschutz e.V." (self-published, 1977), 4, 5.

68. Tompkins, *Better Active*, 156–58.

69. FU Berlin, UA: Gerd Knapienski, "Brokdorf—Übungsplatz für Bürgerinitiativen," *Graswurzelrevolution* 25–26 (December 1976): 4; article on 4–5. Also Bernd Meyerholz, "Legt die Bomben Weg und die Steine," in the same issue of *Graswurzelrevolution*, 6.

70. FU Berlin, UA: Lorenz Kirchner, "Nur Selbstkritik . . .?" *Graswurzelrevolution* 25–26 (December 1976): 7.

71. Alexander Leistner, "Ein Grengänger auf Suche nach Heimat," *Forschungsjournal Soziale Bewegungen* 26, no. 4 (2013): 78–82.

72. FU Berlin, UA: "Kein KKW in Brokdorf!" *Arbeiterkampf* (1 November 1976).

73. Joppke, *Mobilizing*, 103–5.

74. FU Berlin, UA: "Den 19. Februar vorbereiten!" *Arbeiterkampf* (24 January 1977).

75. FU Berlin, UA: MAK, "Zur Politik des KB," *Graswurzelrevolution* 27–28 (1977): 8. This was a submission of a student group from Kassel.

76. The *Sponti* scene was very diverse, and it included groups with beliefs and aspirations grounded in anarchism, grassroots democracy, and non-Marxist–Leninism socialism. Joschka Fischer, who later became German foreign minister, started out as a *Sponti*.

77. FU Berlin, UA: "Den 19. Februar vorbereiten!" *Arbeiterkampf* (24 January 1977); "Großeinsatz der DKP an der Spalterfront," *Arbeiterkampf* (7 February 1977); "30.000 in der Wilster Marsch," *Arbeiterkampf* (21 February 1977).

78. FU Berlin, UA: "KBW dreht durch," *Arbeiterkampf* (21 February 1977); "30.000 in der Wilster Marsch," *Arbeiterkampf* (21 February 1977).

79. Tompkins, *Better Active,* 166–73. On Kalkar, see Andreas Kühn, "Kalkar 1977: Anti-Atomkraft-Bewegung und Polizei im Wandel," *Geschichte im Westen* 22 (2007): 269–89.

80. Joschka Fischer, "Warum eigentlich nicht," *Pflasterstrand* 40 (1978): 88f. Cited and analyzed in Silke Mende, *"Nicht rechts, nicht links, sondern vorn": Eine Geschichte der Gründungsgrünen* (Munich: Oldenbourg Verlag, 2011), 337.

81. FU Berlin, UA: MAK, "Zur Politik des KB," *Graswurzelrevolution* 27–28 (1977): 8–9.

82. Tompkins, *Better Active,* 172–73.

83. This pseudonym was a reference to an Apache tribe and indirectly to the *Stadtindianer* (urban Native Americans), a West German offshoot of the Indiani Metropolitani, far-left Italian protesters loosely associated with anarchism, the political thought of Antonio Negri, and hippy culture.

84. Karrin Hanshew, *Terror and Democracy in West Germany* (Cambridge, UK and New York: Cambridge University Press, 2012), 197–208.

85. See, for example, Klaus Matthiesen's statements in "Jagd auf Aussteiger," *Der Spiegel* (26 January 1981).

86. Jürgen Häusler, *Der Traum wird zum Alptraum: Das Dilemma einer Volkspartei— die SPD im Atomkonflikt* (West Berlin: Sigma, 1988), 123–26.

87. LASH, Abt. 605, Nr. 5509, memorandum from Bundesminister für Wirtschaft III D 1–02 51 00/7, (23 March 1977).

88. LASH, Abt. 605, Nr. 5509, memorandum from Bundesminister für Wirtschaft III D 1–02 51 00/7, (23 March 1977).

89. LASH, Abt. 605, Nr. 5509, letter from Stoltenberg to Schmidt (6 February 1980); letter from Schmidt to Stoltenberg (14 March 1980). On Schmidt's support for the Brokdorf project in a speech on 30 January 1981: LASH, Abt. 605, Nr. 5509, internal document titled "Zeittafel über die Entwicklung des Genehmigungsverfahrens beim Kernkraftwerk Brokdorf."

90. LASH, Abt. 605, Nr. 5509, internal document entitled "Zeittafel über die Entwicklung des Genehmigungsverfahrens beim Kernkraftwerk Brokdorf."

91. "Vom Kopf her," *Der Spiegel* (21 February 1977).

92. "Es gibt doch keine Majestätsbeleidigung, *Der Spiegel* (21 February 1977). Also "Immer nur Ärger," *Der Spiegel* (9 May 1977).

93. Example: LASH, Abt. 605, Nr. 5509, letter of October 1979 from Stoltenberg to Chancellor Helmut Schmidt.

94. Staatsarchiv Hamburg 131-21, 5271, letter from Klose to "friends" (5 January 1981).

95. NDR Archive, *Panorama,* 20 January 1981. The three were Stade, Brunsbüttel, and Krümmel.

96. Jürgen Leinemann, "Oft war es ganz dunkel um mich," *Der Spiegel* (9 February 1981).

97. Staatsarchiv Hamburg 131-21, 5271, letter from Klose to "friends" (5 January 1981).

98. Staatsarchiv Hamburg 622-1/511 200, "Ergaenzungsdrucksache zur Senatsdrucksache Nr. 139," (25 February 1981); Staatsarchiv Hamburg 131-21,

5271, press release by Staatliche Pressestelle Hamburg, "Brokdorf: Beanstandung der HEW-Beteiligung moeglich: Erkennbare Forschritte für alternatives Energie-Konzept" (27 October 1981).

99. "Erst im Jahr 2000," *Der Spiegel* (23 February 1981).

100. "Drunter und drüber," *Der Spiegel* (19 January 1981).

101. "Klose-Rücktritt: Der große Schuschu," *Der Spiegel* (1 June 1981).

102. Horst Bieber, "Brokdorf? Jein danke!" *Die Zeit* (6 February 1981). The author actually uses the term "Glaubenskrieg," which can mean religious war or philosophical war.

103. "Wichtigste Entscheidung seit 20 Jahren," *Der Spiegel* (16 February 1981).

104. NDR Archive, *Panorama*, F 304, 15 November 1976, digitized file.

105. Staatsarchiv Hamburg 131-1 II 11524, memorandum of the government of Hamburg (22 November 1976); memorandum of the government of Hamburg, letter Schulze to Mohrhoff and Poetsch-Heffter (8 December 1976); memorandum dated 17 December 1976.

106. Staatsarchiv Hamburg 131-1 II 11524, letter from the head of the Schleswig-Holstein state chancellery (3 January 1977); letter of Hamburg Senate chancellery to State Secretary of Schleswig-Holstein (29 March 1977).

107. Staatsarchiv Hamburg 131-1 II 11524, *Kirche und Rundfunk*, Nr. 23 (26 March 1977); "Interview zum Sonntag: Was Stoltenberg gegen den NDR unternehmen will" (27 March 1977).

108. Staatsarchiv Hamburg 131-1 II 11524, "Verwaltungsratsvorlage Nr. 50/77 zur 312. Sitzung des Verwaltungsrats am 15. April 1977: Punkt 4 der Tagesordnung." There had long been a standoff between CDU and SPD members on the board.

109. "Wie gebrochen," *Der Spiegel* (15 December 1986).

110. The term refers to people who are very chaotic or who spread chaos, with the implication that they use violence.

111. NDR Archive, Series *Extra III*, program "Aufruf Brokdorf," first aired February 25, 1981, digitized file; Series *Panorama*, F 358 (3 March 1981).

112. "Angriffsobjekt Nr. 1," *Der Spiegel* (29 December 1986).

113. NDR Archive, Series *Panorama* (20 January 1981), digitized file.

114. "Kernenergie: So fälschte Panorama," *Bild am Sonntag* (January 25, 1981).

115. Letter from Stein and Puttfarcken to *Bild am Sonntag* and appendix (4 February 1981).

116. NDR Archive, memorandum from Aust and Biehl (3 February 1981).

117. Frank Bösch, "Politische Macht und gesellschaftliche Gestaltung: Wege zur Einführung des privaten Rundfunks in den 1970er/80er Jahren," in Woyke, *Wandel des Politischen*, 195–214.

118. "Die Zukunft der Atomkraft ist beendet" and "Eine lange Spur von Fragezeichen," *Der Spiegel* (9 April 1979).

119. ZDF Archive, "Schiesst das Zeug doch auf die Sonn: Documentation Gorleben," broadcast on 3 May 1979.

120. Kirsten Alers and Philip Banse, eds., *Unruhiges Hinterland: Portraits aus dem Widerstand im Wendland (in Text und Bild)* (Lüchow: AJB-Verlag, 1997), esp. 12–19, 66–83, and 115–18; Tiggemann, *Achillesferse*, 732.

121. FU Berlin, UA: "Mehr Frauen in die Front der AKW-Gegner!" *Arbeiterkampf* (21 February 1977); "Frauen, lieber heute aktiv, als morgen radioaktiv," *Arbeiterkampf*

(4 April 1977); *Frauen gegen Atomkraftwerke* (Hamburg: Frauenarbeitskreis der Bürgerinitiative Barmbek, 1977).

122. "Anti-Atom-Ikone Marianne Fritzen ist tot: Die vor der Polizeikette? Das war ich," *Tageszeitung-taz* (7 March 2016).

123. Cited in Eva Quistorp, "Frauen in Bürgerinitiativen," in *Die Neue Frauenbewegung in Deutschland: Abschied vom kleinen Unterschied: Eine Quellensammlung*, ed. Ilse Lenz, 2nd ed. (Wiesbaden: VS Verlag für Sozialwissenschaften, Springer Verlag, 2010), 852; article on 851–53.

124. "Die Seele der Grünen: Interview mit Eva Quistorp," *Grünes Gedächtnis* 2010, 18; article on 10–33, retrieved 27 June 2016 from https://www.boell.de/sites/default/files/Jahrbuch_Gruenes_Gedaechtnis_2010.pdf.

125. Jan-Hendrik Schulz, "Die Großdemonstration in Brokdorf am 28. Februar 1981—eine empirische Verlaufsstudie mit Blick auf die Fraktionen der Demonstrierenden und der Polizei," (B.A. thesis, University of Bielefeld, 2007), 53–58, University of Bielefeld, retrieved 9 January 2018 from https://pub.uni-bielefeld.de/person/16907742.

126. Wolfgang Ehmke, ed., *Zwischenschritte: Die Anti-Atomkraft-Bewegung zwischen Gorleben und Wackersdorf* (Cologne: Kölner Volksblatt Verlag, 1987), 65–67; "Frauen gegen Atomkraft: Gebärstreik," *Die Zeit* (11 April 1980).

127. Gudrun Scharmer, "Mein schönes Essen gibts Du an die Leute—Frauen im Landkreis Lüchow-Dannenberg," *Courage* 5 (1976): 6–7. From the clipping collection of the Papier Tiger Archiv, Berlin.

128. Tiggemann, *Achillesferse*, 728–47. FU Berlin, UA, APO-Archiv: S, Sig. 044, Manfred and Niko (no surnames of the authors given), "33 Tage in der Republik Freies Wendland," *Tageszeitung-taz* supplement "Dokumentation: Gorleben" (21 June 1980).

129. FU Berlin, UA, APO-Archiv: S, Sig. 044, Alexander (no surnames of the authors given), "Gegen den Moralismus und den Voluntarismus in der Gewaltdiskussion," *Tageszeitung-taz* supplement "Dokumentation: Gorleben" (21 June 1980). Also Alders and Banse, *Unruhiges Hinterland,* 117–18.

130. FU Berlin, UA, "Widersprüche und Fragen zur Besetzung und Räumung von 1004," *Graswuzelrevolution* 50 (October–November 1980): 23–26.

131. FU Berlin, UA, M.K., "Für und Wider die Gewaltfreiheit," *Graswuzelrevolution* 50 (October–November 1980): 22.

132. NDR Archive, *Nordschau*, 25 February 1981.

133. LASH, Abt. 621, Nr. 564, "Verlaufs- und Erfahrungsbericht der Polizeidirektion Schleswig-Holstein West: Polizeieinsatz Brokdorf, Februar 1981," vol. 1, 5–6. For an example of a portrayal of the nervous mood on television, see NDR Archive, *Tagesthemen*, February 25, 1981, digitized file.

134. NDR Archive, *Nordschau-Magazin*, 27 February 1981.

135. LASH, Abt. 621, Nr. 564, "Verlaufs- und Erfahrungsbericht der Polizeidirektion Schleswig-Holstein West: Polizeieinsatz Brokdorf, Februar 1981," vol. 1, 23.

136. "Politiker sind erleichtert," *Frankfurter Rundschau* (2 March 1981); "Gericht entschied: Demonstration ja, aber nicht in Brokdorf!" *Hamburger Morgenpost* (28 February 1981).

137. Schulz, Die Großdemonstration, esp. 9, 11, 13–14, 16, 25–26, 34–35, 41–42. Regarding the police, see LASH, Abt. 621, Nr. 564, "Verlaufs- und Erfahrungsbericht der Polizeidirektion Schleswig-Holstein West: Polizeieinsatz Brokdorf, Februar 1981," vol. 1, 13–15, 17–18, 22–23.

138. "Ausschreitungen bei insgesamt friedlicher Demonstration gegen Brokdorf," *Süddeutsche Zeitung* (2 March 1981).

139. Countess Marion Dönhoff, "Der Rechtsstaat in Gefahr?" *Die Zeit* (6 March 1981).

140. NDR Archive, *Panorama*, episode 358, 3 March 1981, digitized file. Also NDR Archive, *Extra III*, 25 February 1981, digitized file; *Nordschau-Magazin*, 27 February 1981, digitized file.

141. Mende and Metzger, "Ökopax," 118–34. On the role of the BBU, see Mewes, "A Brief History," 32.

142. Holger Nehring, "Transnationale Netzwerke der bundesdeutschen Friedensbewegung" in *"Entrüstet Euch!" Nuklearkrise, NATO-Doppelbeschluss und Friedensbewegung,* ed. Christoph Becker-Schaum, Philipp Gassert, Martin Klimke, Wilfried Mausbach, and Marianne Zepp (Paderborn: Ferdinand Schöningh, 2012), 219–23; article on 213–28.

143. Sturm, "Polizei und Friedensbewegung," 288.

144. Uekötter, *Deutschland in Grün,* 152–54.

145. Van Buiren, *Die Kernenergie-Kontroverse,* esp. 39–57.

146. Deutsche Kinemathek (Berlin), digitized file of "Loriot: Weihnachten bei den Hoppenstedts," ARD, first broadcast on 7 December 1978.

147. "Abmarsch in den Atomstaat?" cover, *Der Spiegel* (26 March 1979); "Gorleben: Das Zeitalter der Angst? Das nukleare Entsorgungszentrum entscheidet den Kampf um die Kernkraft," *Der Spiegel* (26 March 1979).

148. The scope and representativeness of these three surveys varied greatly. For the 1977 Infratest survey, see "Brauchen Wir Atomkraftwerke? Spiegel-Umfrage über den Bau von Kernkraftwerken," *Der Spiegel* (14 February 1977). Very different results for 1976–1977 (but only for three small regions) can be seen in Battelle-Institut, Einstellungen, A112–13. For the 1981 and 1984 results and analysis, see *Kernenergie und Öffentlichkeit: Ergebnisse einer Befragung von Politikern, Journalisten, Experten und der Bevölkerung* (Allensbach: Institut für Demoskopie, 1984), 3, 5, 8. On the other end of the spectrum, only 16 percent favored abandonment of atomic power and the shutdown of existing nuclear power plants in 1984.

Chapter 6

The Shock of Chernobyl and the Environmentalist Breakthrough in West Germany

On 28 April 1986, a mild, pleasant spring day, Europeans turning on the radio or TV, or perhaps speaking to a friend on the phone, were shocked to learn that a radioactive cloud was crossing Europe. That evening, Soviet media announced that two days earlier, on 26 April, an accident had occurred at the Chernobyl nuclear power station in the Soviet Ukraine.[1] In the worst nuclear power plant accident up to that time, one of the reactors had exploded, releasing radioactive materials into the atmosphere.[2] The fallout was the most intense in the area surrounding the plant, but winds blew the radioactive cloud toward Scandinavia, then the Netherlands, Germany, Poland, and Czechoslovakia.

The prospect of rain was a fearful one, since it would wash the radioactive dust down from the sky and onto towns and cities, forests, and lakes. Rainfall in Bavaria, in fact, badly contaminated the soil. Prone to absorb minerals, mushrooms growing in affected forests became dangerously radioactive. Over the next few days, particles of radioactive iodine, cesium, and strontium spread across wide swaths of Europe and, later, across the globe.

Protest singer and activist Walter Moßmann conveyed his sense of the disaster as a surreal and perverse experience, notable for its lack of a focal point or even a perceptible physical manifestation, in a speech before assembled members of the Green Party in May 1986:

> The cloud is not visible, only a picture that knowledge has created. Nothing to hear, see, smell, taste, touch, nothing; above all, no dirt. This trauma of cleanliness, this nonsense about pollution! What is perverse about radioactivity is that it's clean. And when I look out of the window, I see the cleanest idyllic scene in the world: A cherry tree is in bloom, two sheep are grazing, and children are picking the first

wildflowers. Finally, it's warm and green again, and the wind blows through hair and skirts, but all that isn't supposed to be on this day. Here and there, I catch a few words: everywhere radioactive particles floating in the air, penetrating, concentrating, destroying cells. Then these absurd becquerel[3] units, whose meaning I don't really understand, though I sometimes pompously mouth them, but that doesn't mean much of anything.

That's the eerie thing: No bloody drama, whose significance we could immediately grasp. Here the knife, there the wound—that would make things clear. Here the hammer, there the skull—we could fathom that.[4]

Interviewed fifteen years later by *Der Spiegel* reporters, several prominent German politicians could remember exactly where they were when they first heard about the Chernobyl disaster.[5] For many West Germans, remembering that moment helped transform this piece of history into personal history. Chernobyl became anchored in popular consciousness in a way similar to 9/11 in the United States. Coming at a time when nuclear power had become something of a back-burner issue, Chernobyl rekindled a sense of urgency and won over many who had been unsure of what to think about atomic energy. For years they had been hearing about its dangers—and now the predictions were coming true.

Chernobyl broadened opposition to nuclear power in the Federal Republic. According to a survey conducted in May 1986 by the Emnid Institute on behalf of *Der Spiegel,* the percentage of West Germans who favored construction of new nuclear power plants fell in 1986 to 29 percent from 52 percent in 1980. Twelve percent wanted the Federal Republic to shut down all nuclear power plants immediately, 54 percent after "a transitional period."[6] West Germans poured out into the streets to protest against nuclear power on a scale not seen in France, the Netherlands, or Switzerland.[7]

These fundamental shifts have been downplayed by some observers as the product of "German Angst"—a term hinting at dark corners of the German psyche and self-indulgent philosophical predispositions.[8] Recent work influenced by the history of emotions helps us to place the undeniable fear of the post-Chernobyl period in a different context. This fear was an expression of popular concerns, of desire for health and an intact body—not only for oneself but also for one's children. It was not a helpless fear but a fear articulated as a demand for government action, fear paired with deep frustration and anger over the lack of an adequate response on the part of the political leadership.[9]

In the Federal Republic, this popular outcry was fostered and valorized by growing political mobilization and a rising educational level. The ever-rising sea level of popular participation found outlets in the Greens; the ever-expanding, interconnected peace, anti–nuclear power, ecology, and women's movements; a new generation of street-fighting anarchists (*Autonomen*); citizens' initiatives; but also in a more demanding stance on the part of middle-class, middle-aged moderates, whether or not they were involved in organizations.

Chernobyl Is Everywhere: Power, Persuasion, and Knowledge in West Germany

"Chernobyl Is Everywhere" became the slogan of the day. It encapsulated the frightening realization that there was no escape (at least in West Germany) from the immediate impact of the disaster but also, more generally, from the perils of nuclear power. The upsurge of opposition to nuclear power and the emergence of a new anti–nuclear power consensus were by no means just the result of fear, however. The interplay of power, persuasion, and knowledge were crucial. In West Germany, the resurgent Christian Democratic Party had retaken the national government under the chancellorship of Helmut Kohl (in office 1982–1998). Kohl, along with fellow Christian Democrats heading many state governments, were staunch supporters of nuclear power and atomic power industry exports.[10] They also favored a strong show of state power in the face of political unrest. Just weeks before Chernobyl, two demonstrators died during protests against the building of a nuclear reprocessing plant in Wackersdorf, Bavaria.[11]

There was growing intraparty discord over the NATO double-track decision and nuclear power among Social Democrats. A process of negotiation and accommodation took place between the party establishment and Social Democratic activists.[12] However, Social Democrat–controlled governments were quite capable of employing old guard tactics against demonstrators during post-Chernobyl protests, as will be discussed in this chapter.

The biggest challenge to politics as usual came from the Greens, who helped bring about a breakthrough of environmentalist issues as part of a larger revision of the West German political agenda in this period. The anti–nuclear power movement had served as a kind of political melting pot, bringing together diverse groups to form the Green Party in 1980 and strengthening their bonds. Distrust of government was a central mobilizing force, according to historian Silke Mende. The New Left had long seen the state as the agent of capitalist interests. Since the mid-1970s, far more moderate elements, such as the citizens' initiatives, had become distrustful of local and state governments that seemed willing to bend or even break laws and regulations in order to expand nuclear power.[13]

The Greens attempted to build a new foundation for trust, embracing "grassroots democracy." This principle proved to be controversial, however, generating struggles between the *Fundis,* or "fundamentalists," who saw the party as an alternative to the rigged and corrupted parliamentary system, and the "*Realos,*" or "realists," who wanted to run for election and make the parliamentary system work for their causes. These two factions also clashed over whether violence was justified in political struggles. Riding on a wave of surging environmentalism, Green representatives were first elected to the Bundestag in 1983. In 1985, the Green Party took its place as part of a state government for the first time, as Social Democrats and Greens formed a "red–green coalition" government in Hesse.[14]

State handling of Chernobyl greatly weakened popular trust in the government and the scientific establishment in West Germany. The Christian Democratic leadership's powers of persuasion abandoned them in the crisis. In the words of historian Frank Uekötter, "This [the reaction to Chernobyl] was a confirmation of what the nuclear power movement had been saying all along: No one was prepared for a nuclear catastrophe."[15] The Kohl government—although there was as yet no scientific consensus—initially insisted that the public had nothing to fear from the fallout. However, it instituted border controls on produce trucked in from Eastern Europe. West German border officials checked fruits and vegetables with Geiger counters and saw to it that trucks and cars entering from the East were washed. The government maintained that it was erring on the side of caution to reassure the public. Many West Germans felt unsettled by this contradictory approach.

Adding to the confusion and fear were uncoordinated pronouncements and guidelines of federal, state, and local governments. Christian Democratic leaders tended to downplay the dangers, whereas Social Democratic state and municipal official took the threat to public health more seriously. Standards for maximum allowable levels of radioactivity for milk and other products varied widely from one area to another. In some regions (but not in others), officials advised the public to avoid lettuce. Farmers in Hesse plowed lettuce and other crops under and followed official advice to keep cows in their stalls for a few days, while no such rule existed in Bavaria. Worried parents also heeded government warnings when they forced their bewildered children to keep out of sandboxes or even to avoid playgrounds and parks completely. Municipal government telephone information lines were swamped with calls from disconcerted citizens, and there was a run on iodine tablets. Confusion and anxiety spread, while confidence in the safety of nuclear power plummeted.[16]

Women were quick to realize the implications of radioactive contamination for their children. "Mothers Against Nuclear Power" were known for their grassroots organizing and imaginative events.[17] In Berlin, seventy women came to the first meeting of a new group of nursing mothers concerned about Chernobyl. They formed ties with similar groups across West Germany and marched together in demonstrations against nuclear power. They got information from Dr. Wassermann of the Toxicological Institute in Kiel. Thousands of women flocked to their network. In West Berlin, they experienced rejection, according to activist Anja Röhl: "The Greens accused us of exploiting our children for political purposes, and the feminists accused us of essentializing motherhood."[18] The nursing mothers found greater solidarity in other cities.

Chernobyl also undermined confidence in the scientific establishment, while encouraging the public to become more scientifically informed. Sociologist Ulrich Beck made risk the centerpiece of a rethinking of the relationship between science and society in his enormously influential work *Risk Society: Towards a New Modernity,* which came out shortly after the catastrophe. In this scholarly blockbuster—a rarity—he argued that risk was a central characteristic of modern society. Governments, science, and institutions could not protect society from catastrophic technological failure because these failures are neither predictable nor controllable.

Progress could not eliminate risk, according to Beck, because the threats are the product of the pursuit of progress itself. Science and technology were virtually immune to growing popular criticism and resistant to political control, he argued. He did, however, see a chance for real democratization, a "new modernity."[19] This book, like no other, opened the way for popular participation in major decisions on scientific and technological issues. Increasingly, West Germans saw risk, not as an objectively measurable scientific category but as a qualitative evaluation based on value choices rooted in political, social, and cultural considerations.

The media reflected and promoted this critical approach to risk. On the TV talk show *Drei vor Mitternacht* (Three Minutes to Midnight), experts clashed in June 1986 with counterexperts, and politicians with antinuclear activists.[20] The moderator asked provocatively, "Is there a pact between science, politics and the state? Is 'expert' just a euphemism for 'propagandist?'" She wanted to know why there were no critics of nuclear power on the German Radiation Protection Commission. Ludwig Gerstein, a Christian Democratic spokesman, asserted that the state selected its experts strictly according to professional standards. Jens Scheer, an anti–nuclear power activist and university professor, challenged this claim. He quoted Research

Minister Hans Matthöfer as having told a professor and member of the Jülich nuclear research center advisory board, "You can search for the truth at the university; here you have to promote nuclear power."

Jörg Kuhbier, an environmental official in the Hamburg government and a Social Democrat, pointed out that scientists called on to serve as experts for the government received all sorts of professional opportunities and research money, and so they were able to establish themselves as leading lights of their profession. Opponents of nuclear power, by contrast, did not receive the same professional and official recognition and were not recognized as experts.

The talk show guests also argued about the radiation standards established after Chernobyl. Gerstein asserted that the Radiation Protection Commission had established more stringent standards than those of other countries (such as France and Sweden) because the Federal Republic could afford the luxury of playing it safe. Out of an excess of caution, individual West German states had adopted even stricter rules, which had led to unnecessary confusion and fear.

Till Bastian, head of the German division of International Physicians for the Prevention of Nuclear War, saw things very differently. He pointed out that the maximum allowable amount of radioactive iodine in milk was raised after Chernobyl. This meant that children would be exposed to more than had previously been considered safe, raising the risk of thyroid cancer. Bastian believed that this was done out of consideration for the milk industry. Bastian referred to a calculation that the risk of developing cancer from a radiation exposure of 250–300 millirem[21] per year was one in ten thousand. He asserted that this would mean that six thousand people in the Federal Republic would develop cancer.

The Christian Democratic representative countered that every form of energy had its risks and that one risk had to be weighed against another. Pollution caused by the burning of coal was responsible for the dieback of forests, for example.

Kuhbier argued that science was not corruptible but that it also was not isolated from society. "When a society wants nuclear energy, it finds official organizations and institutions that provide scientific approval for nuclear power." He advocated the idea that diverse points of view should be represented on committees that made decisions concerning standards of radiation exposure. "We should talk about technology and about whether we are willing to accept this level of risk." A member of the audience reacted much more viscerally to the Christian Democrat's justifications of government policies: "Don't you have a heart? Don't you have children?"

The Green Party in Political Debates Concerning Chernobyl

The West German nuclear consensus shattered in the 1970s and was replaced by a new anti–nuclear power consensus that started taking shape in the 1980s. Chernobyl played a role in that shift, but Chernobyl would not have had the effect it did without a successful environmentalist movement or without the Green Party. When Joschka Fischer strode onto the big political stage in his signature sneakers and jeans, it was clear that his party, the Greens, represented an abandonment of politics as usual. Petra Kelly called the Greens an "antiparty." At the same time, the Green Party managed to halt the decline of leftist politics by moving away from the dogmatic, extremist, and violent tendencies of the 1960s and 1970s and by focusing on environmentalism, a consensus-generating topic.

Growing out of the thriving countercultural "alternative scene," the Greens in turn created a new political niche in terms of body language, speech, forms of communication, dress code, and musical idiom, as historians Kathrin Fahlenbrach and Laura Stapane have shown.[22] Their West Berlin Green-affiliated daily newspaper, the *Tageszeitung-taz,* rejected convention in everything, from its lowercase title and odd little interruptions in articles through what appeared to be ad hoc comments of the typesetter to its irreverent tone. Scholar Erik Harms has shown how style and content lined up in the Greens' 1987 election program. They conformed neither to linguistic nor to political norms, presenting goals without any of the usual rationalizations or flourishes, while avoiding any commitment to parliamentary democracy. Not until the 1990s did the Greens evolve into a much more conventional party, working within the system and taking on many of its habits.[23]

Environmentalism became a popular cause in the late 1970s and early 1980s; it was happily embraced by the mainstream parties, particularly the Social Democrats (SPD) and the Free Democrats (FDP). In 1970, Minister of the Interior Hans-Dietrich Genscher (leader of the liberal FDP) initiated West Germany's first program of environmental protection, creating, as historian Uekötter points out, a new term, *Umweltschutz.*[24] Only in 1986, in the wake of Chernobyl, was a ministry for environmental protection created—by Kohl's Christian Democratic government, which was trying to win over environmentally conscious voters. However, the major parties were either proponents of nuclear power or were divided on the subject.

Nuclear power helped to create the Green Party, bringing together fractious elements and winning supporters and votes. Granted, the roots of the Green Party were diverse. It understood itself as a product of the new social movements of the 1970s, and it drew heavily on the "alternative scene." However, as historian Silke Mende has shown, the Greens emerged from a great variety of intellectual traditions and movements: conservative environmentalism (with roots in the Weimar and Nazi eras), the *Lebensreform* (back-to-nature) movement of the late nineteenth and early twentieth centuries, as well as the Left's turn to antiauthoritarianism and grassroots democracy in the 1970s. Blending leftist and rightist thought, the Greens helped redefine progress, rejecting "the fetishism of technological progress" while maintaining a progressive view of society and culture and advocating a rising standard of living for those experiencing material want.[25]

In the late 1970s, as various Green and alternative lists began achieving some success in municipal elections, the Green movement remained quite amorphous. The national Green Party, founded in 1980, was rocked by fundamental debates that sought to overcome deep contradictions. However, delegates to a 1982 party conference committed themselves to a party platform demanding abandonment of nuclear power and a halt to NATO stationing of nuclear weapons in Germany. These remained unifying positions in the party, helping it to overcome bruising innerparty conflicts between the *Realos* and *Fundis*, which were driven not least by disagreements over whether to form coalitions with the Social Democrats.[26]

Women played a much greater role in the Green Party than in any previous German party. Half of all party offices and Green seats in the Bundestag were set aside for women. Activists such as Eva Quistorp, Marianne Fritzen, and Lilo (Lieselotte) Wollny, members of the founding generation of the Green Party, had their roots in citizens' initiatives and the anti–nuclear power and peace movements. They infused the party with ecofeminism. This could entail the belief that women were more caring and in harmony with nature than men, the progenitors of war, violence, and destruction of nature.[27] According to Quistorp, women in the citizens' initiatives saw to it that "non-violence was taken more seriously."[28] However, Petra Kelly simply saw men and women as equals.

For Petra Kelly, the fight against atomic power was part of her identity. She had become convinced that her sister, Grace, had died of cancer at age ten due to radiation exposure. This motivated her to become a peace activist and staunch opponent of nuclear power, as well as a supporter of research on childhood cancer. Her approach to these issues, which in her mind were a matter of deep conviction, was profoundly emotional. Although a Green Party leader and member of the West German parliament, she was a bit of an uncompromising loner. Her rise to media stardom only increased her isolation within the party, which rejected the personalization of politics. However, by putting a face on Green politics, she greatly raised its profile. Abroad, she acted as a sort of ambassador of West German Green politics.[29]

Jutta Ditfurth, one of the most important "ecofundamentalists" and part of the founding generation of the Greens, took an uncompromisingly leftist, anticapitalist line. At a party convention shortly after Chernobyl, she denounced the West German "atomic mafia," which in her opinion was willing to sacrifice the health of children for the sake of the interests of the atomic industry. In her speech, she asserted, "We live in a country that is an active member of the club of international atomic terrorism." She also criticized the Eastern Bloc, which oppressed activists and clung to an "irrational belief in technology."[30]. Although herself a member of the Bundestag, she believed that the Greens could achieve little through parliamentary work. Instead, she advocated direct action, that is, protests of various sorts.

Joschka Fischer and Otto Schily were the "realist" Green counterparts to Petra Kelly and Jutta Ditfurth's fundamentalism. Coming out of the anti-authoritarian but revolutionary K-group of the 1970s, Fischer had become disillusioned with political violence. As an early Green, he believed that his party needed to work within the parliamentary democratic system and form coalitions with the Social Democrats. In 1985, he became minister for the environment and energy in an SPD–Green coalition government in the state of Hesse. Although he had not come out of the anti–atomic power or environmentalist movement, he improved enforcement of environmental standards and induced the Social Democrats to agree not to build any more nuclear power plants in Hesse.

After Chernobyl, he became the go-to person for concerned citizens from across the Federal Republic, as well as a popular Green spokesman in front of TV cameras. The red–green coalition ended when the head of the Hessian government insisted on going forward with the upgrading of a plutonium plant in Hesse. Most Social Democrats were still very resistant to the idea of further coalition building with the Greens, although a new generation—leaders such as Oskar Lafontaine and Gerhard Schröder—were far more amenable.[31]

During a Bundestag debate on 14 May 1986, Chancellor Kohl defended his government's handling of the Chernobyl crisis. Speakers, including the chancellor himself, had to pass through a verbal gauntlet of hecklers from other parties—as was the custom in the Bundestag. Kohl asserted that an accident such as the one in Chernobyl could not take place in the Federal Republic because of the redundant safety systems in West German nuclear power plants. "That's why the nuclear power plants in the Federal Republic are among the safest in the world," he said. "That's what the Soviet Union also claimed!" called out one Green parliamentarian.

Kohl argued that atomic power was cleaner than coal, that it was cheaper, and that it employed many. Abandoning nuclear power would not solve anything because other countries would continue to use nuclear power. Instead, he proposed international agreements regarding safety and accident reporting. He pointed to the successes of his government in promoting economic progress, to which the head of Christian Democratic parliamentarians, Alfred Dregger, called out, "We are world champions!"[32]

In a rebuttal, the leader of the SPD in the Bundestag, Hans-Joachim Vogel, slammed the Kohl government's response to Chernobyl: "The helplessness in this crisis, the contradictory information, assessments and recommendations in these days were appalling. Chancellor [Kohl], you and your government contributed to people's concerns and fear, instead of allaying their concerns and fear." According to Vogel, Minister of the Interior Zimmermann's public statements had consisted of "bored and lethargic irrelevancies." Why had he not coordinated the responses of the national, state, and municipal governments? And why had the chancellor, who was away when the disaster took place, not made reassuring, clarifying statements on television? Vogel felt that the Kohl government had been focusing primarily on ensuring further public support for its nuclear policies—in short, a "propaganda offensive." Vogel went on to point out that if the West German safety standards were among the most stringent in the world, that was because of anti–nuclear power activism.

Signaling a sea change in Social Democratic thinking, Vogel conceded that the SPD had undergone a "torturous" learning process with regard to nuclear power. He proclaimed as a SPD goal the end to expansion of nuclear power and its abandonment after a transitional period, as well as the termination of a couple of projects: the nuclear reprocessing plant in Wackersdorf and the fast-breeder reactor in Kalkar.[33] This represented a remarkable turnabout for a party that had long supported nuclear power.

The Greens demanded the immediate shutdown of all nuclear power plants and termination of the use of nuclear power in West Germany. During the 14 May debate, Green parliamentarian Hannegret Hönes was highly critical of the Kohl government's handling of the Chernobyl crisis, which, she maintained, was no better than the Soviet response. Both, she argued, had engaged in cover-ups in an attempt to save their nuclear programs. She castigated Minister Rita Süssmuth, Minister of Youth, Family, Women, and Health, for defending lax standards regarding radioactive particles in milk. "You can't imagine," Hönes said, "the extent to which you have lost the trust of mothers and concerned fathers during the past few weeks." She asserted that Süssmuth had disappointed many women, who expected female solidarity across party lines.[34]

The Greens increased their portion of the popular vote to 8.3 percent in the January 1987 national elections, up from 5.6 percent in 1983. Perhaps more relevant to Chernobyl was the June 1986 election in the state where Gorleben was located—Lower Saxony—an election that was considered by some to be a "plebiscite on nuclear power." The Greens only marginally increased their presence in this provincial parliament (from 6.5 percent in 1982 to 7.1 percent in 1986), and the Christian Democratic–Free Democratic coalition remained in power. Joschka Fischer was not alone in believing that his party was scaring away voters with radical talk.[35] Many opponents of nuclear power wanted more than radical talk, however. They wanted direct action.

Brokdorf and the Question of Violence: Contradictory Tendencies in Post-Chernobyl Activism

Authorities scrambled to deal with a tremendous rise in militant activity in the wake of Chernobyl. "Brutality has intensified in recent months and weeks, due to Chernobyl, reaching levels never seen before," according to a Hamburg official.[36] Ruud Koopmans's data confirm that, going all the way back to 1965, there had never before been so many protests involving heavy violence in West Germany. At the same time, participation in peaceful antinuclear actions was also surging.[37]

The Green Party was split on whether or not to condone violent protests and, in general, on how to deal with these contradictory tendencies in the movement. In planning a large demonstration against the now-completed Brokdorf nuclear power station, to take place on 7 June 1986, the Greens, together with other organizations, agreed not to exclude any groups from participation. They put together a flyer that declared that all forms of "resistance" were legitimate, including "demonstrations ... blockades, direct action." They agreed that Hamburg protesters would drive to Brokdorf as part of a convoy headed by groups that intended to break through police checkpoints—by force if necessary.[38]

However, many of the approximately 35,000–40,000 participants, particularly those from Schleswig-Holstein, wanted the demonstration to remain nonviolent. One example is a young woman from a small village in Schleswig-Holstein, who cycled to Brokdorf with friends. She had been protesting against nuclear power for years. Chernobyl merely reinforced her views on the Brokdorf nuclear power plant. Like many of her friends, she felt she owed it to her children to do something about it. She spoke with many other cyclists on their way to Brokdorf. "I didn't talk to anyone who was planning to engage in any form of violence, or that was going to the demonstration just for fun," she wrote in a letter to the Minister of the Interior for Schleswig-Holstein. Her purpose was to complain about what happened when she got to the rally, which was held in a field near the nuclear power plant:

> I could see several hundred people, young and old, with serious, recep-
> tive expressions on their faces. Many looked tired after the long walk,
> shivered, ate a little picnic meal ... And suddenly, while the rally was
> still going on, a water cannon and approaching police! No warning!
> People streamed out of the field where the rally was supposed to take
> place, across a ditch and onto the street. A tear gas canister landed next
> to me. I was caught in the crowd, a ditch filled with water on the other
> side of the street. (I don't need to tell you about the effects of tear gas!
> Have you ever experienced it yourself?) I was very afraid![39]

Because they produced few first-person accounts, it is a good deal more difficult to reconstruct the point of view of the *Autonomen*—autonomous leftists, sometimes referred to as anarchists—who participated in this and other anti–nuclear power demonstrations. This radically decentered move-ment sought to destroy capitalism and the state through revolution. It rejected all ideological constraints, including orthodox Marxism–Leninism, as well as "all notions of order and conformity." *Autonomen* created an identifiable milieu rooted in tolerant alternative neighborhoods such as Kreuzberg in Berlin and the Hafenstrasse in Hamburg. Embracing a revolutionary ethos, they engaged in violent confrontations with the police and vandalism, for example destroying shop windows, looting, toppling electrical transmission towers, and setting cars on fire.[40] In 1987, *Autonomen* shot two police officers dead, but actual killings were rare.[41]

Black ski masks, often worn to demonstrations, were part of their signature style, but they were also an attempt to mask their identity and shield themselves from prosecution. From the early 1980s onward, they were a growing presence in anti–nuclear power protests, the fight against the expansion of the Frankfurt airport (Startbahn West), and the squatters' movement. Their methods were rejected by many nonviolent activists, for example during the 1980 "Republic of Wendland" protest in Gorleben. The peace movement rejected any connection with the autonomous movement, as well as what they considered to be violence, in particular physical con-frontations with the police. The *Autonomen* organized their own, separate "antiwar movement," involving confrontations with police during visits to West Germany by top American officials and swearing-in ceremonies of West German army recruits.[42]

There was only limited overlap with the punk scene of that period. Young *Autonomen* of the 1980s were "strongly influenced by the 'No Future' attitude of the time," according to Geronimo, a former activist.[43] However, West German punks had a very ambivalent relationship with the anti–atomic power movement, as historian Jeff Hayton's work shows. Many rebelled against their older brothers and sisters by rejecting that movement, although there were some anti–nuclear power demonstrators among punks, and some punk bands took up this cause.[44]

From the mid-1980s, both the peace movement and the anti–nuclear power movement sought recognition of the blockade as a legitimate, nonviolent form of civil disobedience, involving only passive resistance. Targets of such blockades included US and NATO missile bases as well as the Brokdorf nuclear power plant. Historian Michael Sturm has shown that some state governments, courts, and police forces saw participation in such blockades as prosecutable under criminal law as "coercion" (*Nötigung*), that is, the use of threats of violence to force one's will on others. This lined up with attempts of the Kohl government to undo the previous SPD government's liberalization of aspects of laws on demonstrations—part of a conservative rollback that the Christian Democrats (and their critics) billed as a "spiritual and moral renewal."[45]

However, as Sturm has shown, various factors moderated or cancelled out attempts to crack down on passive resistance and other forms of nonviolent protest. Citizens, some involved in citizens' initiatives called "Citizens Observe Police," as well as the police itself, increasingly used video cameras to monitor police behavior. Participation of prominent people such as Heinrich Böll in peace protests shielded participants to some extent. In addition, a younger generation of police officers was reluctant to intervene physically in blockades. They reported cases of brutality among their ranks, were members of police unions that became part of the peace movement, and were willing to participate in events that brought them into a dialogue with activists. Finally, in the "Brokdorf Decision" of 1985, the courts greatly limited the authorities' ability to forbid peaceful demonstrations, even in cases in which a minority was expected to commit violent acts.[46]

However, these liberalizing factors had a limited effect as far as the anti–nuclear power movement is concerned, as the events of 7 June 1986 and their aftermath demonstrate. Complaints from citizens who had been brutalized poured into officials' offices, as a trove of letters in the Provincial Archive of Schleswig-Holstein plainly show.[47] *Die Zeit* sympathetically recounted the stories told by countless witnesses of police tear gas and water cannon attacks on peaceful demonstrators.[48] Other newspapers were more reluctant to criticize the police,[49] despite the fact that police roughed up a number of reporters. According to journalists speaking out at a panel discussion of members of the German Journalists' Union, the tabloid *Bild,* the *Hamburger Abendblatt* (a local Hamburg paper), and the prominent news agency Deutsche Presse Agentur suppressed articles critical of the police written by their own reporters. Journalists from the NDR broadcast network and the otherwise rather critical *Stern* also felt that there had been internal censorship.[50]

Numerous participants in the 7 June 1986 demonstration brought charges against the police. Since the morning of the 7th, police had been engaged in ongoing battles with the militants at the fence of the nuclear power plant. The peaceful rally took place at some distance in a designated parking lot. The two crowds—the militants and the peaceful demonstrators—were clearly distinguishable, and no one questioned the police's right to use water cannons and tear gas against demonstrators trying to get into the nuclear power plant site by force.

An investigation by the state attorney general's office confirmed, however, that in fact, the police indeed used tear gas and water cannons on peaceful demonstrators—without any provocation and in the middle of a speech—around 2:50 in the afternoon. This was considered an illegal action, one that caused injuries to peaceful demonstrators. Video cameras provided the essential evidence. This is an early example of the way in which widespread video recording of police activity could empower citizens. The police made some of the recordings, which were now becoming common practice in West German police forces, while others were made automatically by surveillance cameras of the company that built the nuclear power plant. Amateur recordings and footage shot by a television crew from the ZDF and ARD networks were also used to ascertain what, exactly, had happened.

Disappointingly, the investigation failed to determine who, precisely, the culprit was. All the police officers in the vicinity at the time denied having fired the tear gas canisters. A member of the water cannon crew admitted only that he had stepped on the lever that sprayed water "by mistake." The responsible supervisors also insisted that they had not given orders to use high-pressure water or tear gas against the nonviolent demonstrators. Government investigators suspected that members of the force were covering for each other. The attorney general, feeling that nothing could be done about this obvious cover-up, determined that no charges could be pressed against any of the officers.[51]

A second serious incident had taken place in connection with the 7 June demonstration on the way to Brokdorf. In the small town of Kleve, at one of many police checkpoints, a violent battle took place between police and militants. These *Autonomen* had been traveling to Brokdorf from Hamburg in large convoys that included buses and motorcycles, their license plate numbers concealed with tape. The police hoped to disarm them at checkpoints on the way. According to a police account, all was quiet in Kleve as a small contingent of officers waited at the checkpoint, when suddenly, seemingly out of nowhere, a convoy appeared, and hundreds of demonstrators jumped out of their vehicles.

Hoping to break through the checkpoint, they attacked police, according to the police report. Their faces covered with balaclavas, many wearing helmets, they came at the police, swinging clubs, throwing stones and Molotov cocktails, and shooting off illumination flares. Up front were young "toughs" that police thought were trained in martial arts, or at least knew how to use their fists. Soon, they started attacking police with slingshots and iron bars. The head of a police unit said the officers feared for their lives. Reinforcements—riot police and SWAT teams (*Spezialeinsatzkommandos*)— were quickly flown in. In the ensuing melee, eighty police officers were injured, some severely, as well as an unknown number of *Autonomen,* and twenty vehicles belonging to militants were destroyed.

The police prevailed.[52] A report in the *Tageszeitung-taz* and an account by autonomist Geronimo confirm that the autonomists had attacked first, hoping to break through the checkpoint and arrive in Brokdorf fully armed. According to Geronimo, the *Autonomen* felt beaten and demoralized. He believed that although the activities of the *Autonomen* had little impact on nuclear power policy, their value lay in having the opportunity to "experiment with different forms of resistance." However, their greatest accomplishment over the next few months was to topple 150 electrical transmission towers[53]—a pointless form of wanton destruction.

Ironically, the most notorious case of police mishandling of anti-Brokdorf activism took place in otherwise liberal and SPD-controlled Hamburg. Unable to get to Brokdorf on 7 June because of the police checkpoints and battles between *Autonomen* and police, many would-be demonstrators from Hamburg turned back and demonstrated in Hamburg the next day. Without warning, police moved in around 12:10 p.m. to surround protesters on the Heiligengeistfeld, a scruffy field where circuses and carnivals were held, stretching from the alternative Karo-Viertel to St. Pauli, an entertainment and red light district. Police claimed that they had no choice because protesters planned to march to downtown Hamburg, where police believed they would riot.[54] They held about four hundred demonstrators for twelve hours and more under degrading circumstances.

This incident blew up into a major nationwide scandal for reasons that become clear from the transcript of hearings conducted soon thereafter by the committee on internal affairs of the Hamburg parliament. Various witnesses—a Protestant minister, a psychologist, a *Stern* photographer—remembered having strolled over to the Heiligengeistfeld to look for friends, join in the protest march about to get under way, or simply see what was going on. All was quiet and peaceful as some thousand demonstrators milled around. A few police officers stood in the distance. Suddenly, a line of police appeared and charged at demonstrators. Panic ensued, but most could not escape because they found themselves facing a second phalanx of police officers. Police pushed and shoved them and beat two men. Surrounded on all sides, they were held in a vise (*Kessel*) that became tighter and tighter as officers slowly pushed them together.[55]

Conditions became very bad over the course of the four or five hours that demonstrators had to stand tightly packed. Police forced protesters to wait for hours to be taken, under police guard, to use the restroom. Both men and women ended up urinating or defecating in public. As the police pushed the detainees into a tighter and tighter circle, it became harder and harder to avoid stepping in the excrement. The detainees were denied food and water for hours on end. In most cases, they were denied the right to make phone calls. Some of the fifteen- and sixteen-year-olds were very worried because they could not call their parents. A few individuals had panic attacks, but most were able to keep calm by holding hands and murmuring words of reassurance to each other.[56]

Police seemed convinced that they had trapped violent demonstrators. A minority probably were, judging from the weapons confiscated by the police.[57] None, however, committed acts of violence during the period in which they were held. The investigative committee estimated that two-thirds of the four hundred demonstrators held until nighttime were activists committed to nonviolent methods. This was not the perception of the police, however, who displayed considerable frustration and rage toward the detainees. When a 33-year-old psychologist tried to speak with the officer in charge, the officer refused, saying, "I don't talk to filth. I only talk to human beings." A witness testified that at one point police suddenly started hitting detainees over the head with truncheons without any provocation, injuring a woman and a young man so badly that they had to be taken away in an ambulance.

After nightfall, police went on the rampage against a line of taxis waiting in a side street, apparently because the taxi drivers signaled their solidarity by flashing their headlights and honking their horns. Police savagely hit the taxis with truncheons, smashing headlights, windshields, and taxi signs and putting dents in the vehicles. Police officers did not testify at the hearings, citing pending criminal charges that they were facing.[58]

A couple of police officers tried to resist the escalating brutality. At one point police started beating their truncheons against their shields, laughing at the fear they saw in the eyes of the detainees. One woman was very afraid that she was about to be hit, but the police officer standing in front of her saw this and whispered to her, "Don't be afraid. I'm not going to hit you." And he kept his word. During the police taxi riot, a different officer refused to go along with the attacks, saying that what his fellow officers were doing was criminal. He was told to "shut up," but he responded that he couldn't remain silent.[59]

The investigative committee both condemned the actions of police and expressed its understanding of the terrible strain police had been under. The Hamburg police, the committee argued, was overwhelmed with the wave of violent protests that had greatly intensified since Chernobyl. There had been a major anti-Brokdorf demonstration in Hamburg on 13 May, culminating in a riot in front of the town hall during which *Autonomen* had thrown stones and paint bombs. A small demonstration had taken place in Brokdorf on 6 June, preceding the big demonstration on 7 June. A survey of articles from the newspaper the *Hamburger Abendblatt* reveals no other antinuclear demonstrations preceding this demonstration, although Hamburg police were complaining in this period about a crime wave. For example, there was a riot between German and Turkish youths, unrelated to Chernobyl, in the St. Pauli section of Hamburg on 25 May, and punks vandalized cars in a parking lot on 2 June.[60]

Many Hamburg police officers were deployed in Brokdorf on 7 June and again faced long hours on duty in Hamburg on the 8th. The day was long, orders were unclear, and rioters taunted police around the perimeter of the Heiligengeistfeld. Police became totally "riled up" and increasingly aggressive. It had become more difficult than ever to distinguish between peaceful and violent protesters. It was understandable that senior police officials wanted to try to nip a riot in the bud by stopping participants at the assembly point where the march was supposed to start, the committee asserted.

However, holding the demonstrators for hours merely on the suspicion that they would join violent protesters was illegal. "If you think through this line of argumentation, it means holding the detainees hostage because of the violent acts of third parties," the report asserted. The committee also roundly condemned violent acts on the part of the police, including vandalism of the taxis. It pointed to major failures in the way the operation was organized, conducted, and overseen. A contributing factor was that the police chief did not oversee the operation because he was on vacation, but rather left it to a very inexperienced deputy police chief who evidently was in over his head. The parliamentary committee also called on peaceful activists to end their solidarity with violent protesters.[61]

In a confidential report, the Hamburg government (the Senat) drew rather different conclusions, expressing approval of police actions but also arguing that the root cause was citizens' concerns stemming from the Chernobyl disaster. In the long term, the report stated, the only real solution would be the abandonment of nuclear power.[62] This represents an astonishing about-face for the Hamburg SPD, which only a few years earlier had defended Hamburg's participation in the Brokdorf project.

Huge numbers of protesters had taken to the streets after Chernobyl, and this activism intensified after the Brokdorf protests of 7–8 June. In Wackersdorf, protests against the planned nuclear reprocessing plant raged on. Bavarian minister-president Franz-Josef Strauß approved the first deployment of CS gas, a particularly aggressive form of tear gas, against demonstrators. Some believed that this caused one protester's fatal asthma attack. As in Brokdorf and Hamburg, peaceful demonstrators declared solidarity with violent militants. In Hamburg, in the days following the Heiligengeistfeld incident, *Autonomen* stormed NDR studios, protesting what they saw as the network's biased coverage. *Autonomen* attacked NDR correspondents, beating them and destroying a camera. NDR journalists took great pains to remain neutral regarding this incident as well as the big Brokdorf demonstration and the detention of the four hundred protesters on the Heiligengeistfeld. At the end of the program, a commentator lamented the rise of violence because it stood in the way of an honest debate about nuclear power. He continued:

> The lesson to be learned from Brokdorf is that whenever there are conflicts of fundamental significance to our state—and nuclear power, yes or no, is such a conflict—there will be know-nothing stone-throwers, a few among thousands of peaceful demonstrators. Neither providing the police with more technology nor restricting the right to demonstrate will help. If we want to eliminate violence, our best bet is to tackle the real cause, for example by abandoning nuclear power.[63]

Conciliatory inclinations were bolstered by the growing number of completely nonviolent demonstrations. For example, on 7 June, the much-watched evening news program *Tagesschau* reported not only on the major, violence-torn demonstrations in Brokdorf and Wackersdorf but also on peaceful demonstrations in Hamm, Würzburg, and Munich.[64] Like the Munich demonstration, a protest in the town of Wilster (near Brokdorf) was primarily for children. One child interviewed in front of TV cameras said, "Nuclear power plants can explode, and then come little pieces, and they can kill." "And who told you that?" asked the reporter. "My mother," responded the child. Was this indoctrination? asked the journalist. One adult demonstrator with children in tow answered, "I don't think it's manipulation. We tell children not to cross the street when the light is red. Is that manipulation?"[65]

Aftermath and Conclusion

The expansion of atomic power came to be regarded as politically impossible in the wake of Chernobyl. The Greens' demands for an immediate shutdown of all nuclear power plants found little resonance. More influential was the Social Democrats' decision in August 1986 to make the abandonment of nuclear power within ten years part of their party platform.[66] Kohl's coalition remained in power after the January 1987 federal election. However, Kohl became a rather lackadaisical and lukewarm supporter of nuclear power and more of a supporter of other kinds of environmentalist policies than many had expected, including the creation of a ministry for environmental protection in 1986, a commitment to taking on a large share of the burden of reducing greenhouse gases in the run-up to the Kyoto Protocol of 1997, and the signing of international agreements on notification in case of nuclear accident, under the aegis of the IAEA, or International Atomic Energy Agency.[67]

It was no longer possible for right-wing politicians such as Franz-Josef Strauß to rally nationwide support for a militarized defense of nuclear power by arguing that antinuclear protesters were terrorists or a vanguard supporting a Soviet takeover.[68] Popular fears were of an entirely different nature in the 1980s than in the 1970s. The third generation of the Red Army Faction was little more than an oddity and a nuisance. With the spread of alternative energy sources and growing energy conservation, no one believed the lights were about to go out. Instead, West German citizens' greatest fear was nuclear war. Peace and anti–nuclear power movements were increasingly converging to form a blended antinuclear movement. The potential use of plutonium in nuclear weapons made the plutonium-producing fast-breeder reactors such as the Kalkar plant, designated as SNR-300, unacceptable to most West Germans. Completed in 1985, it never went into operation.

The Brokdorf nuclear power plant did start producing power for the grid in October 1986, but it was plagued for years by ongoing protests. Just after Chernobyl, there was an incident at the THTR-300 high-temperature thorium reactor in Hamm, resulting in the release of radioactive gases.[69] A December 1987 accident in Biblis A was hushed up for almost a year. A huge scandal enveloped the West German company Transnuklear (a subsidiary of NUKEM) in 1987 because of its negligent handling of nuclear waste. A cloud of scandal and corruption hung over the nuclear industry.

Chernobyl was the catalyst for a new environmentalist consensus, whose foundations had been laid by the anti–nuclear power movement and the Green Party. They had built up too much credibility to be dismissed, even as protests became violent. Police missteps engendered sympathy for protesters. Ultimately, although, it was the loss of trust in the political elite and the scientific establishment, as well as reasoned arguments against nuclear power, that explain the West German public's turn against nuclear power.

Notes

1. Serge Schmemann, "Soviet Announces Nuclear Accident at Electric Plant, *New York Times* (28 April 1986).
2. For details, see Schmid, *Producing Power*.
3. Becquerel are a measure of radioactivity. They cannot be automatically converted into rems and sieverts (measures of the impact of radioactivity on the human body) because of differences in distance from the source of radioactivity, body mass, and type of rays.
4. Archiv Grünes Gedächtnis, B.I.1, Nr. 56, Walter Moßmann, "Rede auf der a.o. Bundesversammlung der Grünen in Hannover," 16.–19.5.1986, 1.
5. Spiegel Online Politik, "15 Jahre Tschernobyl: Wo waren Sie, als der GAU stattfand?" 26 April 2001, retrieved 6 January 2016 from http://www.spiegel.de/politik/deutschland/15-jahre-tschernobyl-wo-waren-sie-als-der-gau-stattfand-a-130413.html.
6. "Neue Mehrheit für den Ausstieg: Spiegel-Umfrage über Tschernobyl und die Deutschen," *Der Spiegel* 20 (12 May 1986).
7. Koopmans, *Democracy from Below*, 204. Katrin Jordan explains the differences in reactions in France and Germany in terms of construction of different narratives in "Ausgestrahlt: Die mediale Debatte um 'Tschernobyl' in der Bundesrepublik und Frankreich 1986/87" (Dr.phil. diss., Humboldt University of Berlin, 2017). I was unable to obtain a copy in time for this publication.
8. For a critical treatment of the term, see Joachim Radkau, "Mythos German Angst: Zum neusten Aufguss einer alten Denunziation der Umweltbewegung," *Blätter für deutsche und internationale Politik* 5 (2011): 73–82.
9. Melanie Arndt, *Tschernobyl: Auswirkungen des Reaktorunfalls auf die Bundesrepublik Deutschland und die DDR* (Erfurt: Landeszentrale für Politische Bildung Thüringen, 2011), Kindle edition, section 4.4.
10. Nico Grasselt, *Die Entzauberung der Energiewende: Politik- und Diskurswandel unter schwarz-gelben Argumentationsmustern* (Wiesbaden: Springer, 2015), 106.
11. "Wackersdorf: Vorzügliches Arrangement," *Der Spiegel* 15 (7 April 1986).
12. Jan Hansen, "Zwischen Staat und Straße: Der Nachrüstungsstreit in der deutschen Sozialdemokratie (1979–1983)," in *Wandel des Politischen: Die Bundesrepublik*

Deutschland während der 1980er Jahre, ed. Meik Woyke (Bonn: Verlag J.H.W. Dietz Nachf, 2013), 521–57, esp. 544.

13. Silke Mende, *"Nicht rechts, nicht links, sondern vorn": Eine Geschichte der Gründungsgrünen* (Munich: Oldenbourg Verlag, 2011), 330–34, 353.

14. On the history of the Greens, see Andrei Markovits and Philip Gorski, *The German Left: Red, Green and Beyond* (New York: Oxford University Press, 1993); Thomas Scharf, *The German Greens: Challenging the Consensus* (Providence: Berg, 1994); E. Gene Frankland and Donald Schoonmaker, *Between Protest and Power: The Green Party in Germany* (Boulder: Westview, 1992); Werner Hülsberg, *The German Greens* (London: Verso, 1988).

15. Uekötter, *Deutschland in Grün*, 162.

16. Arndt, *Tschernobyl*, Kindle edition, sections 4.1 and 4.4; Horst Bieber, "Strahlenstaub über dem schönen Schein: Nach der Katastrophe von Tschernobyl steht die Kernenergie auf dem Prüfstand," *Die Zeit* (9 May 1986).

17. Astrid Kirchhof, "Frauen in der Antiatomkraftbewegung: Am Beispiel der Mütter gegen Atomkraft 1986–1990," *Ariadne: Forum für Frauen- und Geschlechtergeschichte* 62 (2013): 48–57.

18. Anja Röhl, "Wir lassen uns unsere Angst nicht ausreden!" in *Frauen aktiv gegen Atomenergie: wenn aus Wut Visionen werden. 20 Jahre Tschernobyl*, ed. Ulrike Röhr (Wiesbaden: genanet, 2006), 32; article on 31–35. also "Stillende Mütter organisieren sich," *Tageszeitung-taz* (14 May 1986).

19. Beck, *Risikogesellschaft*.

20. Deutsche Kinemathek [German Film Library, Berlin], *Drei vor Mitternacht*, WDR, first shown 24 June 1986. Digitized recording.

21. The rem and the millirem (one-thousandth of a rem), along with the sievert and millisievert, are estimated measures of the impact of radiation on the human body.

22. Kathrin Fahlenbrach and Laura Stapane, "Mediale und visuelle Strategien der Friedensbewegung," in *"Entrüstet Euch!" Nuklearkrise, NATO-Doppelbeschluss und Friedensbewegung*, ed. Christoph Becker-Schaum et al. (Paderborn: Ferdinand Schöningh, 2012), 229–46.

23. Erik Harms, *Der kommunikative Stil der Grünen im historischen Wandel: Eine Überblicksdarstellung am Beispiel dreier Bundestagswahlprogramme* (Frankfurt am Main: Peter Lang, 2008).

24. Uekötter, *Deutschland in Grün*, 119–20.

25. Mende, *Nicht rechts*, 89–91, 132–34, 152–59, 212–13, 239–40, 310–15, 330–39, 356–57, 397–405, 428–34; quotation on 402.

26. Horst Mewes, "A Brief History of the German Green Party," in *The German Greens: Paradox Between Movement and Party*, ed. Margit Mayer and John Ely (Philadelphia, PA: Temple University Press, 1998), 35–37; ch. on 29–48.

27. Mende and Metzger, "Ökopax," 124.

28. Juliane Grüning and Isabelle Faragallah, "Mein ganzes Herz ist mit Tschernobyl verbunden: Zeitzeuginnengespräch mit Eva Quistorp" in *Frauen aktiv gegen Atomenergie: wenn aus Wut Visionen werden: 20 Jahre Tschernobyl*, ed. Ulrike Röhr (Wiesbaden: genanet, 2006), 18.

29. Saskia Richter, *Die Aktivistin: Das Leben der Petra Kelly* (Munich: Deutsche Verlags-Anstalt, 2010).

30. Archiv Grünes Gedächtnis, B.I.1, Nr. 56, Jutta Ditfurth, "Vorschlag für eine Grundsatzerklärung der Grünen," quotations on 3, 4, 5.

31. Paul Hockenos, *Joschka Fischer and the Making of the Berlin Republic: An Alternative History of Postwar Germany* (New York and Oxford: Oxford University Press, 2007), 199–209, 211–13.

32. Archiv Grünes Gedächtnis A. Joschka Fischer, Nr. 12, "Deutscher Bundestag—10. Wahlperiod—215. Sitzung. Bonn, Mittwoch, den 14. Mai 1986," 16523–28; quotations on 16523 and 16527.

33. Ibid., 16528–34; quotations on 16530, 16531, and 16532.

34. Ibid., 16539–42; quotation on 16540.

35. Archiv Grünes Gedächtnis A. Joschka Fischer, Nr. 12, "Von FR, Hans-Helmut Kohl, für Pressestelle Umweltministerium, Herrn Georg Dick." See also "Ein Schiff kann lange brennen," *Der Spiegel* 25 (16 June 1986).

36. Staatsarchiv Hamburg, 131-1 II 8184 Senat der Freien Hansestadt Hamburg, "Protokoll der Sitzung des Innenausschusses am 17. Juni 1986," 7.

37. Koopmans, *Democracy from Below,* 132 and 205.

38. Staatsarchiv Hamburg, 131-1 II 8183 Senat der Freien Hansestadt Hamburg, "Polizeibericht über den Einsatz vom 8. Juni 1986," 3–4; quotations on 4.

39. LASH, Abt. 621, Nr. 632, letter to the Minister of the Interior of Schleswig-Holstein, dated 12 June 1986. The name of the letter writer cannot be revealed due to privacy laws.

40. Geronimo, *Fire and Flames: A History of the German Autonomen Movement,* trans. Gabriel Kuhn (Oakland, CA: PM Press, 2012), 14, 79, 92 (on vandalism); see also George Katsiaficas, introduction to this publication, 1. "Geronimo" is purportedly the pseudonym of a German autonomist who participated in anti–nuclear power demonstrations in the 1980s.

41. Jan Schwarzmeier, "Die Autonomen zwischen Subkultur und sozialer Bewegung" (Dr.phil. diss., University of Göttingen, 1999), 103. According to Schwarzmeier, *Autonomen* across the Federal Republic expressed rejection of this killing.

42. Geronimo, *Fire and Flame,* 90–92, 111–16.

43. Ibid., 79.

44. Jeff Hayton, "Culture from the Slums: Punk Rock, Authenticity and Alternative Culture in East and West Germany" (Ph.D. dissertation, University of Illinois at Urbana-Champaign, 2013), 26, 147–48, 315 fn. 954, 320, 325, 330, 353, 358, 359.

45. On the rhetoric of this rollback, see Peter Hoeres, "Von der 'Tendenzwende' zur 'geistig-moralischen Wende,'" *Vierteljahrshefte für Zeitgeschichte* 61, no. 1 (2013): 93–119.

46. Michael Sturm, "Polizei und Friedensbewegung," in *"Entrüstet Euch!" Nuklearkrise, NATO-Doppelbeschluss und Friedensbewegung,* ed. Christoph Becker-Schaum, Philipp Gassert, Martin Klimke, Wilfried Mausbach, and Marianne Zepp (Paderborn: Ferdinand Schöningh, 2012), 277–93.

47. LASH, Abt. 621, Nr. 632.

48. "'So geht das nicht weiter': Von Brokdorf bis Wackersdorf: Endet der neuerwachte Widerstand gegen die Atomenergie im Teufelskreis der Gewalt?" *Die Zeit* (13 June 1986).

49. "Gruppe von Gewalttätern drückte Brokdorf-Demo ihren Stempel auf," *Flensburger Tageblatt* (9 June 1986); "Politisches 'Wäschewaschen' um Polizeieinsatz in Brokdorf," *Flensburger Tageblatt* (10 June 1986).

50. "Th.J.," "Brokdorf: Auch ein Presse-Desaster," *Tageszeitung-taz* (25 July 1986).

51. LASH Abt. 621, Nr. 635, memorandum of IV 4, signed by Oberstaatsanwalt Schwarz on 13 October 1987.

52. LASH Abt. 621, Nr. 635, report by Police Director Hans-Heinrich Heinsen, "Polizeieinsatz am 07.06.86," 11–13.

53. Geronimo, *Fire and Flame,* 138; quotation on 139.

54. Staatsarchiv Hamburg, 131-1 II 8183 Senat der Freien Hansestadt Hamburg, "Polizeibericht über den Einsatz vom 8. Juni 1986," 8–10.

55. Staatsarchiv Hamburg, 131-1 II 8183 Senat der Freien Hansestadt Hamburg, "Bericht des Innenausschusses über die Vorkommnisse anlässlich und im Zusammenhang mit der Demonstration am Sonntag, dem 8. Juni 1986, auf dem Heiligengeistfeld," 107–10, 142–43.

56. Ibid., 57, 124, 147.

57. On confiscated weapons and age distribution, see Staatsarchiv Hamburg, 131-1 II 8183 Senat der Freien Hansestadt Hamburg, "Polizeibericht über den Einsatz vom 8. Juni 1986," 14, 15.

58. Staatsarchiv Hamburg, 131-1 II 8183 Senat der Freien Hansestadt Hamburg, "Bericht des Innenausschusses über die Vorkommnisse anlässlich und im Zusammenhang mit der Demonstration am Sonntag, dem 8. Juni 1986, auf dem Heiligengeistfeld," 20, 58, 68, 92, 99–101; quotation on 144.

59. Ibid., 9–10; quotations on 95, 120, 101.

60. "Steine gegen Hamburgs Rathaus," *Hamburger Abendblatt* (14 May 1986); "Massenschlägerei auf St. Pauli: Ein Schwerverletzter," *Hamburger Abendblatt* (26 May 1986); "Fischmarkt: Punker zerstachen Reifen von Besucherautos," *Hamburger Abendblatt* (2 June 1986); "Demonstration gegen Brokdorf," *Hamburger Abendblatt* (7 June 1986); "Immer mehr Verbrechen in Hamburg," *Hamburger Abendblatt* (23 May 1986).

61. Staatsarchiv Hamburg, 131-1 II 8183 Senat der Freien Hansestadt Hamburg, "Bericht des Innenausschusses über die Vorkommnisse anlässlich und im Zusammenhang mit der Demonstration am Sonntag, dem 8. Juni 1986, auf dem Heiligengeistfeld," 9–12, 19–20, 25–27, 32; quotation on 19–20. The information on the police chief's vacation is from NDR Archive *Hamburger Journal,* 11 July 1986, on VHS.

62. Staatsarchiv Hamburg, 131-1 II 8183 Senat der Freien Hansestadt Hamburg, "Senatsdrucksache Nr. 898 verteilt am 27.6.1986 für die Senatssitzung am 30.6.1986. Entwurf. Betr.: Abschließenden Stellungsnahme des Senats zu den Vorgängen auf dem Heiligengeistfeld in Hamburg am 8. Juni 1986."

63. NDR Archive, *Hamburger Journal,* 9 June 1986, "Krawalle in Hamburg," on VHS.

64. NDR Archive *Tagesschau,* 7 June 1986, on VHS.

65. NDR Archiv *Extra III,* 13 June 1986, digitized film.

66. Joachim Radkau and Lothar Hahn, *Aufstieg und Fall der deutschen Atomwirtschaft* (Munich: oekom, 2013), 349.

67. Uekötter, *Deutschland in Grün*, 166–67; Nico Grasselt, *Die Entzauberung der Energiewende: Politik- und Diskurswandel unter schwarz-gelben Argumentationsmustern* (Wiesbaden: Springer VS, 2016), 103–5.

68. He did try to make such an argument in 1986. "Wenn der erste auf Demonstranten schießt ..." *Der Spiegel* 30 (21 July 1986).

69. "Umweltfreundlich in Ballungszentren: Hoffnungen und Fehlschläge beim Hochtemperaturreaktor" and "Kernkraft: Funkelnde Augen," *Der Spiegel* (9 June 1986).

Chapter 7

Not Immune to Error

Chernobyl's Impact in the GDR

> Scientists, doctors, engineers, politicians, and military men are, in spite of their expertise, not immune to error, deceit, corruption, carelessness, hunger for power, and vanity. This is why it is vital that dangerous sectors such as the entire nuclear power production cycle be placed under public control.
>
> —Sebastian Pflugbeil, "Pechblende: Der Uranbergbau in der DDR und seine Folgen"

These words of physicist Sebastian Pflugbeil appeared in an East German samizdat[1] publication in 1988. They reflect a skepticism toward elites and concerns about the risks of nuclear power similar to that in the West. However, defended to the hilt by the state, the ruling party, and the secret police, the nuclear power consensus was not seriously questioned in the GDR until Chernobyl. One official claimed, "Doing without [nuclear power] would mean giving up on having a decent life and endangering the world population's food supply."[2] Even after Chernobyl, there were fewer than a thousand environmental activists in the GDR.[3] However, these activists put their careers, their family lives, and their freedom on the line for the sake of anti–nuclear power agitation and other environmentalist causes.

There are indications of broader environmentalist ferment in the GDR. East Germans—aside from those living in the so-called valley of the clueless, an area around Dresden where reception of Western broadcasts was poor—followed reports on Chernobyl on West German television and radio and addressed their concerns to the authorities. These reactions were not merely the results of Western influence, however. Chernobyl mobilized and amplified political and cultural forces already in existence in the GDR.

Some scholars and pundits have argued that state socialism and environmental protection were incompatible and that public discontent over pollution had contributed in major ways to the system's downfall in the GDR.[4] On the basis of its dismal environmental record, historian Frank Uekötter likens the regime of Erich Honecker—leader of the GDR from 1971 to 1989—to "narrow-minded predatory capitalism."[5] However, he, along with historian Tobias Huff, argue that the extreme environmental degradation of the 1980s was not inevitable. They assert that in the 1950s and 1960s, the two Germanys were fairly evenly matched in their plans to protect the environment. However, Honecker largely abandoned environmental protection, making economic growth, accumulation of foreign exchange, and social welfare his top priorities. There was an official, SED-led environmentalist organization in the GDR, but it achieved little.[6]

Another contentious question is how interested the East German population actually was in environmental issues. This is difficult to gauge, given the lack of a free press and the hostility of the SED and the state to that segment of the environmentalist movement not under their control. Activists had difficulties reaching the public. Rudolf Bahro and Robert Havemann published books about the environment and environmentalism in 1980 in West Germany, but these were unavailable in the GDR.[7] In 1982, the Council of Ministers declared all environmental data a state secret.

The West German Green movement encouraged East German activism, most spectacularly in a peace demonstration by Petra Kelly and other Greens on 11 May 1983 on East Berlin's Alexanderplatz. The flow of environmentalist literature from the West into the GDR certainly helped activists formulate and defend their agenda and spread awareness of ecological problems. For example, the East Berlin Ecology Library acquired a large collection of Western publications from the West German Green Party. West German television and radio also helped disseminate information about the Western ecology movement and environmental problems, although they largely ignored the situation in the GDR.

However, as historian Astrid Kirchhof has pointed out, profound differences in activist cultures in the two Germanys attenuated the impact of transfers of knowledge and ideas and impeded collaboration.[8] East German peace and ecology activists were able to attain a measure of autonomy from state control because the Protestant Church[9] and, to a lesser extent, the Catholic Church, provided them with institutional backing, protection, and places to meet. Most West German Greens found these religious ties unappealing and had little understanding for what they saw as the antisocialist orientation of the East German activists. Currents in the Eastern Bloc also helped mold GDR activism. Mikhail Gorbachev's Perestroika policies and increasingly hands-off attitude toward Eastern Europe encouraged those who sought change in the GDR, as did the Solidarity movement in neighboring Poland.

This chapter looks, first, at the inner organization, values, culture, and evolution of the intertwined peace, ecology, human rights, and anti–nuclear power strands of East German activism. The lifestyle, ethos, ideology, and religious orientation of activists help explain the strengths and weaknesses of the movement. A second theme of this chapter is the outward appeal of the movement. Earlier chapters of this book tried to assess how the West German anti–nuclear power movement was seen by the broader society and to explain why it gained acceptance over time. Given the embryonic nature of East German civil society in the second half of the 1980s, we have little more than scattered hints regarding interactions between activists and the East German public. What we do know is that, faced with a system that defined itself in terms of science and progress, activists turned to science to refute official claims. Science played an even greater role in sparking opposition to nuclear power than in the West. However, popular science accounts critical of nuclear power were seldom available in libraries and bookstores.[10]

Peace, Environmentalism, and Other Roots of Anti–Nuclear Power Activism

In the GDR, peace, ecology, human rights, and anti–nuclear power activism grew out of contacts with the West but also out of homegrown subcultures of dissidence and nonconformism. Two Marxists, Robert Havemann and Rudolf Bahro, were among the most important critics of the SED dictatorship and modern industrial society. Havemann was expelled from the SED and dismissed from his university professorship in 1963 and was placed under house arrest in 1976. Bahro could not publish his masterwork, *The Alternative,* in the GDR and was convicted of treason in 1978 and deported to West Germany in 1979.

The SED felt threatened by critical works by East German artists, writers, and musicians, who fed an increasingly rebellious youth culture. Because of his lack of obedience to the party, leftist East German protest and folk singer Wolf Biermann was expelled from the SED and forbidden to perform in public—temporarily in 1963, permanently in 1965. In 1976, Biermann was allowed to give a concert in West Germany but was not allowed to return to the GDR afterward, and his citizenship was revoked.[11] In December 1965, Erich Honecker, then a prominent member of the Central Committee, declared before that body, "Our German Democratic Republic is a clean state. In it, there are unchanging standards of ethics, morals, decency, and good behavior." Honecker blamed "rapes and manifestations of rowdiness, . . . unruliness at school and in the workplace . . . [and] drunkenness" on rock music, which he considered a form of Western decadence.[12] In the party press, fans were likened to the Hitler Youth, and idle youth were equated with criminals. The People's Police (Volkspolizei) forced longhaired boys and men to have their tresses cut off by police barbers. Most rock bands were banned, and a few fans were sent off to work camps.

The second half of the 1960s was characterized by a tense intergenerational conflict and blooming of youth culture in the shadow of threatened reprisals, culminating in 1968, the year of the crushing of the Prague Spring. Although the legacy of the 1960s was a bitter one in the GDR, popular legends were created in that era that brought about changes in the self-image of East Germans. Trotskyist and Maoist groups proliferated in the GDR in 1970s but were soon crushed. Youth culture and intellectual opposition were forces that the Communist regime was never able to fully tame, although it certainly tried to do so.[13]

Critics of the system increasingly turned to the Protestant Church for protection. Church–state dialog in the 1970s led to agreements under which the Church pledged its loyalty to the state as a "Church in Socialism," and the state promised noninterference in religious observance.[14] This made it possible for the Church to shield groups that met in church buildings, under the auspices of the Church. There was deep disagreement within the Church. Some believed that as Christians they were called on to stand up for human rights, decry injustice, and promote peace, even if this meant challenging the authority of the SED. Others, motivated by fear of endangering Church–state rapprochement, as well as by loyalty to the SED, believed the Church needed to show "solidarity with the state."[15] One such Church official was Manfred Stolpe, who signed on as unofficial secret police informant. He claimed he was trying to influence the Ministry of State Security to the benefit of the Church.[16] Activists who got into trouble with the secret police could not count on the support of these SED loyalists.[17]

Peace was the theme that drew masses to the Church. In the 1970s, the Church began to establish its credibility and relevance among young people, whether or not they were religious. Church-run outreach initiatives focused on creating safe spaces where young people could develop a deeper sense of self.[18] When the SED undertook measures widely seen as a militarization of society, for example by introducing paramilitary instruction into schools (in 1978), young people and their parents turned to the Church for moral support and assistance. Some hoped for a religious exemption that would allow them to perform their military service as unarmed "construction soldiers." Based on an unlikely marriage of scripture[19] and a Soviet sculpture of 1959,[20] the image of "swords into plowshares" became the compelling symbol of a popular peace movement that grew out of the Church's youth ministry.

An SED crackdown virtually guaranteed that the movement would escape SED control. For example, the SED forbade the "swords into plowshares" patches that many young people wore on their clothing. This only contributed to the popularity of peace-themed church services, vigils, events, and marches, according to historian, theologian, and former GDR activist Ehrhart Neubert. Pastor Rainer Eppelmann became a major figure in the Church-led peace movement when he, together with Robert Havemann, wrote the Berlin Appeal, an open letter advocating nuclear disarmament in East and West and the demilitarization and neutralization of Germany. Eppelmann was jailed for only three days but the Church leadership denied him any form of support, disciplined him, and strongly advised Church members not to become signatories.[21]

Havemann and Eppelmann's collaboration was seen as a joining of forces between Christians and Marxists. The same phenomenon was taking place on the local level. Referring to non-Christian activists, former SED member and militant activist Vera Lengsfeld (born in 1952) wrote, "Some had seldom or never been in a church in their lives. For them, it was a problem that they had to meet on church premises. Similarly, a few Christians had problems working together with atheists and Marxists." However, she asserts, they learned to be tolerant, and the risks and dangers of their work bound them together in communities based on mutual respect and loyalty.[22] Far-left activist Wolfgang Rüddenklau also perceived the relationship between Christian and leftist activists as harmonious.[23] This cultural and political context helps explain why this environmentalist movement was quite different in character from the West German Green movement.

Sebastian Pflugbeil, who disseminated his views concerning the dangers of atomic power in Church circles, is of the opinion that there was no ecology "movement" as such in the GDR, only "diffuse groups of committed people concerned with specific issues, and diffuse circles of friends of these committed people." He asserts that these groups were unstructured but were always connected with a Protestant congregation because by the late 1980s that was the only legal way they could meet.[24] Lengfeld writes, "The independent [i.e., not state-controlled] peace, ecology, and human rights movement of the GDR ... is lacking in the classic characteristics of a political opposition: no program, no real structure, no discernible operating procedures, no leaders in the traditional sense."[25] Many ecological initiatives in the GDR were deeply rooted in local problems, for example runoff from an industrial-scale hog farm in one rural area.[26]

There were, however, centers of Church-led activism. A small Protestant institution, the Kirchliches Forschungsheim Wittenberg (Church Home for Research in Wittenberg) put together a traveling environmentalist exhibition that attracted some twenty thousand visitors in 1979–1980. Its work inspired the founding of what became one of the most active environmentalist groups in the GDR, the Ecological Working Group of the Dresden Church District, in 1980. This group wrote a brochure, "Die Erde ist zu retten" (The Earth Can Be Saved), which only circulated in Church circles in samizdat form.

The Stasi forced the cancellation of an exhibit based on this booklet in 1983. The head of the Church Home for Research, Hans-Peter Gensichen (born in 1943), was very careful not to provoke oppression. For example, he advised activists participating in the Church environmentalist meeting in 1984 to avoid confrontation with the state.[27] In 1983, the Church Home for Research began publishing an environmentalist magazine, *Briefe zur Orientierung im Konflikt Mensch-Erde* (Letters of Orientation in the Conflict between Humans and the Earth), whose circulation grew from four hundred to four thousand. Its publications, exhibits, talks, and discussion groups inspired decentralized protests, for example against forest dieback.[28]

The Stasi saw these activities as potentially subversive and tried to block them whenever possible. A 1985 secret police report claimed that the purpose of environmentalist activism in the GDR was "to attack the environmentalist policies of the socialist state, to slander, to discredit, and to create a platform for a coalition of oppositional and discontented forces."[29] The Stasi carefully monitored environmentalist groups, planted informants in their midst, and sought to control them through a policy of "subversion, intimidation, isolation and decimation."[30] The Stasi claimed in internal documents to have infiltrated every oppositional group in East Berlin.[31]

Vera Lengsfeld, whose memoir provides insight into the emotional lives of GDR activists, discovered at age seventeen that her father was a Stasi agent. In younger years, she conformed to expectations, joining the SED, only to go on to become a leading opponent of the system.[32] Fully aware of the ways of the Stasi, she was nonetheless traumatized when agents searched her apartment in the 1980s, leaving everything a shambles. This experience haunted her for quite a while. Far worse, the authorities made life difficult for her son, for example preventing her from taking him, at age thirteen, on a vacation that would have meant a great deal to him, stopping them at the last minute at the airport. Lengsfeld contends that the Stasi carefully studied dissidents' private lives with the aim of destroying whatever it was that each individual treasured most, be it the chance to study at the university, their career, their friendships, their family, or their ability to travel.[33] Stasi documents as well as studies on the Stasi confirm her analysis.[34]

However, ecology activists found creative ways to circumvent repressive measures. In the early 1980s, tree plantings and bicycle protests emerged as signature forms of environmentalist activism in Church circles. Despite their innocuous nature, tree plantings were forbidden in some towns. Activists also organized ecological bicycle tours. "The bicycle became a symbol," writes *Umweltblätter* editor Rüddenklau. In particular, white ribbons were tied to handlebars to symbolize protest. This, as well as the carrying of signs, was forbidden by the authorities.[35] However, according to Lengsfeld, activists occasionally were able to get around such restrictions. On one occasion cyclists dressed colorfully to signal that they were participating in a demonstration. Police told them they had to keep a "safety distance" between each other, in hopes that they would be indistinguishable from random cyclists. The demonstrating cyclists happily obeyed, thus disrupting traffic on the very busy Schönhauser Allee in East Berlin.[36]

Anticonsumerism was also a hallmark of the ecology movement in the GDR. Tobias Huff argues that religiously motivated self-denial did not resonate with the majority of East Germans, who longed to emulate West German consumerism.[37] However, the activists' emphasis on personal commitment, responsibility, and authenticity not only conformed with Protestant teachings but also Western alternative culture, which held considerable emotional appeal for at least some young East Germans. For example, at one of the first meetings of the group concerned with the hog farm in 1986, members discussed the causes of ecological problems, "lack of responsibility, indifference, consumption of meat, lifestyles," asking themselves what they could do to overcome these problems: "What new approaches can we take to change our lifestyle and attitudes?" Early meetings centered on consciousness raising. Activities at a small, local Ecology Day in Knau in 1987 included "spinning, herbs, tea tips … [and] the exchange of militaristic toy [for other toys]."[38]

At a Protestant Church congress for Thuringia in 1988, there were discussion groups on nutrition, "spiritual health," "reasons to believe," "communicative meditation," and "alternative medicine." At evening church services, traditional religious elements were interspersed with "witnessing" (i.e., speaking the truth) of nuclear power, child rearing, and pollution from factory farms.[39] Many of these elements, such as vegetarianism, campaigns against militaristic toys, alternative medicine, could be found in Western alternative culture. In the GDR, it was difficult but far from impossible to pursue such a lifestyle, which was intimately connected with the search for a more authentic outlook and emotional life.[40]

Music helped created a sense of identity independent of "real existing socialism" and supplied the emotional backdrop to oppositional activities. Concerts brought disparate groups together, giving them a sense of purpose. The music initially associated with East German ecology and peace activism was inspired by the Western protest and folk music tradition.[41] Pastor Rainer Eppelmann allowed East German blues musician Günter (Holly) Holwas to perform at "blues services," which became popular in the first half of the 1980s. In the second half of the decade, however, punk music became increasingly popular in oppositional circles, marking a generational shift. As historian Jeff Hayton has pointed out, punk concerts, which often took place in churches, became a crucial site for dissemination of activist messages. The punk scene also provided a vector for forbidden contacts with Western and Polish punks.[42]

Throughout the GDR activist community, an inner-directed, self-reflective, and community-building orientation predominated. At times, social contacts and activism blurred together. Activists met in private homes for birthday celebrations, literary readings, and meetings of discussion groups.[43] Lengsfeld often cooked extra food at dinnertime because people were constantly dropping by her home. She writes, "The majority of the peace and ecology groups did not conceive of themselves primarily as political groups but as an attempt to develop their own forms of living together." Participants in the peace and ecology movements had to learn to overcome shyness and openly express their opinions, something they had not learned to do in school or in public. "It was an intellectual liberation from constraints that we had grown up with and that in adulthood still held us back and deformed us."[44]

East Berlin far-left activists were more outward directed but not necessarily in a politicized way, seeing themselves as resisting "the subservient ethos of the 'little people.'"[45] However elitist this statement may sound, it reflects a kind of self-assertion that had long been lacking in East German society. Was this the result of exposure to Western culture? Perhaps in part. However, GDR activists did not necessarily identify with West German activists, some of whom seemed unable to comprehend the human rights abuses the East Germans faced.[46] Chernobyl was to stimulate and broaden these fragile beginnings of a civil society and to make nuclear power a major issue among GDR environmentalists in a way that it had never been before.

Chernobyl in the GDR

East Germans first heard that some sort of nuclear accident had occurred from Western media on 28 April 1986. This was the result of Soviet secretiveness. According to historian Melanie Arndt, the GDR government first learned of the accident from the IAEA, which informed the East German office in charge of nuclear safety, the SAAS, of high radioactivity readings in Scandinavia.[47] Not until 14 May 1986 did Mikhail Gorbachev appear on TV to tell viewers about the Chernobyl accident. GDR newspapers only briefly alluded to Chernobyl initially.[48] On 2 May *Neues Deutschland,* the official party newspaper of the SED, ran an article that confirmed the basic facts of the disaster, although without mentioning that the Soviet authorities had waited twenty-four hours before evacuating the residents of the nearby town of Pripyat, resulting in many receiving dangerous levels of radiation exposure.

The East German article reassured citizens that "the health of the citizens of our state and nature" were not in any danger, nor had they been at any time during the crisis. The article's authors, Karl Lanius and Günter Flach, were leading East German physicists. They relied on the ultimate authority of science in seeking to reassure readers. Close to the beginning of the article came a technical explanation of how the reactor type of the sort used in Chernobyl functioned, mentioning that the reactors in use in the GDR were water moderated, unlike the graphite-moderated Chernobyl reactor that had exploded. The authors seemed convinced that average people would support nuclear power if they understood the science behind it.

Affirming the universal nature of science, the article pointed to Soviet and East German commitment to international nuclear safety standards as established by the IAEA. In fact, Chernobyl cemented the internationalization of the GDR's safety regime, as discussed in Chapter 2. The authors of the article accused Western media of engaging in "scaremongering" in an attempt to discredit the Soviet Union and to distract the public from Soviet disarmament initiatives in the face of the US SDI (Strategic Defense Initiative, or "Star Wars") program. They also made use of tried and true arguments that went back to the 1950s, notably the socialist version of "atoms for peace," the universality of technological progress but also the superiority of (East) German technology, as evidenced by its development of a new reactor safety system.[49]

Put in charge of monitoring radiation levels in food, the SAAS publicly insisted that milk and all foods were safe, even for small children,[50] although some foods were found to contain levels of radiation that exceeded GDR standards. SAAS and the government assumed that short-term exposure to levels of radiation higher than those allowed for nuclear plant workers was not harmful, even to children. Although unaware of this, customers refused to buy lettuce and other rare fruits and vegetables that suddenly appeared because they could no longer be sold in West Germany. These foods were then used for school and daycare center lunches.[51] Some parents of such children were strong-armed into giving permission to personnel to serve their children milk and vegetables.[52]

Maintaining production levels, not wasting food, and not unsetting the public were the officials' top priorities. For example, in the fall after the disaster, SAAS advised the ministry in charge of agriculture that it should see to it that contaminated animal feed harvested in May and June was mixed with uncontaminated grain to reduce cesium levels in milk. In order to win over the ministry, SAAS noted, "These are, above all, organizational and operative measures that will not diminish output and will not involve the population."[53] To its credit, the SED did allow measurements of radioactivity to be published, and there was more media coverage of Chernobyl in the GDR than in other Eastern Bloc countries. East German media coverage very much downplayed the possible aftereffects of the fallout in the GDR, however.[54]

Nevertheless, East Germans expressed a good deal of concern. One mother in Karl-Marx-Stadt wrote of her worries about her children's health in a diary entry that was later published in a women's magazine: "I can almost visualize radioactivity, the little particles that cling to my skin, to my daughter's skin ... maybe she won't be able to bear children!"[55] In the days after the disaster, citizens inundated the SAAS with questions about the veracity of East German media accounts and about the health impact of the radioactive cloud.[56]

The Stasi held on to 243 queries submitted to the government by residents of the East Berlin area. These citizens wanted to know whether someone who had traveled to Eastern Europe or the Soviet Union in that period needed to be examined by a doctor. Was it safe to travel to the East? Was it safe to consume certain foods? Some complained that the GDR media were not covering the crisis well enough. Others used a Geiger counter to detect radioactive contamination on people's clothing and reported these measurements.[57] District health officials, called together for big meetings in Leipzig and Berlin, were united in their belief that the Chernobyl disaster posed no danger to people in the GDR. The Ministry of Health advised these health officials to try to reassure citizens who came to them with concerns and to give them a sense of "safety and security" but also to take down detailed information about them. Only if they could not be dissuaded were they to be given an appointment with a physician specializing in radiation medicine.[58]

Reports streaming into the secret police reflected sharp divides in the population. Some felt that Western media coverage was propaganda that was being used to defame the Soviet Union. Others, however, distrusted GDR and Soviet reports and relied on Western media for information during the crisis. Stasi informants had heard people expressing doubts concerning the safety of Soviet-made reactors in the GDR, and nuclear power in general.[59] Prominent novelist Christa Wolf pointed to the significance of Chernobyl for East Germans in her novel *Accident/A Day's News,* published in 1987.[60]

Organized opposition to nuclear power emerged for the first time in the GDR after Chernobyl. On 25 May 1986, the synod, or Church council, of the Protestant Church in the GDR adopted a resolution expressing misgivings regarding nuclear power and calling for the development of alternative energy sources.[61] On May Day—the most important socialist holiday— members of a congregation in the Lichtenberg section of East Berlin sent a letter to the Council of Ministers and the Soviet ambassador, asking for a shutdown of GDR nuclear power plants. On 9 June, other activists started a petition addressed to the Volkskammer, the powerless East German parliament, asking it to allow East German citizens the chance to vote on whether or not to continue using nuclear power.

In May and June, peace movement members submitted an anti–atomic power "citizens' letter" (*Eingabe*)—the most important channel of communication between rulers and ruled in the GDR.[62] A Berlin group called Doctors for Peace pointed in another citizens' letter to the risks of nuclear power, protested GDR media policies and criticized the sale of fruits and vegetables harvested around the time of Chernobyl to Eastern Bloc countries because Western countries would not buy them. A great variety of similar actions took place across the GDR.[63] Punk bands also protested against nuclear power and helped distribute petitions and information in the wake of Chernobyl.[64] A hand-painted poster (Illustration 7.1) equating nuclear power with death, symbolized by a cross, probably from this period, disappeared into a Stasi file.

Members of the East German peace and human rights movement started a petition titled "Chernobyl Has an Impact Everywhere," a variation on the West Germany anti–nuclear power slogan, "Chernobyl is everywhere." After collecting 141 signatures, they presented it to the Council of Ministers and ADN (Allgemeiner Deutscher Nachrichtendienst), the East German press agency, on 4 June, 1986. This was only the start of this campaign. When activist Vera Lengsfeld drove around the GDR collecting signatures, she was welcomed in many homes. "It was good to know that there were like-minded people everywhere," she later wrote.[65]

The Stasi saw this activism as subversive and potentially inimical to SED control over society. The Stasi was convinced that the West German Green Party, the allied West Berlin party Alternative List, and the West Berlin newspaper *Tageszeitung-taz* were the main instigators of this "hostile and negative"

Illustration 7.1 Anonymous, hand-painted anti–nuclear power poster, GDR, confiscated by the secret police. BStU (Stasi Archive), MfS BV Berlin Abt. XX 5242 Bd. 2–6, 53. Courtesy of the BStU.

upsurge in anti–nuclear power activity in the GDR.[66] By contrast, the GDR leadership and press sought confirmation of the rationality of their policies among supporters of nuclear power in the Federal Republic and even seem to have developed a kind of understanding for the West German Christian Democrats, who proclaimed that the fallout from Chernobyl was harmless.[67]

The Protestant Church synod in September 1986 drew the attention of the SED and the Stasi. No fewer than twenty-eight secret police informants were present at that gathering, many of them Church officials.[68] This level of surveillance and manipulation of Church events was not unusual. The Stasi hoped to bolster the position of the "realistic" elements in the Church, who wanted to continue to cooperate with the state and avoid confrontation. However, representatives of the Western press and diplomatic corps were also present, perhaps temporarily strengthening the position of the more critical church members. These independent-minded elements rankled the Stasi, which complained about critics' "distinctive, exaggerated self-confidence and craving for recognition, as well as their overestimation of the role of the Protestant Church in public life." Rather contradictorily, the secret police reported that they did not carry the day because they were "not well prepared, ineffectual, behaving in a passive, hesitant way."

However true this may be, they helped to maintain the Church's focus on societal issues, pressing forward with demands regarding the peace movement, the autonomy of the Church, the rights of Christians who did not want to bear arms, and nuclear power.[69] The synod specifically addressed nuclear power in a resolution pointing to human error as a problem in nuclear safety systems that Christians should be concerned about. It called for a worldwide moratorium on the construction of new nuclear power plants and "drastic energy conservation."[70] Critical church members clearly did not pose a danger to the SED. It is remarkable, however, that they persevered in the face of pressure to conform ideologically and fear concerning their own careers and families. They seem in part to have been inspired by Gorbachev.[71] Because they were able to maintain the respect of many East Germans unhappy with SED policies, they were able to reinforce the role of the Church as the principal guardian of a fledgling civil society.

Developing Rational Arguments

While the Church provided a unifying institutional framework for the disparate activist groups, activists began to look elsewhere for a unifying cognitive framework. They were ever less inclined to treat religion or even ethics as the foundation of their critique of nuclear power. This doubtlessly had to do with the ideological diversity of East Germans engaged in this cause. According to one activist, they fell into three categories: a small group of social conservatives who saw nuclear power as a threat to "unborn life"; a tiny minority of "leftist–Green Christians"; and the majority, who came out of the East German ecology movement, had been mobilized by Chernobyl, and were pragmatic in outlook.[72] The foundational documents that circulated in samizdat form reflect the attitudes of this third group, specifically a rational analysis of nuclear power from scientific, economic, and historical perspectives.

These trends can be seen in the talks delivered at the Third Berlin Ecology Seminar, which took place 28–30 November 1986 in the Zion Church and was organized by the Ecology Library. The talks were the basis for the hectographed samizdat, "Rotting Reactors: Nuclear Power in the GDR— Continued Use or Abandonment?," excerpts of which appeared from time to time in later Church publications.[73] A detailed explanation of how the Chernobyl accident was thought to have occurred began the volume. It was followed by a discussion of alternative energy sources and ways of improving the energy efficiency of fossil fuels.

The next section called into question economic arguments in favor of nuclear energy, employing highly differentiating and sophisticated arguments grounded in economic analysis, backed up by quantitative data. The authors, economist Thomas Klein and activist Wolfgang Wolf,[74] rejected claims that society had to choose between coal and nuclear energy. Renewable energy sources represented an alternative, they insisted. Nuclear power only appeared to be a cheaper energy source because so much money had already been pumped into it. The authors maintained that in both capitalist and Communist countries, false ways of thinking about nuclear energy had been promoted by economic and political elites. The paper recommended a comprehensive program of energy conservation and energy efficiency. Throughout, it cited West German and US studies.[75]

Other sections of "Rotten Reactors" made use of a variety of kinds of arguments. One author, who began by arguing that a belief in God (although not necessarily the Christian God) was a necessary element in a respect for nature, mainly focused on technical and scientific arguments regarding nuclear power.[76] The last paper called for "resistance" against the "atomic state" (a term used by Robert Jungk) and for a fight against "inhumane societies," referring to both East and West Germany. "We are not fighting for powdered milk but against the atomic mafia, against patterns of thought that accept the risks of destroying our environment and the world around us, and [in fact] plan on it." The talk ended with a radical motto commonly used among protesters in West Germany: "Destroy what is destroying you!"[77] The various authors seem to have represent different factions—economic experts, Christians, and leftist radicals. What was missing in "Rotten Reactors" was a strong scientific foundation. A final section addressed this deficiency.

A major confrontation between activists and the East German scientific establishment, reported on in the appendix of "Rotten Reactors," lay bare the unequal positions of the two sides in the debate over nuclear power. Activists had asked to meet with nuclear power experts from SAAS. After canceling two meetings, one at the last minute, scientists finally met with some forty peace and ecology activists on 9 February 1987. The scientists were accompanied by a group of members of the local chapter of the state-run women's organization, who, in the words of one activist "cursed at us and mocked us continually from the sidelines, shouting that the event should end soon." The environmentalists present also thought that ID checks and screeching speakers were part of an attempt to intimidate them. They were able to identify as untrue claims made by the SAAS representatives that the problems of permanent disposal of radioactive waste and decontamination of decommissioned nuclear power plants had been solved by the GDR. SAAS claims that the East German system of pressurized rooms within the reactor building was at least as good as Western containment domes was met with skepticism.

However, activists also blamed themselves for the failure of this attempt to establish a dialogue because they felt that they had not been "properly prepared." They did not have a sufficient understanding of the scientific and technical fundamentals to be able to critically evaluate the "confusing profusion of facts, some familiar, some not" that overwhelmed them. How could it be that—as the experts claimed—radioactivity in the body was less harmful than exposure to atmospheric radioactivity? Was it true that during the decontamination process at Three Mile Island, workers were told to take off their radiation suits?[78] The counterexperts' lack of mastery of the facts hampered their ability to take on the experts. Ethical objections to nuclear power were not enough in the minds of those present at the meeting. This experience convinced activists of the importance of scientific arguments.

Fighting Science with Science

According to Sebastian Pflugbeil, five people—two Protestant ministers, two medical doctors, and one physicist (himself)—provided the scientific foundations for anti–nuclear power activism in the GDR. However, by his account their relationship with science was complex. He wrote in an email to the author of the present study, "Early on, we discussed the reality that science is a good deal more subjective, corrupt, subservient, and intentionally false than the average citizen might imagine. And in scientific fields related to ecology and nuclear technologies, there is far more science for sale than science that is honest and accurate. That itself was an important topic [of our conversations]."[79] These five activists sought to make what they considered to be scientifically sound research results accessible to the East German public.

In researching the comprehensive study of 1989, *Energie und Umwelt* (Energy and Environment),[80] Pflugbeil and his coauthor, Joachim Listing, a biomedical researcher, took care to avoid political problems by citing Eastern Bloc studies. They did not put their names on the study, which was published by the Protestant Church, marked "for internal use only."[81] Ties with the Church protected them and other activists from persecution, but these studies had little or no religious content. The basic structure of their arguments was scientific in nature.

Pflugbeil gave talks in Protestant Churches and wrote "Letters to my Godson," in which he explained, in epistolary form, what he considered the most important aspects of the debate concerning nuclear power.[82] In the "Letters," he expressed a great deal of concern about the health impact of nuclear power. Radiation protection, he explained, was largely based on long-term studies of victims of the Hiroshima and Nagasaki bombings. These studies underestimated the impact of radiation exposure because medical doctors had often listed the cause of death on death certificates as something other than cancer to spare the grieving families from social ostracism. RERF (Radiation Effects Research Foundation), a US–Japanese research organization, took up the study of the Japanese bombing victims after the disbanding of the US Atomic Bomb Casualty Commission (ABCC) in 1975. Undertaking a reanalysis of the ABCC data on behalf of RERF, Dale Preston and Donald Pierce calculated that the expected incidence of additional cancer deaths among a million people exposed to one hundred millisieverts of radiation was not a thousand as previously thought but fifteen to seventeen thousand.[83]

This meant that standards of radiation exposure were overly optimistic and should be revised, concluded Pflugbeil. He explained that there are many causes of cancer, including smoking, exposure to chemicals, and poor eating habits. It is impossible to determine in an individual case whether radiation was the cause of a particular cancer case. However, statistically, it could be shown that radiation exposure increased the overall cancer risk. He was even able to point to studies, one conducted under the aegis of Alice Stewart, the British expert on radiation exposure and childhood cancer, that pointed to a correlation between regional variations in natural background radiation and childhood cancer deaths.[84]

Another important study on the impact of nuclear power was Michael Beleites's 65-page samizdat, "Pechblende" (Pitchblende, a term for uranium ore), which uncovered problematic aspects of the history of uranium mining in the GDR.[85] Not a scientist, he received help from Pflugbeil in tracking down scientific and technical literature on the subject. A Soviet-run uranium mining enterprise since 1946, "Wismut" became one of the most important sources of uranium in the Eastern Bloc. Every aspect of the operation was top secret. Beleites's study, the fruit of years of research, revealed to East

German readers for the first time the serious health consequences of this mining operation, both for miners, many of whom developed lung cancer, and for residents of the area, located in Saxony and Thuringia, in the south of the GDR. He began by explaining in considerable detail the Wismut mining operations and the impact of exposure to radioactivity on the human body.

Beleites had to rely in some cases on decidedly crude measures of contamination, for example the miners' exposure to radioactive dust and radon. He was not privy to Geiger counter or dosimeter measurements, but a geologist had told him that a device, presumably a Geiger counter, had emitted rapid clicks in the mines that in some spots had changed to an urgent whistle. Contaminated water was pumped out of the mines and dumped in the river, or it seeped into the groundwater—presumably aquifers that supplied water to the population. Air vents spread radioactive particles and radon onto farmlands.

Sludge from the mines was dumped outdoors. It got into the air and water and was used until 1980 as a building material. Beleites's account is filled with chilling details: miners' children who played with "pretty" rocks from the mines; residents of the town of Gera who picked the abundant mushrooms that grew in the contaminated waste sites; and apparent cancer clusters in the areas around the mines. The SED dealt with the problems in cynical ways, paying miners high wages and continuing an older tradition of providing them with tax-free hard liquor. What could be done? Beleites advocated both nuclear disarmament and the abandonment of nuclear power.[86]

It is difficult to gauge the impact of "Pechblende." In theory, this samizdat was "only for official use within the Church." The number of copies in circulation is unknown. To the extent that they were able to get their hands on it, East Germans must have found it to be a revelation. Much has been published about environmental devastation in the GDR since 1990, but far less was known before then. In 1982, the East German Council of Ministers had declared all environmental data to be a state secret. Uekötter asserts that this was "probably the moment when the SED leadership lost control over the ecology debate."[87]

"Pechblende" is generally considered to have given a highly accurate account of environmental problems caused by Wismut, although many details did not come to light until after 1990. In a secret report, a Stasi informant—presumably someone with a scientific background—praised as highly accurate those sections of "Pechblende" that discussed the scientific fundamentals concerning the health effects of radiation and the technical explanations of uranium mining. He or she did find the sections on the ecological and public health impact of Wismut "speculative and biased."

However, Beleites had to rely on speculation, this report pointed out, because little information was available to GDR citizens. The informant noted "that in the GDR too little information about such sensitive topics is made available."[88]

Lacking hard data on the GDR, Beleites cited West German studies on pollution caused by uranium mines in the Federal Republic, as well as cases relating to uranium mining in the United States.[89] Using the greatly expanded body of information on Wismut that has become available since 1990, recent studies have judged "Pechblende" to have been a valuable pioneering work.[90]

Another Church samizdat publication was "... nicht das letzte Wort: Kernenergie in der Diskussion" by Joachim Krause, who had a university degree in chemistry but turned to theology and became a Protestant minister.[91] It sought to acquaint readers with the scientific and technical fundamentals of nuclear power and its health impacts. Nuclear power had been developed with a "conscientiousness unrivaled in the history of industrialization" yet remained risky.[92] He accurately explained how nuclear power plants and safety systems functioned. He also demonstrated familiarity with debates in the West and a firm command of the mathematics behind risk calculation, such as was found in WASH-1400 (the 1975 US study) and a 1979 study of the West German Association for Reactor Safety (Gesellschaft für Reaktorsicherheit, or GRS).[93]

Krause also pointed to major kinds of risk that these studies did not take into account, such as the possibilities of an aircraft crash onto a nuclear power plant, earthquakes, sabotage, and mechanical failures of various sorts. He argued that errors in safety systems only became known when those errors unexpectedly occurred and that estimates of human and mechanical errors were not based on hard data. He also repeated the criticism that if there was a one in ten thousand chance of a reactor core meltdown in any given year in any given reactor, that meant that, on average, a meltdown could be expected every twenty-five years, given that there were four hundred reactors in the world at that time. He pointed out, however, that this did not imply a firm prediction of the precise timeframe in which such an accident would actually occur.[94] This samizdat did a good job of conveying some of the complexity of debates in the West.

The Crackdown on the Ecology Library

The Ecology Library, housed in the parish hall of the Zion Church in East Berlin, provided the East German public access to scientific information and environmentalist and ecological publications that had previously been unobtainable in the GDR, at least for average citizens. It had become the center of East German ecological activism and perhaps of the opposition as a whole.

In East Berlin, the oppositional scene was particularly radical. Far-left activists involved in the Ecology Library considered themselves anarchists and opponents of Stalinism, capitalism, and fascism.[95] Wolfgang Rüddenklau, former editor of the East German samizdat publication, *Umweltblätter* (Ecology Newsletter), writes of the views then prevalent in these circles, "Society was divided into those who dominated and those who were victimized. Domination not only destroyed humans but also nature. The reconciliation of humans and nature could only take place in a society without domination."[96] According to Hayton, punks contributed to this and other environmentalist publications and were an integral part of this scene.[97] A group of leftist radicals broke with the Church to form the "Church from below," which engaged in overtly political activities.[98]

This milieu was swept up in a state crackdown on the Ecology Library in late 1987. Police stormed the church during the night of 24–25 November 1987, hoping to catch printers at work on an issue of *Grenzfall* (Border Case), an illegal samizdat publication of the banned "Initiative for Human Rights." The planned raid failed when the Stasi was double-crossed by its secret informant, a *Grenzfall* editor, according to one version of the story. So, instead, they arrested seven members of the staff of *Ecology Newsletter*, which was a legal publication and which was in fact being printed in the middle of the night. Secret police agents confiscated four printing presses, two of them over fifty years old, one a gift of West German Greens.[99]

Four staff members were arrested and charged with forming an "association for the purpose of illegal activities." In protest, supporters of the library held a candlelight vigil around the clock at the Zion Church. On 27 November, police beat a protester, pulled him into their vehicle, and disappeared with him. Two days later, demonstrators marched a mile in silence, carrying candles from the Elias Church to the Zion Church—a rare display of public protest.[100]

The organizers hoped that the vigil would serve as a powerful symbol and would win public support, which it in fact did. Visible from the street and filling the church and the entryway were flowers and burning candles. These were, despite Christian associations, part of a broadly familiar visual culture of public commemoration. People from the neighborhood around the Church brought protesters food and hot beverages, and many flocked to the services held every evening.[101] Dissident Bärbel Bohley got in touch with former East German Roland Jahn, who now lived in West Berlin. Jahn in turn contacted the Western media, which pointed out that this was the first direct attack on the Church in the GDR since the Stalin era.[102]

West German politicians protested against the arrests, and the Western press reported extensively on the vigil. Many East Germans heard about the Ecology Library for the first time from Western TV and radio reports, adding to its presence in public consciousness.[103] When the popular newspaper *Junge Welt*[104] equated the participants in the vigil with "rowdies with a fascistic vocabulary and weapons," Lengsfeld wrote a retort and, when it was not published, initiated a lawsuit against the editors. She contacted a reporter of the leftist West Berlin newspaper *Tageszeitung-taz,* who published a report on her actions. Lengsfeld had misgivings about this move, which went against the habitual caution of East German activists regarding contacts with Western media.[105]

The activists won this confrontation, gaining media attention in the West and—via Western radio and TV—in the East. The SED had to back down, setting free the four people who were arrested and eventually dropping the charges against them. Given this success, the Church was little inclined to put pressure on the opposition to scale back its activities, especially since the Church was anxious to remain relevant in a largely secular society. The ranks of activists swelled, and they became increasingly bolder. No matter how riddled with Stasi informants, the peace and ecology groups continued to expand and deepen their activities.

Some activists broke away from the Ecology Library to form the Green–environmentalist network Arche in 1988. A cofounder, Carlo Jordan (born in 1951) rejected the Ecology Library's Marxism. Arche was more political, was more interested in collaborating with other groups, and had closer ties to the West than its predecessor. The West German Greens smuggled a printing machine into the GDR for Arche.[106] Thus, repression only served to strengthen the links between environmentalism, oppositional politics, criticism of the scientific establishment, and respect for science as a cognitive system.

Conclusion

Although skeptical of scientists, East German activists genuinely embraced science. GDR counterexperts cited scientific studies on radiation exposure and on technical risks associated with nuclear power, and their arguments were rooted in science. Beliefs that were specifically religious or esoteric in nature became peripheral. However, as was the case with their West German counterparts, the East German counterexperts challenged the validity of some of what was normally understood to be legitimate science on the grounds that mainstream scientists were bending to political pressure and putting their own careers first. Their criticism was ethical and political in nature. "I am outraged over the culpable among the experts, and I am saddened by all those who know [what is going on] but don't say anything and don't do anything," wrote Pflugbeil. However, himself a scientist, he did not see an easy path forward, writing his godson, "The public's increasing loss of trust in 'experts' may have consequences that the two of us may not like."[107]

Critical views on nuclear power could be disseminated through Church channels. Many saw Christian beliefs, activism, and science as quite compatible. Mobilized by Chernobyl, science, the Church, and youth culture provided the foundations for anti–nuclear power activism. Thanks to Church protection, some activists, such as Beleites, were harassed but not imprisoned,[108] but Lengsfeld—who broke with the Church hierarchy and the political system in a fundamental way—was incarcerated and then deported to West Germany.

Despite these ties to the Church, there were undercurrents of independent mindedness. Youth culture and Marxist dissidence tested the boundaries of the safety zone created by the Church, not always with the best of results. Collective experience of oppression generated fear but also a sense of solidarity among the disparate groups that made up the activist scene, including supporters of human rights, peace, environmentalism, and feminism. They bridged the divides between religious and Marxist viewpoints, as well as the hippy and punk generations. By 1988, GDR environmentalists were becoming increasingly political, in the sense that they were trying to build organizations, but there is little evidence that they rejected socialism. Lengsfeld writes, "We did not want to overthrow the system; we unintentionally destabilized it from below."[109] Whatever their intentions, they helped lay the foundations for the peaceful revolution of 1989–1990.

The efforts of anti–nuclear power activists in the GDR also had an impact on popular opinion. According to a 1990 Emnid[110] survey, a similar percentage of East and West Germans favored abandonment of nuclear power—either immediately (12 and 16 percent, respectively) or gradually (46 and 49 percent, respectively).[111] East and West Germans arrived at these views via somewhat different paths. In the GDR, anti–nuclear power activism grew out of a comprehensive peace, human rights, and environmentalist movement. Although some of the basic texts of the movement were Western in origin, many were not. East and West German anti–nuclear power activists did, however, share some characteristics: opposition to a growth philosophy that they feared could bring about environmental disaster, distrust of experts, and the embrace of scientific arguments.

Notes

1. Photocopy machines were widespread in the West at that time but virtually inaccessible in the GDR due to the authorities' attempts to prevent the circulation of texts that were not officially sanctioned. Samizdat publications were mimeographed, reproduced by hectograph, or simply typed up with carbon copies. Those that bore the label "Only for use within the Church" were usually tolerated by the authorities but could not go beyond the bounds of what was loosely construed as the religious sphere. Such samizdats were theoretically not available to the public but are known to have circulated from hand to hand.

2. Robert Havemann Gesellschaft, TH 2/5a 122, handwritten notes of discussions of GDR officials at ISES Solar World Congress 1987 in Hamburg.

3. Uekötter, *Deutschland in Grün*, 186.

4. Ilko-Sascha Kowalczuk, *Endspiel: Die Revolution von 1989 in der DDR*, 2nd ed. (Munich: Beck, 2009), 126–27.

5. Uekötter, *Deutschland in Grün*, 184.

6. Tobias Huff, *Natur und Industrie im Sozialismus: Eine Umweltgeschichte der DDR* (Göttingen: Vandenhoeck & Ruprecht, 2015).

7. Rudolf Bahro, *Elemente einer neuen Politik: Zum Verhältnis von Ökologie und Sozialismus* (West Berlin: Olle und Wolter, 1980); Robert Havemann, *Morgen: Die Industriegesellschaft am Scheideweg: Kritik und reale Utopie* (Munich: Piper, 1980). Havemann was living in the GDR at the time. Bahro had been deported to the West in 1979.

8. Astrid Kirchhof, "'For a Decent Quality of Life:' Environmental Groups in East and West Berlin," *Journal of Urban History* 41, no. 4 (2015): 625–46.

9. In Germany, the Protestant Church was and is a mainline denomination that encompasses Lutheran, Reformed (Calvinist in origin), and mixed Lutheran–Reformed strains. It was divided into separate federations for East and West Germany but shared a history, theological foundations, and personal ties.

10. Email from Sebastian Pflugbeil to Dolores Augustine, 31 March 2015.

11. Ehrhart Neubert, *Geschichte der Opposition in der DDR 1949–1989*, 2nd ed. (Berlin: Ch. Links Verlag, 1998), 142–62. On Havemann, also see Dirk Draheim, Hartmut Hecht, and Dieter Hoffmann, eds., *Robert Havemann: Dokumente eines Lebens* (Berlin: Links Verlag, 1991); Dieter Hoffmann, "Robert Havemann: Antifascist, Communist, Dissident," in *Science under Socialism: East Germany in Comparative Perspective*, eds. Kristie Macrakis and Dieter Hoffmann (Cambridge, MA: Harvard University Press, 1999), 269–85.

12. Ulrich Mählert, *Kleine Geschichte der DDR*, 7th ed. (Munich: Verlag C.H. Beck, 2010), 107.

13. Neubert, *Geschichte*, 163–68, 201–08, 214–48. On Trotskyites and Maoists, also see Wolfgang Rüddenklau, *Störenfried: DDR-Opposition 1986–1989: Mit Texten aus den "Umweltblättern"* (Berlin: BasisDruck Verlag, 1992), 21. Some actually had contact with the West German KPD and KBW.

14. Robert Goeckel, *The Lutheran Church and the East German State: Political Conflict and Change under Ulbricht and Honecker* (Ithaca: Cornell University Press, 1990), 202–29.

15. Neubert, *Geschichte*, 248–84.
16. Dolores Augustine, "The Impact of Two Reunification-Era Debates on the East German Sense of Identity," *German Studies Review* 27 (Fall 2004): 563–78.
17. For a local example, see Jan Schönfelder, *Mit Gott gegen Gülle: Die Umweltgruppe Knau/Dittersdorf 1986–1991: Eine regionale Protestbewegung in der DDR* (Rudolstadt: Hain-Verl., 2000), 100–103.
18. Sebastian Bonk, Florian Key, and Peer Pasternack, "Die Offene Arbeit in den Evangelischen Kirchen der DDR: Fallbeispiel Halle-Neustadt," in *Wissensregion Sachsen-Anhalt: Hochschule, Bildung und Wissenschaft: Die Expertisen aus Wittenberg*, ed. Peer Pasternack (Leipzig: Akademische Verlagsanstalt, 2014), 203–7.
19. Micah 4:3.
20. In 1959, the Soviet Union donated a sculpture by socialist realistic sculptor Yevgeny Vuchetich titled "Let Us Beat Swords into Plowshares" to the United Nations.
21. Neubert, *Geschichte*, 374–84, 389–411.
22. Vera Lengsfeld, *Ich wollte frei sein: Die Mauer, die Stasi, die Revolution* (Munich: F.A. Herbig, 2011), 113–16, 119; quotation on 119.
23. Rüddenklau, *Störenfried*, 68.
24. Email from Sebastian Pflugbeil to Dolores Augustine, 31 March 2015.
25. Lengsfeld, *Ich wollte*, 119.
26. For example, see Schönfelder, *Mit Gott gegen Gülle*, 33–37, 89–98.
27. BStU MfS ZAIG Nr. 20671, 4.
28. Huff, *Natur und Industrie*, 324–32; Rüddenklau, *Störenfried*, 43–47.
29. BStU MfS-HA XX Nr. 23556, 1.
30. BstU MfS-HA XX Nr. 23830, 5.
31. BstU MfS BV Berlin, Abt. XX Nr. 7194, 60.
32. "Mein Vater war beim MfS—Stasikinder," YouTube video, four parts, 44 min. (total), directed by Thomas Grimm and Ruth Hoffmann (Berlin: Zeitzeugen TV, 2012), DVD, retrieved 21 August 2016 Part 1 from https://www.youtube.com/watch?v=Y8xI8cLqauI; Part 2: https://www.youtube.com/watch?v=DVMk4fvlyTk; Part 3: https://www.youtube.com/watch?v=nEH6zR-vVPA; Part 4: https://www.youtube.com/watch?v=8NzsyqVHXzc.
33. Lengsfeld, *Ich wollte*, 122–23, 131, 133, 167–68.
34. Jens Gieseke, *The History of the Stasi: East Germany's Secret Police, 1945–1990*, trans. David Burnett (New York: Berghahn Books, 2014); Jens Gieseke, "The Stasi and East German Society: Some Remarks on Current Research," *GHI Bulletin Supplement* 9 (2014), 59–72. Retrieved 21 August 2016 from http://www.ghi-dc.org/fileadmin/user_upload/GHI_Washington/Publications/Supplements/Supplement_9/bu-supp9_059.pdf.
35. Rüddenklau, *Störenfried*, 47.
36. Lengsfeld, *Ich wollte*, 120.
37. Huff, *Natur und Industrie*, 338–39.
38. Schönfelder, *Mit Gott gegen Gülle*, 39, 44.
39. Schönfelder, *Mit Gott gegen Gülle*, 62, 66.
40. Example in Renate Hürtgen, *Ausreise per Antrag: Der lange Weg nach drüben: Eine Studie über Herrschaft und Alltag in der DDR-Provinz* (Göttingen: Vandenhoeck & Ruprecht, 2014), 91–93.

41. Some of the important musicians include Stefan Krawczyk, Freya Klier, Gerhard Schöne, Ingo Barz, and Bettina Wegner.

42. Jeff Hayton, "Culture from the Slums: Punk Rock, Authenticity and Alternative Culture in East and West Germany" (Ph.D. dissertation, University of Illinois at Urbana-Champaign, 2013), 484–93.

43. Ulrike Poppe, "'Der Weg ist das Ziel.' Zum Selbstverständnis und der politischen Rolle oppositioneller Gruppen der achtziger Jahre," in *Zwischen Selbstbehauptung und Anpassung: Formen des Widerstandes und der Opposition in der DDR*, ed. Ulrike Poppe, Rainer Eckert, and Ilko-Sascha Kowalczuk, 244–72 (Berlin: Ch. Links Verlag, 1995), 256–57.

44. Lengsfeld, *Ich wollte*, 116, 117, 159–61.

45. Rüddenklau, *Störenfried*, 25.

46. On the lack of understanding shown by West German leftist Protestants in the 1960s and 1970s, see Gisela Sommer, *Grenzüberschreitungen: Evangelische Studentengemeinde in der DDR und BRD: Geschichte-Verhältnis-Zusammenarbeit in zwei deutschen Staaten* (Stuttgart: Alektor-Verlag, 1984), 93–96, 101, 120, 123, 142, and 153. On the Western peace movement's insensitivity, see Holger Nehring, "Transnationale Netzwerke der bundesdeutschen Friedensbewegung," in *"Entrüstet Euch!" Nuklearkrise, NATO-Doppelbeschluss und Friedensbewegung*, ed. Christoph Becker-Schaum, Philipp Gassert, Martin Klimke, Wilfried Mausbach, and Marianne Zepp, 213–28 (Paderborn: Ferdinand Schöningh, 2012), 218.

47. Arndt, *Tschernobyl*, section 5.1.

48. Deutsches Rundfunkarchiv, Staatliches Komitee für Rundfunk. Redaktion Monitor, "DLF [Deutschlandfunk] 12.05h. vom 30.4.1986."

49. Karl Lanius and Günter Flach, "Sicherheit—oberstes Prinzip bei der friedlichen Nutzung des Atoms zum Wohle der Menschheit," *Neues Deutschland* (2 May 1986).

50. Werner Finkemeier, "Meßdaten in der DDR zeigten keine erhöhte Gesundheitsgefährdung," *Unsere Zeit* (11 June 1986). *Unsere Zeit* was a publication of the West German DKP, a socialist publication friendly to the SED.

51. Arndt, *Tschernobyl*, section 5.1. On grocery store customers not buying fruits and vegetables, see BStU MfS-ZAIG Nr. 14617, 7.

52. BStU MfS-ZAIG Nr. 14617, 2–6.

53. SAPMO/Barch DC/20/4991, 35–36. The memo is dated 25 November 1986. The document did not mention specific maximum exposure levels beyond which fodder had to be disposed of and in fact did not mention disposal.

54. Arndt, *Tschernobyl*, section 5.1.

55. Robert Havemann Gesellschaft, PS 68/2 7, Susi Franke, "Tagebuchaufzeichnungen vom Frühjahr 1986," *Lila Band* 1987 Nr. 2.

56. BStU MfS HA VII Nr. 1333, 152.

57. BStU MfS BV Bln AKG 3364 (passim); MfS BV Bln AKG 3482 (passim); MfS BV Bln AKG 3496.

58. BStU MfS-ZAIG Nr. 14617, 2–6, 11.

59. BStU MfS HA VII Nr. 1333, 257–61.

60. Christa Wolf, *Accident/A Day's News*, trans. Heike Schwarzbauer and Rick Takvorian (New York: Farrar, Straus & Giroux, 1989). First German edition:

Christa Wolf, *Störfall: Nachrichten eines Tages* (Berlin/Weimar: Aufbau-Verlag, 1987). The Chernobyl disaster figures large in the novel. The novel is a meditation on the impact of technology on the modern world but also on what it is to be human, to experience the small events of daily life.

61. Andreas Passarge, "Die Herausforderung unserer Zeit," *Arche Nova* (26 April 1989), 3. *Arche Nova* was a samizdat publication of the Protestant Church in the GDR. Source: BstU MfS BV Bln Abt. XX Nr. 7192, 4.

62. BStU MfS-HA XX/AKG Nr. 6400, 1–6. See also Henrik Eberle, ed., *Mit sozialistischem Gruss!: Parteiinterne Hausmitteilungen, Briefe, Akten und Intrigen aus der Ulbricht-Zeit* (Berlin: Schwarzkopf & Schwarzkopf, 1998).

63. Robert Havemann Gesellschaft, PS 107/12 2–5; BstU MfS HA XX/AKG Nr. 1150, 46–50.

64. Hayton, "Culture from the Slums," 484.

65. Lengsfeld, *Ich wollte,* 160. Also BStU MfS-HA XX/AKG Nr. 6400, 1–6.

66. BStU MfS-HA XX/AKG Nr. 6400, 1–6.

67. For an example of a Western expert criticizing recommendations to avoid sandboxes, grass, and swimming pools, see Deutsches Rundfunkarchiv, OVC 7413, series "Aktuelle Kamera," Episode "Pressekonferenz in Moskau zur Havarie in *Tschernobyl,*" 6 May 1986. For a discussion of West German Christian Democrats, see BStU MfS HA VII Nr. 1333, 270–71.

68. BStU MfS-HA XX/4 Nr. 3285, 1–3.

69. BStU MfS-HA XX/4 Nr. 3285, 21–48; quotations on 46.

70. Vollrad Kuhn, "Die Energieproblematik in Synodenbeschlüssen der Ev. Kirchen in der DDR und in den Papieren der 2. Vollversammlung für Gerechtigkeit, Frieden und Bewahrung der Schöpfung," *Arche Nova* (26 April 1989): 39–43; quotation on 40. Source: BstU MfS BV Bln Abt. XX Nr. 7192, 40–44.

71. An unnamed source, evidently a Church official who was also a Stasi informant, wrote in a report to the Stasi dated 24 September 1986, "It is politically sensational how the churches have made themselves part of Gorbachev's program." BStU Mfs-HA XX/4 Nr. 3285, 169.

72. Robert Havemann-Gesellschaft, TH 2/5a 73, Gunnar Schubert, "Atomi NO: Gedanken zum Anti-Atom-Protest in der DDR," Samizdat dated 16 July 1988.

73. Robert Havemann-Gesellschaft, PS 74 1–60, "Morsche Meiler: Atomkraft in der DDR—Weiternutzen oder Ausstieg? Reader zum 3. Berliner Ökologieseminar (28.–30. November 1986)," samizdat publication, 1986/1987.

74. Identified in Thomas Klein, *"Frieden und Gerechtigkeit!" Die Politisierung der Unabhängigen Friedensbewegung in Ost-Berlin während der 80er Jahre* (Cologne: Böhlau Verlag, 2007), 266.

75. Robert Havemann-Gesellschaft, PS 74 27–43, "Wirtschaftlichkeitsdenken in der Energiepolitik," in "Morsche Meiler."

76. Robert Havemann-Gesellschaft, PS 74 50–53, "Referent am Sonntag: Philosophische Betrachtungen der Umwelt- und Energieproblematik und Resumé," in "Morsche Meiler."

77. Robert Havemann-Gesellschaft, PS 74 54–55, "Mein Thema ist das Problem—Wie nach Tschernobyl leben?," in "Morsche Meiler."

78. Debates concerning the impact of radioactive particles lodged in the body are ongoing. For an expert's explanation of the debates, pitched to a nonexpert audience, see https://www.youtube.com/watch?v=l9l3nPs75lg (retrieved 1 September 2017). The speaker is Arnie Gundersen, a long-time nuclear engineer turned nuclear industry critic. However, events such as the polonium poisoning of a critic of the Russian government demonstrate that even minute amounts of such radioactive compounds can be fatal. See Alan Cowell, "Radiation Poisoning Killed Ex-Russian Spy," *New York Times* (24 November 2006). This is particularly true of alpha radiation. The story about the protective suits is almost certainly apocryphal. See Stuart Diamond, "Key Cleanup Step Planned for a 3-Mile Island Reactor," *New York Times* (19 July 1974).

79. Email from Sebastian Pflugbeil to Dolores Augustine, 31 March 2015.

80. Bund der Evangelischen Kirchen in der DDR, *Energie und Umwelt: Für die Berücksichtigung von Gerechtigkeit, Frieden und Schöpfungsverantwortung bei der Lösung von Energieproblemen in der DDR: Erarbeitet vom Unterausschuss "Energie" des Ausschusses "Kirche und Gesellschaft" im Auftrag der Konferenz der Evangelischen Kirchenleitungen in der DDR* (East Berlin: Bund der Evangelischen Kirchen in der DDR, 1989).

81. Email from Sebastian Pflugbeil to Dolores Augustine, 31 March 2015.

82. Robert Havemann-Gesellschaft, TH 2/5b 84–95, Sebastian Pflugbeil, "Briefe an meinen Patensohn," Samizdat dated 2 March 1988.

83. This reclassification of deaths was possible because autopsy reports were available. See Dale Preston and Donald Pierce, "The Effect of Changes in Dosimetry on Cancer Mortality Risk Estimates in the Atomic Bomb Survivors," Technical Report RERF TR 9–87. Retrieved 10 September 2016 from http://www.rerf.jp/library/archives_e/tr1987.htm. See also Richard Sposto, Dale Preston, Yukiko Shimizu, and Kiyohiko Mabuchi, "The Consultant's Forum: The Effect of Diagnostic Misclassification on Non-Cancer and Cancer Mortality Dose Response in A-Bomb Survivors," *Biometrics* 48 (June 1992): 615–17.

84. George Kneale and Alice Stewart, "Childhood Cancers in the U.K. and Their Relation to Background Radiation," *Radiation and Health* 16 (1987): 203–20.

85. Michael Beleites, "Pechblende: Der Uranbergbau in der DDR und seine Folgen," samizdat publication, unnumbered page, BStU, HA XVIII Nr. 18237, 82–151.

86. Beleites, "Pechblende," 15–24, 28–34, 40–43, 47, 55 (cited according to the samizdat).

87. Uekötter, *Deutschland im Grün,* 184; see also 185–86.

88. BStU, HA XVIII Nr. 18237, 152–53; quotation on 152.

89. Beleites, "Pechblende," 24–25, 29 (cited according to the samizdat).

90. Rainer Karlsch and Rudolf Boch, "Die Geschichte des Uranbergbaus der Wismut: Forschungsstand und neue Erkenntnisse," in *Uranbergbau im Kalten Krieg: Die Wismut im sowjetischen Atomkomplex,* ed. Rudolf Boch and Rainer Karlsch, (Berlin: Ch. Links Verlag, 2011), 12; article on 9–15; see, in the same volume, Manuel Schramm, "Strahlenschutz im Uranbergbau. DDR und Bundesrepublik Deutschland im Vergleich (1945–1990)," 308; chapter on 271–328.

91. Joachim Krause, "... nicht das letzte Wort: Kernenergie in der Diskussion," Samizdat publication, Kirchliches Forschungsheim Wittenberg, 1987, BstU, HA XVIII Nr. 18237, 16–54.

92. Ibid., 31.

93. Anton Bayer, *Deutsche Risikostudie Kernkraftwerke*. Vol. 8: *Unfallfolgenrechnung und Risikoergebnisse* (Cologne: Verlag TÜV Rhineland, 1981).

94. Joachim Krause, "... nicht das letzte Wort: Kernenergie in der Diskussion," samizdat publication, Kirchliches Forschungsheim Wittenberg, 1987, BstU, HA XVIII Nr. 18237, 29–33.

95. Dietmar Wolf, "Die Berliner Umwelt-Bibliothek: Links, anarchistisch und auch immer ein wenig chaotisch," retrieved 20 August 2016 from http://umwelt-bibliothek.de/umwelt-bibliothek.html.

96. Rüddenklau, *Störenfried,* 68.

97. Hayton, "Culture from the Slums," 484.

98. Neubert, *Geschichte,* 686–90.

99. "MfS-Aktion gegen die Umwelt-Bibliothek," ed., Bundeszentrale für politische Bildung and Robert-Havemann-Gesellschaft e.V., most recently edited September 2008, retrieved 29 March 2015 from www.jugendopposition.de/index.php?id=203.

100. BStU Archiv der Zentralstelle, MfS-ZAIG Nr. 20660, 30–39, 52.

101. Lengsfeld, *Ich wollte,* 186–87.

102. "Stasi und DDR-Opposition. Zoff um Zion," Spiegel Online (26 November 2007), retrieved 22 August 2016 from http://www.spiegel.de/einestages/stasi-und-ddr-opposition-a-948898.html.

103. "MfS-Aktion gegen die Umwelt-Bibliothek," ed. Bundeszentrale für politische Bildung and Robert-Havemann-Gesellschaft e.V., most recently edited September 2008, retrieved 29 March 2015 from www.jugendopposition.de/index.php?id=203.

104. *Junge Welt* had the highest circulation of East German newspapers and was the official organ of the Communist youth organization, the Freie Deutsche Jugend (FDJ, or Free German Youth).

105. Lengsfeld, *Ich wollte,* 164, 188–90; quotation on 188.

106. Huff, *Natur und Industrie,* 370–71.

107. Robert Havemann-Gesellschaft, TH 2/5b 84–95, Sebastian Pflugbeil, "Briefe an meinen Patensohn," samizdat dated 2 March 1988.

108. Michael Beleites, *Untergrund: Ein Konflikt mit der Stasi in der Uran-Provinz* (Berlin: BasisDruck, 1991); Lengsfeld, *Ich wollte.*

109. Lengsfeld, *Ich wollte,* 117.

110. Emnid is a German opinion polling organization.

111. "Den Neuen fehlt Selbstvertrauen," *Der Spiegel* 46 (12 November 1990).

Chapter 8

Abandoning Nuclear Power—Or Not?

Wildly gyrating walls, trembling skyscrapers, children hiding under desks, terrified businessmen in dark suits scrambling for safety, a deep, dull roar—these were some of the first images and sounds of the Japanese earthquake of 11 March 2011 that went around the globe via television and internet. Massive fires broke out, buildings collapsed, and highways buckled in the strongest earthquake ever to hit Japan, 9.0 on the Richter scale. It was followed by about a hundred aftershocks, many of them as strong as a powerful earthquake. Far more devastating was the tsunami that hit the northeastern coast of Japan.

Residents standing on a tall building recorded the devastation as a tsunami rushed into their town, its ghastly black waters pouring onto a street, carrying away cars in a great flood that filled streets, flipped boats, and carried away buildings. Scenes of utter devastation were widespread as the tsunami inundated about two hundred square miles, damaging or destroying some 375,000 buildings and a quarter of a million automobiles and killing twenty thousand people.[1] Waves that were 42 feet high flooded the Fukushima Daiichi Nuclear Power Plant. The destruction of backup power generators caused a loss of reactor cooling, resulting in three core meltdowns, a hydrogen explosion, and release of radioactive materials into the environment.[2]

The German reaction was swift. On 12 March, sixty thousand protesters chanting, "*Abschalten!*" (Shut it down!) formed a 28-mile–long human chain that reached from the nuclear power plant Neckarwestheim to Stuttgart, the capital of Baden-Württemberg.[3] On 30 May 2011, the conservative and heretofore pro–nuclear power government of Angela Merkel announced a plan to abandon nuclear power by 2022. Italy, Belgium, and Switzerland took similar steps in the wake of Fukushima, although Belgium later backtracked.

In their attempts to explain this uncompromising German reaction to Fukushima, some scholars have clung to psychological explanations ultimately rooted in conceptions of "German fear." For example, Klaus-Dieter Maubach asserts that the impact of the earthquake and tsunami reminded Germans of wartime scenes of bombed-out cities, particularly Hiroshima and Nagasaki. Although the destruction was not caused by the Fukushima core meltdown, he argues, the wider disaster became conflated and "emotionally linked" in the minds of Germans with the nuclear power plant accident.[4] While there may be some truth to such explanations (although hard evidence is lacking), they overlook the impact of decades of environmental activism and public debate, which continued after German reunification in 1990. The first major issue was the discovery of major safety problems with East German nuclear power plants and the decision to shut them down.

Anti–nuclear power activism had been on the wane but was reactivated by massive protests against the transportation of nuclear waste in dry cask storage, known under the brand name Castor (an acronym for "cask for storage and transport of radioactive material") to Gorleben, starting in 1995. More quietly, a professionalization of environmentalism was taking place, particularly within the Green Party and institutions such as the Öko-Institut (Ecology Institute) in Freiburg. Renewable energy came to enjoy government support, entrepreneurial enthusiasm, and growing popularity even before Fukushima. A phaseout of nuclear power had been under discussion for years but suddenly won widespread support after Fukushima. Most recently, Germany has not been immune to doubts as to whether renewables will cover future energy needs. In the age of global warming, nuclear power has won friends across the globe. The last section of this chapter addresses the question of whether the German rejection of nuclear power is irreversible.

A Radioactive Issue: Nuclear Safety Problems in the Former East

Environmental degradation, particularly air pollution and toxic waste, emerged as a major topic in protests and public discussions in the tumultuous period leading up to the fall of the Berlin Wall on 9 November 1989 and the reform period thereafter. Nuclear power was one of the issues that had animated environmental activists in the Communist period, and it continued to do so in the period of democratization. Sebastian Pflugbeil, a long-time critic of nuclear power, brought this issue into the New Forum, which was founded by a group of dissidents in September 1989 and which became a broad-based movement demanding democratic rights. The transitional government of Hans Modrow, in power from 13 November 1989 to 12 April 1990, began to tackle the ecological crisis. Dominated by the SED elite, that government was, however, strongly committed to the GDR's nuclear power program.

A visit to the Greifswald nuclear power station by the West German minister for environmental protection in January 1990 made clear, however, that the GDR was going to have to work closely with the Federal Republic and establish a new basis for trust. On 22 January the Modrow government released a list of nuclear incidents and accidents that had taken place in the Greifswald station over the previous decades and announced the formation of an investigative commission, to be run by the West German Society for Reactor Safety (GRS), with participation of the West German technical inspection organization TÜV, the Materials Testing Office in Stuttgart, SAAS (the East German State Office for Nuclear Safety and Radiation Protection), and the socialist combine for nuclear power plants.[5]

Media outcry was swift and effective. Revealing a long a list of nuclear incidents—a "chronicle of horror"—at Greifswald, Der Spiegel reported that the plant was "little more than a pile of scrap metal." Worst of all, the article asserted, was the cover-up, orchestrated by the SED and the Stasi. The article quoted employees as referring to Greifswald as "Chernobyl North" and calling its construction "endlessly shoddy."[6] A cover article called for its shutdown, while lambasting the "arrogance ... of nuclear power supporters and propagandists," including those in the West.[7] These articles attracted a great deal of attention, garnering the praise of the left-liberal West German newspaper Frankfurter Rundschau but the opprobrium of the East German weekly Wochenpost, which accused the West of trying to force the GDR into bankruptcy in hopes of buying up its assets for nothing. The head of the socialist combine for nuclear power plants, Reiner Lehmann, asserted in the Wochenpost that the articles in Der Spiegel were riddled with errors and misrepresentations.[8]

On 3 February, ten thousand demonstrators from East and West Germany converged on Gorleben to protest against nuclear power plants in both Germanys, and on 11 March, eight thousand protested against the construction of the atomic power plant in Stendal, East Germany. However, some East German environmentalists had mixed feelings regarding atomic power because they saw the reduction of sulfur dioxide emissions of coal-burning power plants and open-pit mining of lignite as higher priorities.[9]

Under political pressure, the Modrow government asked eight representatives of the political opposition who took part in round table negotiations with the government to join the government as ministers without portfolio in February 1990. Among them was Pflugbeil, who immediately used his position to go on an inspection tour of the Greifswald nuclear power station. He was greeted by protesting employees, who came out in support of their plant. He now had access to SAAS reports critical of conditions in Greifswald, which he used to argue in favor of shutting down reactor units 1–4. Ministers from the round table demanded that a second investigative commission on Greifswald be set up. It was dominated by scientists opposed to nuclear power. In its final report, it recommended the complete decommissioning of all Greifswald reactor units, while the commission headed by the GRS only favored shutting down units 1–4. According to Felix Matthes of the Eco Institute, nuclear power was the first "integrated East–West German political arena," with supporters and opponents in both German states.[10]

In the period after German reunification, which took place on 3 October 1990, the West German energy industry, along with the federal government, proved unwilling to take the risk of propping up the East German nuclear power industry because this could cause political problems for the West German program and would have entailed considerable expense.[11] The entire Greifswald nuclear power station was decommissioned and torn down. The nuclear power plant in Stendal never went into operation.[12]

Political opposition to nuclear power may have played a role in convincing political and industrial leaders that it was not worth saving the East German nuclear power plants. However, this triumph was widely interpreted as resulting from the failure of Soviet and East German technology. Soon, nuclear power was being discussed as a climate-friendly technology that would help combat global warming and air pollution.[13] The anti–nuclear power movement fell into a funk. In 1991, one activist told a journalist from the magazine *Atom*, "The people who have been fighting against nuclear energy seem to be at the end of their strength."[14]

Gorleben Redux

In 1995, the anti–nuclear power movement came roaring back to life with protests against the transportation of nuclear waste to Gorleben. Massive protests against the building of a nuclear reprocessing plant in Gorleben had caused Minister-President Ernst Albrecht to give up on the project in 1979. However, over time, it became clear that this was nothing more than a tactical step backward. Industry went ahead with two temporary storage facilities for nuclear waste in Gorleben and continued with exploratory work on salt mines, where, it was hoped, highly radioactive spent nuclear fuel could be stored permanently—as long as a million years. In 1989, the West

German energy company Veba formed a partnership with a French energy company, Cogema, allowing West German nuclear power plants to send spent nuclear fuel assemblies to a nuclear reprocessing plant in La Hague, France. These were then to be shipped to the Gorleben storage facility. The first such shipment arrived in 1995, the last in 2011.[15]

The Gorleben movement of the 1990s and early twenty-first century was a world apart from the "Free Republic of Wendland," a joyous outpouring of utopianism made concrete in a village built by protesters in 1980. The Gorleben movement became radicalized in the 1980s, as activists resorted to sabotage, bomb threats, and violent confrontations with police.[16] In the 1990s, the Environmentalist Citizens' Initiative of Lüchow-Dannenberg (Bürgerinitiative Umweltschutz Lüchow-Dannenberg) was the most prominent of a sprawling network of anti-Castor activist groups. They organized blockades of streets, highways, and rail lines, following the peace movement's model of nonviolent civil disobedience. Ten thousand protesters participated in a peaceful rally in Dannenberg on 4 May 1996 and twenty thousand in a demonstration in Lüneburg on 1 March 1997. Farmers organized tractor demonstrations in March 1979, March 1997, and March 2001 and drove tractors in larger demonstrations, for example on 7 November 2008.[17]

However, autonomists (anarchists), along with a growing number of "peaceful" activists, engaged in acts of sabotage aimed at the state railway, the Bundesbahn. Pointing to injuries to railway employees, *Der Spiegel* quoted the head of the Bundesbahn as calling these saboteurs "political terrorists."[18] Some forms of protest did, however, recapture the imaginative quality of the Republic of Wendland phase. Blockade participants made music and sang protest songs. Some were older songs from the 1970s but punk and "beat-boxing" were also part of the protesters' repertoire. A video of the May 2003 protests made by the video group of Environmentalist Citizens' Initiatives of Lüchow-Dannenberg showed a kind of folk festival that included games and demonstrations for children, including a model railroad with figurines of protesters that snapped onto the tracks, stopping the little model train.[19] Inspired by the 2003 British film *Calendar Girls,* male and female farmers of the Farmers' Emergency Association of Lüchow-Dannenberg (Bäuerliche Notgemeinschaft Lüchow-Dannenberg) posed nude for a calendar, whose sales proceeds they hoped to use to help finance their cause.[20]

The Social Democratic governments of Gerhard Schröder (in office in Lower Saxony 1990–1998), Gerhard Glogowski (1998–1999), and Sigmar Gabriel (1999–2003) and the Christian Democratic–led governments of Christian Wulff (2003–2010) ordered massive, militarized police mobilizations to counter attempts to stop Castor deliveries. Nineteen thousand police officers were deployed in May 1996, twenty thousand in February 1997, thirteen thousand in November 2003, and sixteen thousand in November 2010. In February 1997, the police union began complaining that police were being exposed to dangerous levels of radiation, a fear that the government claimed was unfounded. By 2003, the cost of defending a Castor delivery had risen to thirty million euros, according to *Der Spiegel*. In 2010, the police union and the German Taxpayers' Federation (Bund der Steuerzahler e.V.) called on industry to help defray the enormous costs. Groaning under the weight of this expense, the government of Lower Saxony requested a reimbursement from federal funds.[21]

Scholar–activist Felix Kolb points out that the anti-Castor movement created a narrative that came across well in the media: the Gorleben movement aimed to bring down the entire German nuclear power program by hitting it at its most vulnerable spot; the transportation of nuclear waste to Gorleben entailed an unacceptable level of risk of accidents; temporary storage of spent nuclear fuel in Gorleben was untested with regard to possible dangers such as earthquakes and aircraft crashes, while permanent storage in salt domes was even more questionable; and although individual attempts to stop Castor transports might fail, the constant pressure would eventually achieve its goal. A series of revelations of safety problems served to underline and validate this narrative. In May 1998, it was revealed that the nuclear industry had for years failed to disclose radiation leaks from the Castor containers. Additional factors greatly increased the impact of the Gorleben protests.

The movement was extremely well organized, thanks to the Environmentalist Citizens' Initiative of Lüchow-Dannenberg as well as international organizations such as Greenpeace and Robin Wood, and enjoyed the wholehearted support of the Green Party and *Tageszeitung-taz*.[22] The older generation of activists who had participated in the building of the Free Republic of Wendland in 1980—for example Marianne Fritzen, Undine von Blottnitz, and Rebecca Harms—were still involved, providing examples of lifelong dedication to the cause.[23] Intergenerational continuity was crucial to the success of the Gorleben movement of the 1990s and early twenty-first century. The anti–nuclear power movement had become rooted in German political culture in a way that was not seen in countries such as the United States and France. As a result, the German movement was about to achieve its uppermost goal—the official abandonment of nuclear power. However, the movement's more utopian goals were being overtaken by the emergence of a green economy and a convergence of antipolitics and conventional politics.

The Greening of Germany and the Professionalization of Environmentalism

The weaving of ecology into the fabric of the German economy and political system had started decades before. The basic technologies of solar and wind power were developed long before their commercial development. The president of Harvard, James Bryant Conant, predicted in 1951 that solar power would become the dominant energy source by the end of the twentieth century.[24] The energy crisis of the 1970s sparked some interest in renewables among scientists, political leaders, and industry in the United States and Western Europe, but research, development, and production were very limited in scope. Nonetheless, in the 1960s and 1970s, improvements to photovoltaic technology—used in solar panels and solar cells—prepared the way for mass use.[25] Carol Hager points to the 1981 founding of the Fraunhofer Institute for Solar Energy Systems in Freiburg by Adolf Goetzberger as one of several examples of the crucially important transition from activism to innovation.[26]

In West Germany, the anti–nuclear power movement played a decisive role in the push to develop renewable energy sources because it generated demand and put pressure on government to subsidize research. Founded in 1977 by activists from the Wyhl movement, the Eco Institute of Freiburg (Öko-Institut Freiburg), proposed in 1980 that West Germany work toward a future abandonment of nuclear power and oil through energy conservation, energy efficiency,[27] and development of renewable energy sources, coining the now-ubiquitous term *Energiewende* (energy transition).[28]

Concern about global warming, greenhouse gases, and the use of fossil fuels burst onto the world stage with the worldwide sustainable development discourse of the 1980s, as historian Miina Kaarkoski points out. Under Chancellor Helmut Kohl, Germany took on the role of climate policy pioneer with the 1990 announcement that the Federal Republic would reduce its carbon dioxide emissions by at least 25 percent by 2005, a pledge that German representatives repeated at the United Nations Earth Summit in Rio de Janeiro in 1992. Kohl's Christian Democrats (CDU/CSU) and their coalition partners, the FDP, proposed replacing fossil fuels with an "energy mix" that would include nuclear power. This put them at odds with opponents of nuclear energy, who only slowly came to embrace the term "energy transition," as Kaarkoski has shown.[29] Ironically, it was the Christian Democrats, in power until 1998, who initiated the development of renewable energy sources on the national level.

In Germany, the transition to alternative energy sources was initiated by a 1990 law that required big energy corporations to buy up energy produced by small-scale renewable energy producers at a set, above market rate. Corporations tried to fight this highly contentious law and its successor, the Renewable Energy Law of 2000. Instead of developing these new technologies themselves, they pressed politicians and the courts to eliminate subsidies for renewables.[30] It was "tinkerers and idealists" who established large numbers of medium-sized businesses that produced wind turbines and set up wind farms in the 1990s, according to Maubach. Employing people who were "highly motivated and innovative," the wind industry spawned technological improvements, lowered costs, and created jobs.[31]

Solar power and, to a certain extent, biomass power also promoted a decentralization of energy production. Solar power experienced a breakthrough in Germany in 1999–2003, when low-interest loans induced many citizens to install solar panels on their roofs. Demand for loans outstripped availability, and the program was discontinued. However, mass production led to sinking costs, and research was ramped up with the help of state subsidies. Solar power proved surprisingly popular in this sun-deprived country because it was accessible to individual homeowners, farmers, and communities. Volunteers set up and ran community solar power plants. Farmers participated in the growth of biomass power, which made use of manure, grass, plant waste, and energy crops such as corn.[32]

There has been considerable disagreement concerning the social impact of this turn to alternative energy sources. Social Democratic politician Hermann Scheer believed that the development of renewable energy empowered the people.[33] A study by anthropologist Jennifer Carlson on the impact of the renewable energy boom in a small town in northern Germany suggests, however, that propertied male farmers were the primary beneficiaries.[34] Large corporations have also entered the market, building large-scale wind farms, both on land and offshore. Critics of wind power argue that it destroys the German landscape and is too noisy and expensive.[35] Citizens complain that they are being excluded from the siting of wind turbines, but are finding new ways to make their voices heard.[36] There has been a fair amount of public criticism of biomass energy due to the lack of biodiversity it causes, the need for insecticide, and the unpopularity of large biogas plants. Around 2006–2007, the media in Germany as well as in other countries increasingly criticized the importation of biomass products because they crowd out food production in the Global South.[37]

However, from the perspective of nuclear power, the turn to renewable energy sources has been a success, laying the foundations for an economy with a declining need for nuclear power. According to the Working Group on Energy Balances (Arbeitsgemeinschaft Energiebilanzen e.V.), the percentage of power generated in Germany by renewables rose from 3.4 percent in 2000 to 18.6 percent in 2016, while that of nuclear power fell from 34.7 percent to 18.9 percent over the same period.[38]

At the same time that renewable energy was experiencing a breakthrough but also becoming a "normal industry,"[39] the Green Party, along with other NGOs and media associated with the anti–nuclear power movement, were also entering the mainstream. As historian Jens Ivo Engels has shown, the Eco Institute evolved from an activist organization into a think tank and ecological consultant to both corporations and the state, accommodating itself to their needs but at the same time attaining much greater influence. Essentially, the counterexperts came to be accepted as experts.[40]

Professionalization also took place in the Green Party. When Alliance '90/the Green Party[41] joined a coalition government headed by German chancellor Gerhard Schröder in 1998, it definitively turned its back on its earlier identity as an "antiparty" and as part of the new social movements. Joschka Fischer's rise to the offices of vice-chancellor and foreign minister marked the triumph of his strategy of accommodation and coalition. Fischer faced criticism for his participation in revolutionary activism in the 1960s and 1970s but prevailed as a new member of the establishment.[42] However, with regard to nuclear power, the Greens found themselves drawn into a process of negotiation and compromise more protracted and difficult than expected.

Nuclear Realpolitik before Fukushima

On the face of it, an ecological consensus had emerged in reunited Germany, thanks in no small part to the prominence of consensus-building issues such as forest dieback due to acid rain (*Waldsterben*).[43] The future of nuclear power nonetheless continued to be a fraught topic. The Greens made the abandonment of nuclear power a condition of their participation in the "Red–Green coalition" with the Social Democratic Party, which governed Germany from 1998 to 2005. However, Chancellor Gerhard Schröder did not, in fact, pursue the rapid phasing out of nuclear power. He was concerned that if nuclear power plants were shut down quickly, jobs would be lost and the courts would order the government to make compensation payments for this loss of private property to the energy industry, which had open-ended operating permits for nuclear power plants. In talks behind closed doors with the big energy and utility companies, the Schröder government agreed not to shut down the nuclear power plants for thirty-five years. In exchange, the corporations agreed not to ask for compensation payments.

When the Greens balked, a public tug-of-war ensued. The minister-president of Bavaria, Edmund Stoiber, a member of the pronuclear Christian Socialist Party, threatened that Bavaria would veto any legislation that called for a shutdown of nuclear power plants in *less* than thirty-five years.[44] Compromises were reached in 2000 and 2002, but they diverged from the demands of the anti–nuclear power movement. No new nuclear power plants were to be built; those already in existence were to be shut down after, on average, thirty-two years in operation; and investigation of the salt domes in Gorleben as a possible permanent repository for nuclear waste was to be suspended. The new agreements set limits on the amount of power that each nuclear power plant could generate, but more could be produced at one plant if less was produced at another.[45] Reluctantly, Green minister for the environment, Jürgen Trittin, a member of the Green Party, acquiesced. This led to a rift between the anti–nuclear power movement and the Greens, and not a few Green Party members left the party.[46]

Scholars disagree in their assessment of Schröder's nuclear power policies. One political scientist describes Schröder's style of governing as similar to Konrad Adenauer's inasmuch as both liked to negotiate with lobbyists, ignoring their ministers.[47] In the 1970s, he had headed the Social Democratic youth organization, the Jusos (Young Socialists) at a time when it was distinctly leftist and militant in orientation. As chancellor, however, he treated political activism as irrelevant and did not give environmentalist organizations a voice in negotiations. Instead, he plotted a pragmatic, centrist course for his party similar to that of Tony Blair's New Labour and Anthony Giddens's "third way."[48] However, Schröder did manage to achieve what had previously been unattainable: a plan for the eventual abandonment of nuclear power. Political scientist Karlheinz Niclauß views this as "a personal negotiating success" for Schröder.[49]

Schröder's successor, Angela Merkel, was not expected to continue the push to shut down Germany's nuclear power program. Quite the contrary: she ran in 2005 on a party platform that promised to postpone the shutdown of nuclear power plants for years.[50] Christian Democrats viewed atomic energy as a stopgap technology that would help fill Germany's energy needs until enough renewable energy was available, presumably in decades.[51] Christian Democrats had been arguing since at least 1991 that Germany could only reduce its carbon emissions if it continued to use nuclear power.[52] Merkel could not put this policy into practice in 2005–2009, when her party and its coalition partner, the Christian Socialists, formed a "grand coalition" government with the Social Democrats. During this period, Schröder's nuclear policies remained in force.[53]

The Atomic Forum (a pronuclear lobbying organization) hired a communications agency to run a pro–nuclear power campaign in the run-up to the 2009 election. Shedding its Social Democratic coalition partners after the 2009 elections, the Merkel government—now an alliance of Christian Democrats, Christian Socialists, and Liberals (FDP)—was able in 2010 to overturn Schröder's policies, postponing the phaseout of atomic power and restarting exploratory work on the Gorleben salt domes.[54] Activists protested the government's nuclear policies in a series of demonstrations in 2010. It is ironic that it was precisely this probusiness, pronuclear Christian Democratic government that oversaw Germany's abandonment of nuclear power.

"Fukushima Is Everywhere"

Just as 9/11 had destroyed the West's sense of invulnerability, Fukushima[55] showed the technological mastery of nuclear power to be an illusion, asserted journalist Michael Bauchmüller in an opinion piece in the *Süddeutsche Zeitung* shortly after the disaster.[56] The world watched in mounting apprehension as the catastrophe unfolded over the course of hours, days, and weeks. The earthquake, hitting Fukushima at 2:46 p.m. on 11 March 2011, caused the reactors to automatically shut down. However, the temperature in three reactors remained high due to decay heat caused by radioactivity. Fifty minutes later, the tsunami arrived. The seawall proved not to be high enough to keep out the waters of the nearly fifty-foot waves. This was only one of many lines of defense that failed, exposing the inadequacy of the plant's safety features. The cooling systems and control panels shut down due to a widespread blackout caused by the earthquake. Unfortunately, the backup generators were located in basements, which flooded. The building housing a second set of backup generators also flooded, causing them to fail.

TEPCO (Tokyo Electric Power Company) attempted to restore power, but without success. As a result, the cooling water boiled off, leaving the reactor cores uncovered. Firefighters' attempts to inject water into the reactors were initially largely ineffectual. The nuclear fuel became so overheated that core meltdowns took place in units 1, 2, and 3. The molten fuel melted through the reactor core, but the concrete floors below held—at least initially. Overheating caused zirconium to react with water, producing hydrogen that escaped through damaged seals into the reactor buildings. Hydrogen explosions took place in units 1, 2, and 3 on 12, 14, and 15 March, causing the spread of cesium and other radioactive substances into the atmosphere and across the surrounding areas. Radioactive water also leaked into the ocean. The Fukushima disaster eventually was, like Chernobyl, given a rating of 7 on the INES scale, although the repercussions were considerably less severe.[57]

A sense of uncertainty and anxiety dominated news coverage of the crisis. Would this nuclear accident remain in the Three Mile Island range, or could it blow up into a second Chernobyl?[58] German experts expressed considerable concern regarding the possible impact of the disaster on human health in Japan. Sebastian Pflugbeil, the president of the Society for Radiation Protection and earlier critic of the East German nuclear power program, said in a television interview on 16 March that a radioactive cloud could seriously endanger the health of the inhabitants of Tokyo but that little could be done because such a large city could not be evacuated. But even a proponent of nuclear power such as Peter Jacob, the director of the Institute for Radiation Protection of the Helmholtz Center of Munich, was reluctant to appear to be overly optimistic. He admitted that Tokyo's population could suffer serious health consequences if the worst-case scenario—massive release of radioactive materials into the atmosphere—came to pass in Fukushima.[59]

Despite the constant stream of updates available on the internet[60] and extensive reporting on television and in the print media, solid information was hard to come by. Television viewers saw videos of the cloud of smoke streaming out of reactor buildings and firefighters spraying great arcs of water on the affected reactors.[61] However, what was happening in the interior of the reactors was unclear. Had a core meltdown taken place? The Japanese media said yes, while the Japanese government initially denied it. "We don't know what to make of it because contradictory reports are circulating," said one German television journalist reporting from Japan. His on-air colleague back in the studio in Germany expressed her anxiety bordering on anger through body language and facial expressions.[62]

Der Spiegel strongly criticized the information politics of the Japanese government and TEPCO, attributing it to "the typical Japanese habit of trying to sweep accidents under the carpet—out of shame and a false sense of group loyalty or corporate spirit."[63] Soon, it became clear that TEPCO had been ignoring safety issues in Fukushima for years and that close ties between government and industry in Japan had fatally weakened the nuclear regulatory system.[64] In the case of Chernobyl, proponents of nuclear power, including future chancellor Angela Merkel, had argued with some success that the 1986 disaster was the result of the poor state of Soviet technology.[65] Why, then, was it not possible in 2011 to blame Fukushima on the specific shortcomings of Japanese society? For one, decades of scandals and cover-ups, most recently of radiation leaks of Castor transports to Gorleben, had undermined public trust in the responsible elites.

Another important factor was that the German political leadership and technical experts fully recognized Japan as a technologically advanced industrial society and as a democracy and drew the conclusion that such an accident could take place in Germany. This thought is evident in Chancellor Merkel's comment, "Before Fukushima, I accepted the residual risks connected with nuclear power because I was convinced that in a technologically advanced country with high safety standards, [these risks] would not materialize, as far as is humanly possible to judge."[66] This statement is emblematic of the return of risk to the forefront of public debate.[67] "Residual risk"—originally a technical term connoting unforeseen risks outside of normal risk calculation—had become a central concept of the anti–nuclear power movement.

The double punch of the most violent earthquake ever recorded in Japan and the highest tsunami ever known to have hit the Fukushima coastline was a classic case of a residual risk or, in colloquial English, a "black swan"—an event so seemingly improbable as to be thought impossible. Obviously, earthquakes and tsunamis were not the problem in Germany. But what if a terrorist attack on a nuclear power plant took place? asked *Die Zeit*.[68] In addition, there was always the unknowable factor of limitations to human insight. In a front-page opinion piece in *Frankfurter Allgemeine Zeitung*, Berthold Kohler wrote, "As with Three Mile Island and Chernobyl, *afterwards* there were many explanations in the Japanese case for how the disaster occurred and how it could have been prevented. The fact of the matter is that these accidents occurred, despite comforting probability calculations, in a time span far shorter than [the predicted] thousands of years."[69]

Seventy percent of Germans surveyed by Infratest on 14 March 2011 believed that an accident such as the Fukushima disaster could take place in Germany. An equal number were in favor of a complete shutdown of German nuclear power plants. Not only long-time activists but also "completely normal citizens" demonstrated against nuclear power in an estimated 450 German towns and cities on 14 March, a workday evening.[70] Vigils, human chains, picketing in front of the chancellor's office, and large demonstrations in big cities continued for weeks.

Angela Merkel moved swiftly and decisively. On 14 March she announced the temporary shutdown of seven older nuclear power plants, along with the mishap-plagued plant Krümmel. Roundly criticized by the other parties, Merkel was under a great deal of political pressure.[71] In elections on 27 March in Baden-Württemberg, Merkel's Christian Democratic Party had its worst showing in that state since 1952, receiving only 39 percent of the vote, while the Greens achieved at all-time high of 24.2 percent of the vote. For the first time, a member of the Green Party became minister-president of a German state, heading a coalition with the Social Democrats.

Merkel's government set up an ethics panel that, along with the Reactor Safety Commission, was to advise the government on how to proceed with nuclear power. It was headed by the most important proponent of environmentalism in the CDU, Klaus Töpfer, and included representatives of most major parties, academia, churches, and nuclear industry but excluded environmentalist organizations and the far-Left party Die Linke, or "The Left." This panel recommended the abandonment of nuclear power, advice that Merkel followed, despite the finding of the Reactor Safety Commission that Germany's nuclear power plants were safe. Announcing her decision to permanently close eight nuclear power plants and to phase out the others by 2022, Merkel explained, "Fukushima changed my stance on nuclear power ... [since it revealed that] even in a high-tech country such as Japan the risks of nuclear power cannot be reliably contained."[72]

This was a tremendous about-face for Merkel, who had been a supporter of nuclear power for years, as well as for the Christian Democratic and Christian Socialist Parties, the great majority of whose members supported her decision.[73] The Bundestag voted in favor of Merkel's proposed phaseout of nuclear power 513 to 79.[74] Merkel's personal shock over the devastation of the Fukushima plant may have played something of a role in this political reinvention. Above all, however, it was a response to public opinion. A *Frankfurter Allgemeine* commentator wrote that it would have been "political suicide" not to follow this path.[75]

The law met with a very positive popular response. According to a large-scale survey, Germans became more concerned about the environment and more risk averse after Fukushima, especially if they lived near a nuclear power plant. A follow-up survey showed that the decision to phase out nuclear power very much relieved their concerns. This effect was particularly strong among women who supported the Green Party.[76]

Although a fair degree of consensus was achieved regarding nuclear power in 2011, there were still major pockets of resistance. Not surprisingly, the big energy corporations were strongly opposed to an abandonment of nuclear power. Their lobbyists immediately started a campaign against the new policy, warning that consumers would be paying more for electricity and that blackouts could be expected and asserting that Germany would be reliant on nuclear energy in any case because the country would have to import power from France.[77] Some voices in the conservative press also expressed skepticism regarding the nuclear phaseout. Andreas Mihm wrote in the *Frankfurter Allgemeine,* "The real questions are where the energy that keeps our society going will come from, and whether it can be produced safely and reliably. Anyone who cannot guarantee this and shuts down nuclear power plants without having a substitute [energy source] is acting irresponsibly."[78] *Die Welt,* a high-circulation conservative newspaper, took Merkel to task for what it held to be an ill-considered and rash policy change.[79]

Merkel, however, saw Germany as blazing a trail by developing the energy sources of the future. She compared this "Herculean" project to the reunification of Germany. Federal minister of the environment Norbert Röttgen saw this program as an investment in the future.[80] In a sense, the German abandonment of nuclear power was a "Nixon in China" phenomenon, that is, the achievement of a long-time left-liberal goal made possible by its embrace by a conservative government. To recognize this is not in any way to diminish the role that Greens and Social Democrats, not to mention millions of activists and concerned citizens, played in making this profound policy change possible.

This is not the end of the story, of course, because climate change has ushered in a new chapter in debates about nuclear power.

Nuclear Power in the Age of Global Warming

With the emergence of global warming as the leading environmental issue, worldwide support for nuclear power has grown, and Germany has been no exception. In the *Frankfurter Allgemeine Zeitung,* political scientist Reinhard Wolf pointed in 2015 to the prognosis of the Intergovernmental Panel on Climate Change (IPCC) that by 2020, 75–250 million Africans will confront water shortages, while a billion Asians will be in the same situation by 2050—due to climate change. Wolf cites studies according to which nuclear power—even if we accept some of the higher estimates of deaths due to Chernobyl—has been responsible for far fewer deaths than coal.[81] According to the World Health Organization (WHO), climate change is causing about 150,000 deaths per year worldwide.[82] WHO estimates that annually an additional seven million people die prematurely due to air pollution—a part of which is caused by coal.[83]

Under the Kyoto Agreement of 1997, Germany pledged to reduce its greenhouse gas emissions by 21 percent by 2012, using 1990 levels as the base line. It achieved that goal in 2007, thanks partly to the growth of renewables and other improvements and partly to German reunification and the shutdown of many pollution-producing East German factories. Germany voluntarily declared a series of goals in efforts to reduce 1990 levels of greenhouse emissions: 40 percent by 2020, 55 percent by 2030, 70 percent by 2040, and 80–95 percent by 2050.[84] Recent developments cast doubt on the ability of Germany to reach these goals. According to Federal Ministry of the

Environment statistics, greenhouse gas emissions sank 27.7 percent between 1990 and 2014 but increased in 2012 and 2013, declined in 2014 but again increased—slightly—in 2015 and 2016.[85]

Renewable energy cannot, at the moment, solve the problem. The two renewable energy sources with the greatest growth potential are solar and wind energy.[86] The problem is that production fluctuates tremendously, surging when the sun shines and the wind blows but falling precipitously at night or when the air is still. Large-scale storage of energy, for example in batteries, is not yet economically feasible. Therefore, power plants that run on natural gas or coal are needed as a backup. However, coal plants cannot be started up and shut down as needed, so they are run at full capacity at all times.[87] In 2000, 23.8 percent of electricity was produced with black coal, 26.6 percent with lignite (brown coal). By 2016, black coal fell to 18.8 percent of electricity generation, while brown coal rose to account for 28.1 percent of the total.

The percentage for natural gas fell from 10.4 percent in 2010 and 2011 to 7.8–7.9 in 2014 and 2015.[88] Energy corporations shut down gas-powered plants before coal-powered plants because coal was distinctly less expensive than gas. The mandatory cap-and-trade system in Germany was supposed to make coal more expensive, but prices of CO_2 allowances have been so low that they have not really served their purpose.[89] However, use of natural gas recovered in 2016, rising to 10.3 percent.[90] Natural gas was thought of as a clean energy source until 2015.[91] Quite recently, however, scientists have discovered that large amounts of methane are released during natural gas production, making it harmful to the environment.[92] Presumably, the data of the Federal Environment Agency do not consider this new evaluation of the environmental impact of natural gas, thus understating the problem of release of greenhouse gases through use of gas and coal in power production.

The transition to renewable energy has also raised utility customers' energy bills. Energy prices, determined through trading on the European Energy Exchange in Leipzig, decline precipitously when large amounts of wind or solar power are fed into the energy network. At times, in fact, prices actually became negative, that is, suppliers have to *pay* to get rid of the power they produce.[93] This is a major problem for coal-fired plants, which are often run 24/7, but not for the renewables because they enjoy fixed prices. The cost for this de facto subsidy of renewable energy is passed on to consumers in the form of higher electricity bills.[94] Between 2002 and 2015, electricity prices paid by households in Germany doubled. Energy companies turned off nearly 350,000 customers' electricity because they did not pay their bills.[95] In 2016, the Merkel government attempted to cut subsidies to renewable energy and slow the growth of onshore wind power.[96] Offshore wind farms are a more dependable energy source and therefore produce energy more economically, because ocean wind currents are more constant.

The big energy corporations feel threatened by the new energy regime. Werner Wenning, a member of the board of directors of energy corporation Eon,[97] complained to the *Frankfurter Allgemeine Zeitung* in December 2015, "There isn't another branch of industry that has been given such a rough deal politically. So inconsistent. So threatening to its livelihood."[98] In 1998, Germany liberalized its energy markets, ending the monopoly position of energy companies in conformity with European Union rules.[99] The big corporations felt they were at a disadvantage because as the owners and operators of the power grid they had to invest in its expansion and improvement to accommodate the wind power producers, among others. Moreover, they were obligated to use all renewable energy supplied.[100]

A big part of the problem was that they continued to rely on coal and natural gas and moved too slowly into the renewable energy sector. As a result, Eon and RWE[101]—two of the biggest energy companies in Germany—experienced considerable losses in 2014 and 2015. In 2016, Eon and RWE adopted similar strategies, each splitting into a conventional energy producer and a company that focuses on renewable energy production. Corporate management has expressed optimism over the future of their respective corporations.[102]

All the legal claims connected with the shutdown of nuclear power plants have been settled. In December 2016, the German Federal Constitutional Court ruled that Eon and RWE must be compensated but did not set a concrete figure. Experts believe that the state will probably have to pay about 2.5 billion euros. This ruling establishes that the 2011 decision to shut down nuclear power plants is constitutional.[103] On the other hand, the German government successfully negotiated a settlement with these major conglomerates over the allocation of expenses and risks connected with the decommissioning and disassembling the nuclear power plants, as well as with the disposal and storage of nuclear waste.[104] On 3 July 2017, these corporations paid 24 billion euros in exchange for the state assuming the full costs—whatever they may turn out to be—of permanent disposal of nuclear waste now sitting in temporary storage in Gorleben and elsewhere.[105]

Consumers have also put up resistance to certain aspects of the energy transition. Wind power, produced mainly in the north, where ocean winds are strong, has to be moved to other parts of Germany. However, citizens along planned routes of this transmission network, called Südlink or Sued. link (South Link), have resisted the building of high-power lines through their communities. Tennet TSO, the corporation carrying out this billion-euro project involving thousands of power masts along five hundred miles, has been trying to negotiate with these communities and the citizens' initiatives representing them.[106]

The Christian Socialist–governed Bavarian state government threatened to block the extension of *Südlink* into Bavaria, arousing suspicions that Christian Socialists—long-time ardent supporters of atomic power—were trying to torpedo the transition.[107] In July 2015, the Bavarian minister-president agreed to a compromise: far fewer lines would be built in Bavaria. The energy company was fairly successful in appeasing residents by promising to place some of the transmission lines underground.[108]

Nonetheless, some observers have expressed skepticism as to whether Germany will succeed in its transition from nuclear to renewable energy in the long run.[109] Most believe that it would be politically unthinkable and very difficult from a logistical point of view for Germany to backtrack on its abandonment of nuclear power. The aim of most critics of the German phaseout of nuclear power seems to be to prevent Germany from becoming a model to other nations.[110] Like previous generations of nuclear energy proponents, they argue that atomic power is clean, inexpensive, and potentially safe, or that it has the potential to become so.

Some German nuclear experts, such as Antonio Hurtado, head of the Institute for Energy Technology at the Technical University of Dresden, believe that safe nuclear power plants could be built, nuclear waste greatly reduced, and future accidents avoided, truly making nuclear power a clean energy source. However, as the weekly business news magazine *Wirtschaftswoche* makes clear, skepticism regarding such claims abounds in Germany. Christoph Pistner of the Eco Institute in Darmstadt is not impressed with the "fourth-generation" projects touted by nuclear scientists: "Much of what sounds exciting in theory fails in the end because the technological implementation is far more complicated than had been realized, and the costs of development and construction are usually grossly underestimated."[111]

Is nuclear power really inexpensive? As far as costs go, economists and ethicists point out that nuclear power was always heavily subsidized, both directly and indirectly. The energy bills paid by consumers did not in the past reflect the costs of nuclear waste disposal. Instead, these costs were shouldered by taxpayers or were left for future generations to pay. This is an example of an "externality"—a cost to society resulting from the production of a good or service and not paid by the producer or the consumer but left for the larger society (or posterity) to pay. Ethicist Kristin Shrader-Frechette points out that many of the US studies that show nuclear power to be a low-cost energy source were commissioned by the nuclear industry and that their estimates of costs unjustifiably excluded such externalities, resulting in an underestimation of the costs of nuclear power by a factor of six.[112] At present, in fact, renewables are producing such a glut of power that they are making nuclear power unprofitable worldwide.[113]

Germany can only turn its back on both fossil fuels and nuclear power with the help of technological breakthroughs. Scientists and engineers such as Eicke Weber, director of the Fraunhofer Institute for Solar Energy Systems, are developing technologies that greatly improve the performance of renewable energy. These include expansion of the power transmission network, adaptation of the grid to fluctuating energy loads, and development of methods of energy storage that make it possible to deal with fluctuations in renewable energy production.[114]

German energy efficiency policies have also been very effective, particularly with regard to building standards.[115] Fuel economy standards, technical improvements, and high taxes on gasoline have led to improved fuel economy of automobiles. Moreover, Germany has deployed a mixture of regulations, economic incentives, research and development efforts, and public education to improve energy efficiency in a wide range of areas, including industry, appliances, lighting, and public transportation.[116] Environmentalist organizations and the media have kept up pressure on the German government to continue to improve energy efficiency.[117]

However, Germany's achievements in the areas of renewable energy and energy efficiency have only just made up for the loss of nuclear energy production, primarily because of increased automobile use. Moreover, the Volkswagen scandal, which spread to other automakers as well, revealed that German cars were producing far more greenhouse gas emissions and air pollution than had long been claimed. This major setback makes it appear quite unlikely that Germany will be able to achieve its climate goals for the next few decades. It is far too early to judge what the long-term impact of abandonment of nuclear power will be.

Conclusion

Two converging, but in some ways contradictory, forces brought about the decision to abandon nuclear power: citizens' activism, on the one hand, and environmentalism's professionalization and fuller integration into the capitalist system, on the other. The latter process was multifaceted. The Federal Republic's political leadership saw to it that East German nuclear power plants were evaluated according to Western standards. Scientific experts and corporate management took charge and dispatched the ailing GDR nuclear infrastructure with relative haste. The professionalization of the Green

Party and the triumph of the pragmatic politicians such as Joschka Fischer made Gerhard Schröder's Red–Green coalition government of 1998–2005 possible.

However, Schröder proved himself to be so flexible in negotiations with industry that the nuclear phaseout hardly met activists' expectations. Nonetheless, the agreement to shut down all nuclear power plants over a 32-year period set an important precedent. Renewable energy was another area where environmentally conscious citizens seized the opportunity to help build a greener economy. Some were eventually pushed aside by energy corporations and agrobusiness. Nonetheless, the development of wind, solar, biomass, and other forms of renewable energy made the abandonment of nuclear power conceivable. Professionalization and what might be called the entrepreneurial turn in environmentalism are only part of the story, however.

The anti–nuclear power movement also set the stage for Merkel's 2011 decision. In Germany, activism did not die in the 1990s. The massive protests against the Castor transports to Gorleben did not just give the movement focus and a public voice. The Gorleben protesters found the neuralgic point in the nuclear power system and exerted so much pressure on it that the state and energy industry had to respond. Anti–nuclear power protests were more intense in Germany than in any other country in the 1990s. By the end, the costs of defending the nuclear waste deliveries to Gorleben were putting an intolerable strain on the system.

Once again, the movement's strategy of disruption proved to be highly successful and was buttressed by a high degree of organization and a compelling narrative that mainstream media communicated to the German public. In addition, the movement's strategy of passive resistance—upheld by the great majority of protesters—gave it a political respectability that won over a growing number of Germans. The context was important as well: Communism had just fallen, and terrorism was not a major issue in the 1990s, although that changed with 9/11. Thus, society felt more tolerant toward protesters at the time.

Fukushima was one disaster too many for Germans. The failure of nuclear safeguards in a nation with a reputation for technological prowess like Japan caused many Germans, along with their chancellor, to fear that such a disaster could occur in Germany. Fukushima had a greater impact in Germany than elsewhere because of the long history of anti–nuclear power activism. Germany has certainly gone further than any other country in imagining a postnuclear future. Two social scientists have called Germany a "laboratory for the energy mix of the future."[118] Germany's strides in the development of renewable energy and in energy conservation and energy efficiency have not, however, been able to counterbalance auto emissions, a problem that was long hiding in plain sight but became manifest with the Volkswagen scandal. The expansion of renewables has also proven more difficult than originally thought.

The anti–nuclear power movement has faded—a victim of its own success. The Greens, now a party like any other, garnered only 8.9 percent of the national vote in 2017 and show little potential for further growth. The chances for a resurrection of the leftist agenda of the movement's mainstream seem dim. Nonetheless, it is remarkable that the anti–nuclear power movement has lasted as long as it has. As late as 2016, nuclear power still brought Germans onto the street—not in the numbers seen in times past but in far greater numbers than in other countries.[119] Historians have often overlooked the longevity of the movement. If it was a reaction to the problems of one generation or a reflection of the preoccupations of one era, it certainly grew to be far more.

The thrust of the movement was twofold. The first is popular resistance to the rule of experts and distrust of a state allied with elites. The second is human health, understood in scientific terms. The central demand of the anti–nuclear power movement was that the people have a major voice in deciding what technological risks they consider unacceptable. But popular empowerment makes little sense in an era in which the experts, bureaucrats, and professional politicians have seemingly accepted all the demands of the protesters. This professionalization of the anti–nuclear power movement and its integration into the capitalist system put its ultimate goal—the abandonment of nuclear power—within reach but appeared to condemn the movement to irrelevance. This actually leaves the energy future of Germany an open question.

Notes

1. Harsh Gupta and Vineet Gahalaut, *Three Great Tsunamis: Lisbon (1755), Sumatra-Andaman (2004) and Japan (2011)* (Dordrecht and New York: Springer, 2013), 53–60. For an example of German TV coverage, see the ZDF program, *Heute Journal,* 12 March 2011, retrieved 31 October 2017 from https://www.youtube.com/watch?v=24iY6QKa73o.
2. Shinya Mizokami and Yuji Kumagai, "Event Sequence of the Fukushima Daiichi Accident," in *Reflections on the Fukushima Daiichi Nuclear Accident,* ed. Joonhong Ahn, Cathryn Carson, Mikael Jensen, Kohta Juraku, Shinya Nagasaki, and Satoru Tanaka (Cham: Springer, 2015), 21–50.
3. N24 (a commercial television network), *Nachrichten,* 12 March 2011. Retrieved 30 October 2017 from https://www.youtube.com/watch?v=0Rc3jjP9EIw.
4. Klaus-Dieter Maubach, *Energiewende: Wege zu einer Bezahlbaren Energieversorgung,* 2nd ed. (Wiesbaden: Springer Fachmedien, 2014), 13.

5. Felix Christian Matthes, *Stromwirtschaft und deutsche Einheit: Eine Fallstudie zur Transformation der Elektrizitätswirtschaft in Ost-Deutschland* (Berlin: Matthes; Norderstedt: Libri, 2000), 215–50.

6. "Zeitbombe Greifswald," *Der Spiegel*, special issue 2: *162 Tage Deutsche Geschichte: Das halbe Jahr der gewaltlosen Revolution* (1 February 1990). Also see "DDR/ Kernkraft: Alles tot," *Der Spiegel* 4 (22 January 1990).

7. "Hausmitteilung Betr.: Titel," *Der Spiegel* 5 (29 January 1990).

8. "Nach 17 Jahren Schweigen: Zur Situation im Kernkraftwerk Greifswald," *Wochenpost*, 9 March 1990; "'Greifswald' stand mehrmals vor der Katastrophe," *Frankfurter Rundschau*, 29 January 1990.

9. Wolfgang Ehmke, "Zwischen Braunkohle und Atomstrom," *Deutsche Volkszeitung Die Tat*, 16 March 1990.

10. Matthes, *Stromwirtschaft*, 250–51; quotation on 273. On the recommendations for decommissioning of the Greifswald nuclear power plant, also see Abele and Hamper, "Kernenergiepolitik der DDR," 75–76.

11. Matthes, *Stromwirtschaft*, 377.

12. Schönherr, "Die ersten vier Blöcke," 303–5.]

13. Miina Kaarkoski, "'*Energiemix*' versus '*Energiewende*': Competing Conceptualisations of Nuclear Energy Policy in the German Parliamentary Debates of 1991–2001" (Ph.D. diss., University of Jyväskylä, Finland, 2016), 81–87, 109–17, 131–34, retrieved 31 October 2017 from https://jyx.jyu.fi/dspace/handle/123456789/50961.

14. Cited according to "Phönix aus der Asche," *Der Spiegel* 44 (28 October 1991).

15. Anselm Tiggemann, *Die "Achillesferse" der Kernenergie in der Bundesrepublik Deutschland: Zur Kernenergiekontroverse und Geschichte der nuklearen Entsorgung von den Anfängen bis Gorleben 1955 bis 1985* (Lauf an der Pegnitz: Europaforum-Verlag, 2004), 747, 769–70; Laufs, *Reaktorsicherheit*, 979.

16. Tiggemann, *Achillesferse*, 771.

17. Felix Kolb, "Der Castor-Konflikt: Das Comeback der Anti-AKW Bewegung," *Forschungsjournal Neue Sozial Bewegungen* 10, no. 3 (1997): 18–20; article on 16–29, retrieved 31 October 2017 from http://forschungsjournal.de/jahrgaenge. Also "Castor-Transport: 10.000 Atomkraftgegner protestieren," *Spiegel* Online Politik, 25 March 2001, retrieved 27 June 2016 from http://www.spiegel.de/politik/deutschland/castor-transport-10-000-atomkraftgegner-protestieren-a-124674.html; "Protest gegen Castor-Transport: Erste Festnahmen im Wendland," *Spiegel* Online Politik, 8 November 2008, retrieved 27 June 2016 from http://www.spiegel.de/politik/deutschland/protest-gegen-castor-transport-erste-festnahmen-im-wendland-a-589240.html.

18. "Demonstranten: Hau weg den Scheiß," *Der Spiegel* 45 (4 November 1996).

19. Video-Gruppe der Bürgerinitiative Umweltschutz Lüchow-Dannenberg e.V., "Aufstehen und Widersetzen—Castor, Gorleben," retrieved 31 October 2017 from https://www.youtube.com/watch?v=s4If4-90yIQ.

20. Carola Meyer und Melanie Thun, "Gorleben—Der Aufstand der Bauern," NDR, first shown on 17 May 2010, retrieved 31 October 2017 from https://www.youtube.com/watch?v=-Ozgz6ZyAtI.

21. "Umwelt: Schiet in de Kiste," *Der Spiegel* 28 (11 July 1994); "Atommüll: Endlager Sibirien?" *Der Spiegel* 20 (13 May 1996); "Atomenergie: Mehr als ein Bauernkrieg," *Der Spiegel* 9 (24 February 1997); "Castor-Transport: Lohnender Einsatz, " *Der Spiegel* 46 (10 November 2003); "Castor-Transport: Polizeigewerkschaft will Atomkonzerne zur Kasse bitten," *Spiegel* Online Politik, 5 November 2010, retrieved 27 June 2016 http://www.spiegel.de/politik/deutschland/castor-transport-polizeigewerkschaft-will-atomkonzerne-zur-kasse-bitten-a-727372.html. On police fears of radiation exposure, see "Übertriebene Fürsorge," *Der Spiegel* 9 (24 February 1997).

22. Kolb, "Castor-Konflikt," 23–25; Kaarkoski, *Energiemix,* 144–58; "Atommüll: Der strahlende Castor," *Der Spiegel* 22 (25 May 1998).

23. "Der 1. Castortransport nach Gorleben 25.4.1995," retrieved 31 October 2017 from https://www.youtube.com/watch?v=L7WVYo8VexY; Matthias Gebauer, "Undine von Blottnitz: Kämpferin bis zum Ende," *Spiegel* Online Politik, 5 March 2001, retrieved 27 June 2016 from http://www.spiegel.de/politik/deutschland/undine-von-blottnitz-kaempferin-bis-zum-ende-a-120932.html; "Castoren am Ziel: Anti-Atom-Aktivisten feiern historische Blockade," *Spiegel* Online Politik, 9 November 2010, retrieved 27 June 2016 from http://www.spiegel.de/politik/deutschland/castoren-am-ziel-anti-atom-aktivisten-feiern-historische-blockade-a-728053.html.

24. Linus Pauling, "Nuclear Fission Is Not the Answer," *Bulletin of the Atomic Scientists* 33, no. 3 (1977): 66; "Sonnen-Energie/Wissenschaft: Nur bei schönem Wetter," *Der Spiegel* 41 (10 October 1951).

25. John Perlin, *Let It Shine: The 6,000-Year Story of Solar Energy* (Novato, CA: New World Library, 2013), 353–54 and 430–36.

26. Carol Hager, "The Grassroots Origins of the German Energy Transition," in *Germany's Energy Transition: A Comparative Perspective*, ed. Carol Hager and Christoph Stefes (New York: Palgrave Macmillan, 2016), 14–15; article on 1–26.

27. Energy conservation is reduction in energy use, while energy efficiency is more efficient use of energy, e.g., through energy-efficient machines.

28. Florentin Krause, Hartmut Bossel, and Karl-Friedrich Müller-Reißmann, *Energiewende—Wachstum und Wohlstand ohne Erdöl und Uran* (Frankfurt am Main: S. Fischer, 1980).

29. Kaarkoski, "*Energiemix,*" 106–15.

30. "In die falsche Tasche gegriffen," *Der Spiegel* 22 (29 May 1995).

31. Maubach, *Energiewende,* 35–45, 59–60; quotations on 43.

32. Elke Bruns, Dörte Ohlhorst, Bernd Wenzel, and Johann Köppel, *Renewable Energies in Germany's Electricity Market: A Biography of the Innovation Process* (Dordrecht: Springer, 2011), 110–11, 193–97, 201, 204, 209.

33. Hermann Scheer, *Energy Autonomy: The Economic, Social & Technological Case for Renewable Energy* (New York: Routledge, 2006).

34. Jennifer Carlson, "Unruly Energies: Provocations of Renewable Energy Development in a Northern German Village" (Ph.D. diss., University of Texas at Austin, 2014). She conducted her field work in 2010–2011.

35. Otfried Wolfrum, *Windkraft: Eine Alternative, die keine ist* (Frankfurt am Main: Zweitausendeins, 1997); Otfried Wolfrum, "Der große Schwindel Windstrom:

Deutschland leistet sich hohe Subventionen für einen energiepolitischen Irrweg," *Frankfurter Allgemeine Zeitung* (24 October 1996). also Gerd Rosenkranz, "Windkraft: Sturmlauf gegen den Ökostrom," *Spiegel* Online Politik, 26 August 2003, retrieved 28 June 2016 from http://www.spiegel.de/politik/deutschland/windkraft-sturmlauf-gegen-den-oekostrom-a-262875.html.

36. Andrea Bues and Ludger Gailing, "Energy Transitions and Power: Between Governmentality and Depoliticization," in *Conceptualizing Germany's Energy Transition: Institutions, Materiality, Power, Space*, ed. Ludger Gailing and Timothy Moss (New York: Palgrave Macmillan, 2016), 69–91.

37. Bruns et al., *Renewable Energies*, 139–40.

38. Arbeitsgemeinschaft Energiebilanzen e.V., "Auswertungstabellen zur Energiebilanz für die Bundesrepublik Deutschland 1990 bis 2016," Table 4.1, retrieved 30 October 2017 from http://www.ag-energiebilanzen.de/10-0-auswertungstabellen.html.

39. Maubach, *Energiewende*, 60.

40. Jens Ivo Engels, "'Inkorporierung' und 'Normalisierung' einer Protestbewegung am Beispiel der westdeutschen Umweltproteste in den 1980er Jahren," *Moving the Social* 40 (2008): 94–97; article on 81–100; retrieved 31 October 2017 from http://moving-the-social.ub.rub.de/index.php/Moving_the_social/issue/view/53.

41. This merger of the Green Party with Alliance '90, an East German ecological party, had little impact on the Green Party's orientation.

42. Paul Hockenos, *Joschka Fischer and the Making of the Berlin Republic: An Alternative History of Postwar Germany* (New York: Oxford University Press, 2008).

43. Scholars now view the West German discourse of the 1980s concerning forest dieback primarily as a cultural and political phenomenon lacking in a solid scientific foundation. Birgit Metzger, *"Erst stirbt der Wald, dann Du!" Das Waldsterben als westdeutsches Politikum (1978–1986)* (Frankfurt am Main: Campus Verlag, 2015); Engels, "Inkorporierung," 85–94; Uekötter, *Deutschland im Grün*, ch. 8.

44. Under the German constitution, the *Bundesrat*, the upper house of parliament, consisting of representatives of the German states, can veto legislation on matters under the administrative, financial, or organizational authority of the states.

45. Sabine Weiland, *Politik der Ideen: Nachhaltige Entwicklung in Deutschland, Großbritannien und den USA* (Wiesbaden: VS Verlag für Sozialwissenschaften, 2007), 250–53; Kaarkoski, *Energiemix*, 120–34; Dominique Beaucaire, "Atomkonsens, Laufzeitverlängerung und Atomausstieg: Eine Policy-Analyse der deutschen Atompolitik von 2000 bis 2011" (Diplom thesis, Freie Universität Berlin, 2012).

46. Wolfgang Rüdig, "Germany," in *Green Parties in National Governments*, ed. Ferdinand Müller-Rommel and Thomas Poguntke (London and New York: Routledge, 2002), 97–98; ch. on 78–111.

47. Karl-Rudolf Korte, "Solutions for the Decision Dilemma: Political Styles of Germany's Chancellors," *German Politics* 9, no. 1 (2000), 12; article on 1–22.

48. Anthony Giddens, *The Third Way: The Renewal of Social Democracy* (Cambridge, UK: Polity, 1998). Giddens postulated that the conventional Left–Right political divide was breaking down, opening the way for a political "third way."

49. Karlheinz Niclauß, *Kanzlerdemokratie: Von Konrad Adenauer bis Angela Merkel* (Wiesbaden: Springer VS, 2015), 330; also see 265–333. For a scholarly biography of Schröder, see Gregor Schöllgen, *Gerhard Schröder: Die Biographie* (Munich: Deutsche Verlags-Anstalt, 2015).

50. "Protest: Greenpeace-Aktivisten belagern Tagungsort der Union," *Spiegel Online Politik*, 11 July 2005, retrieved 1 July 2016 from http://www.spiegel.de/politik/deutschland/protest-greenpeace-aktivisten-belagern-tagungsort-der-union-a-364682.html.

51. Falk Illing, *Energiepolitik in Deutschland: Die energiepolitischen Massnahmen der Bundesregierung 1949–2015*, 2nd ed. (Baden-Baden: Nomos, 2016), 255.

52. Kaarkoski, *Energiemix*, 109–17, 131–34.

53. Grasselt, *Entzauberung*, 122.

54. Beaucaire, "Atomkonsens," 45–47, 55, 57.

55. "Fukushima ist Überall," *Der Spiegel* 11 (14 March 2011). The article title is a play on the anti–nuclear power slogan, "Chernobyl Is Everywhere."

56. Michael Bauchmüller, "Der GAU im Wohnzimmer," *Süddeutsche Zeitung* (15 March 2011).

57. That was due partly to luck: the winds turned and blew most of the radioactive cloud out to sea. Thanks to better plant construction, there was no raging fire as in Chernobyl. In addition, authorities acted quickly to evacuate the population from the affected area. In hindsight, critics have claimed that this evacuation was unnecessary or even harmful but at the time it was thought to be a necessary precaution. Failure to evacuate caused many civilian deaths in Chernobyl in 1986. See Mizokami, "Event Sequence," 22–49; Martin Fackler, "Report Finds Japan Underestimated Tsunami Danger," *New York Times* (1 June 2011); IAEA, "The Fukushima Daiichi Accident: Report by the Director General," retrieved 31 October 2017 from http://www-pub.iaea.org/MTCD/Publications/PDF/Pub1710-ReportByTheDG-Web.pdf. See also comments of Japanese ex-premier: Wieland Wagner, "Ex-Premier Kan über Fukushima-Katastrophe: 'Die Frage war, ob Japan untergeht,'" *Spiegel Online Politik*, 9 October 2015, retrieved 22 July 2016.

58. Jeanne Rubner, "Wahnsinn mit Methode: Was Man aus den Katastrophen in Harrisburg und Tschernobyl für das jetztige Drama folgern kann," *Süddeutsche Zeitung* (15 March 2011).

59. *Tagesschau*, ARD, 16 and 17 March 2011, retrieved 31 October 2017 from https://www.youtube.com/watch?v=NJhGxhQV0fg.

60. Bauchmüller, "Der GAU." For an example, see "Havariertes AKW: Video zeigt massive Zerstörung in Fukushima I," *Spiegel Online Wissenschaft*, 17 March 2011, retrieved 13 July 2016 from http://www.spiegel.de/wissenschaft/mensch/havariertes-akw-video-zeigt-massive-zerstoerung-in-fukushima-i-a-751574.html.

61. *Tagesschau*, ARD, 19 March 2011, retrieved 31 October 2017 from https://www.youtube.com/watch?v=laClhezRQlA.

62. "Heute spezial," ZDF, 12 March 2011, retrieved 31 October 2017 from https://www.youtube.com/watch?v=oNlLaRGr3JA.

63. Wieland Wagner, "Problematic Public Relations: Japanese Leaders Leave People in the Dark," *Spiegel Online International*, 15 March 2011, retrieved 12 July 2016

from http://www.spiegel.de/international/world/problematic-public-relations-jap anese-leaders-leave-people-in-the-dark-a-751109.html. *Der Spiegel* published dozens of articles highly critical of Japanese safety.

64. "Japans Atomgigant Tepco: Filz, Vetternwirtschaft, Fukushima," *Spiegel* Online Panorama, 31 March 2011, retrieved 11 January 2018 from http://www.spiegel.de/ panorama/japans-atomgigant-tepco-filz-vetternwirtschaft-fukushima-a-754098. html.

65. Kaarkoski, *Energiemix,* 78–80.

66. "Regierungserklärung zur Energiewende: 'Eine Herkulesaufgabe,'" *Frankfurter Allgemeine Zeitung* (9 June 2011). Retrieved 13 July 2016 from http://www.faz. net/aktuell/politik/energiepolitik/regierungserklaerung-zur-energiewende-eine-herkulesaufgabe-1657503.html.

67. On the revival of risk as a central concept in debates about energy policy and nuclear power, see Ingeborg Lauvhjell, "Die Atomkraftdebatte in Deutschland: Eine Diskursanalyse über die Repräsentation der Atomkraft in der deutschen Öffentlichkeit" (M.A. thesis, University of Oslo, 2012), 60–61.

68. Gero von Randow and Hans Schuh, "AKW-Sicherheit: Im Blindflug," *Die Zeit* (17 March 2011).

69. Berthold Kohler, "Noch eine Häutung," *Frankfurter Allgemeine Zeitung* (31 May 2011). For years, supporters of nuclear power had been using a predictive model based on WASH-1400 (a reactor safety study conducted by experts working for the US Nuclear Regulatory Commission), according to which a serious nuclear power plant accident was likely to occur no more than every 10,000 years. Laufs, *Reaktorsicherheit,* 663–64.

70. Christoph Twickel, "Deutsches Anti-AKW-Gefühl: Im Land der Mahnbürger," *Spiegel* Online Kultur, 15 March 2011.

71. *Tagesschau,* ARD, 15 March 2011, retrieved 31 October 2017 from https://www. youtube.com/watch?v=2hzQyvK1qaE.

72. Beaucaire, "Atomkonsens," 73–74; see also 71–72.

73. Günter Bannas and Peter Carstens, "Merkels Agenda 2022," *Frankfurter Allgemeine Zeitung* (31 May 2011); Albert Schäffer, "Ein Mann als Volkspartei," *Frankfurter Allgemeine Zeitung* (31 May 2011).

74. Deutscher Bundestag, "117. Sitzung des Deutschen Bundestages am Donnerstag, 30.Juni 2011: Endgültiges Ergebnis der Namentlichen Abstimmung Nr. 1," retrieved 31 October 2017 from https://web.archive.org/web/20110812123738/ http://www.bundestag.de/bundestag/plenum/abstimmung/20110630_17_6070. pdf.

75. Kohler, "Noch eine Häutung."

76. Jan Goebel, Christian Krekel, Tim Tiefenbach, and Nicolas R. Ziebarth, "How Natural Disasters Can Affect Environmental Concerns, Risk Aversion, and Even Politics: Evidence from Fukushima and Three European Countries," *Journal of Population Economics* 28 (2015): 1150–53, 1155–61; article on 1137–80.

77. Gisela Lang, "Meiler-Moratorium: Greenpeace kritisiert Blackout-Drohung der Atomlobby," *Spiegel* Online Wirtschaft, 4 April 2011, retrieved 15 July 2016 from http://www.spiegel.de/wirtschaft/soziales/meiler-moratorium-greenpeace-kritisiert-blackout-drohung-der-atomlobby-a-754940.html.

78. Andreas Mihm, "Kosten der Wende," *Frankfurter Allgemeine Zeitung* (31 May 2011).

79. Jennifer Rose Linehan, "Fukushima to Fessenheim: France and Germany's Policy Responses to the Fukushima Nuclear Disaster, 11 March to 18 June 2011" (M.A. thesis, Carleton University, Ottawa, 2013), 117.

80. "Regierungserklärung zur Energiewende: 'Eine Herkulesaufgabe,'" *Frankfurter Allgemeine Zeitung* (9 June 2011).

81. Reinhard Wolf, "Deutschland, gutes Klimaland?" *Frankfurter Allgemeine Zeitung* (10 August 2015).

82. World Health Organization, Health and Environment Linkages Initiative, "Climate Change," retrieved 31 October 2017 from http://www.who.int/heli/risks/climate/climatechange/en/.

83. World Health Organization, "7 million premature deaths annually linked to air pollution," retrieved 31 October 2017 from http://www.who.int/mediacentre/news/releases/2014/air-pollution/en/.

84. Umwelt Bundesamt, "Minderungsziele für Treibhausgase," retrieved 31 October 2017 from http://www.umweltbundesamt.de/daten/klimawandel/treibhausgas-emissionen-in-deutschland.

85. Umwelt Bundesamt, "Treibhausgas-Emissionen in Deutschland," retrieved 30 October 2017 from http://www.umweltbundesamt.de/daten/klimawandel/treibhausgas-emissionen-in-deutschland. Emissions are not measured in the air but are calculated based on reporting of known sources of emissions. Umwelt Bundesamt, "Wie funktioniert die Berichterstattung," retrieved 31 October 2017 from http://www.umweltbundesamt.de/themen/klima-energie/klimaschutz-ene rgiepolitik-in-deutschland/treibhausgas-emissionen/wie-funktioniert-die-berichte rstattung. Thus, e.g., the Volkswagen diesel automobile emissions cover-up obscured whatever contribution diesel automobiles made to emissions.

86. Water power is a traditional renewable energy source but it has reached its growth potential. Biomass energy is limited by various problems. The factory farms that produce manure devastate the environment. And the use of corn and other food crops drove up food prices in nations of the Global South in recent years, leading to popular outcry.

87. Illing, *Energiepolitik*, 274, 289, 312–14; Wilfried Köplin, "Die Energiewende erfolgreich gestalten: Mit der Industrie und mit dem Markt," in *Energiewende zwischen Klimaschutz und Atomausstieg—Lösungen in die Umsetzung tragen: 18. Internationale Sommerakademie St. Marienthal,* ed. Fritz Brickwedde and Dirk Schötz, 58–64 (Berlin: Erich Schmidt Verlag, 2013), 63.

88. Arbeitsgemeinschaft Energiebilanzen e.V., "Auswertungstabellen zur Energiebilanz für die Bundesrepublik Deutschland 1990 bis 2016," Table 4.1, retrieved 30 October 2017 from http://www.ag-energiebilanzen.de/10-0-auswertungstabellen.html.

89. Cap and trade is an emissions trading system in which CO_2 emissions are capped but companies can buy and sell certificates, called CO_2 allowances (*CO2-Zertifikate*), which allow the holder to produce additional CO_2 emissions. The burning of coal produces large CO_2 emissions, necessitating the purchase of CO_2 allowances and adding to its cost as an energy source. On the low prices of CO_2 allowances, see

Jan Wittenbrink, "Emissionszertifikate: Klima-Handel," *Die Zeit* 51 (17 December 2015).

90. Arbeitsgemeinschaft Energiebilanzen e.V., "Auswertungstabellen zur Energiebilanz für die Bundesrepublik Deutschland 1990 bis 2016," Table 4.1, retrieved 30 October 2017 from http://www.ag-energiebilanzen.de/10-0-auswertungstabellen. html.

91. Frank Drieschner, "Schmutziger Irrtum," *Die Zeit* 50 (11 December 2014).

92. John Schwartz, "Methane Leaks in Natural-Gas Supply Chain Far Exceed Estimates, Study Says," *New York Times* (18 August 2015).

93. Maubach, *Energiewende,* 272–73.

94. "Energiewende wird teurer," *Süddeutsche Zeitung* (20 May 2016).

95. "Bundesnetzagentur: Rund 350.000 Haushalten wurde der Strom gesperrt," *Spiegel* Online Wirtschaft, 15 November 2015, retrieved 18 July 2016 from http://www.spiegel.de/wirtschaft/service/strom-350-000-haushalte-mit-stromsp erre-a-1062889.html.

96. "Bund will Förderung von Windrädern reduzieren," *Zeit* Online, 11 May 2016, retrieved 21 July 2016 from http://www.zeit.de/wirtschaft/2016-05/ erneuerbare-energien-windkraft-foerderung-kuerzung.

97. Also written E.ON, from the English word "eon." The following energy companies fused to form Eon in 2000: VEBA, VIAG, PreussenElektra, and Bayernwerk. The takeover of Ruhrgas followed in 2003.

98. Georg Meck, "Werner Wenning Eon-Aufsichtsratschef: Der Staat stiehlt sich beim Atomausstieg davon," *Frankfurter Allgemeine Zeitung* (26 December 2015).

99. Maubach, *Energiewende,* 61, 70.

100. Ibid., 69–70, 72–73, 85–93, 108.

101. Formerly *Rheinisch-Westfälisches Elektrizitätswerk AG,* or Rhine-Westphalia Power Company. RWE merged with VEW (Vereinigte Elektrizitätswerke Westfalen, or United Electricity Works of Westphalia) in 2000 and took over Energy Resources Holding and part of Essent (a Netherlands corporation).

102. "Eon drängt auf rasche Klärung beim Atomthema," *Manager Magazin* (8 June 2016); Rolf Schraa, "Eon und RWE spalten sich auf—wer macht es besser?" *Manager Magazin* (8 June 2016); "Dividende gestrichen: RWE-Krise wird für Kommunen zum Albtraum," *Spiegel* Online Wissenschaft, 17 February 2016, retrieved 18 July 2016 from http://www.spiegel.de/wirtschaft/unternehmen/rwe-streicht-dividende-und-schockiert-kommunen-a-1077834.html.

103. "Staat muss RWE, Eon und Co. entschädigen," *Manager Magazin,* 6 December 2016, retrieved 30 October 2017 from http://www.manager-magazin.de/ unternehmen/energie/atomausstieg-staat-muss-rwe-und-eon-entschaedigen-a-1124608.html.

104. "Bundesverfassungsgericht Atomkonzerne pochen auf Milliarden-Schadensersatz," *Frankfurter Allgemeine Zeitung* (15 March 2016). On RWE's decision to permanently abandon nuclear power, see "Kurswende: Neuer RWE-Chef will keine Atomkraftwerke mehr bauen," *Spiegel* Online Wirtschaft, 18 June 2012, retrieved 22 July 2016 from http://www.spiegel.de/wirtschaft/unternehmen/ neuer-chef-von-rwe-will-keine-atomkraftwerke-mehr-bauen-a-839433.html. There have been dozens of other court cases as well. E.g., see "Keine Entschädigung

fuer E.on," *Zeit* Online, 4 July 2016, retrieved 24 July 2016 from http://www.zeit. de/wirtschaft/2016-07/atomausstieg-e-on-atomkraft-schadenersatz-klage.

105. "Wie überweist man 24 Milliarden Euro?" *Manager Magazin*, 3 July 2017, retrieved 30 October 2017 from http://www.manager-magazin.de/unternehmen/energie/atomkonzerne-rwe-und-eon-kaufen-sich-von-endlagerrisiko-frei-a-1155433.html.

106. Katja Scherer, "Streit um neue Stromtrassen: Aber bitte nicht hier!" *Spiegel* Online Wirtschaft, 20 July 2014, retrieved 17 July 2016 from http://www.spiegel.de/wirtschaft/soziales/energiewende-gemeinden-wehren-sich-gegen-die-geplanten-stromtrassen-a-981615.html.

107. "Seehofer hält neue Stromtrasse für unnötig," *Süddeutsche Zeitung* (14 March 2014).

108. "Das ist ein Null-Ergebnis," *Süddeutsche Zeitung* (2 July 2015).

109. Maubach, *Energiewende*, 290–92.

110. See, e.g., Ernest Moniz, "Why We Still Need Nuclear Power: Making Clean Energy Safe and Affordable," *Foreign Affairs* 90, no. 6 (2011), 83–94.

111. For the hopes of a nuclear power proponent, see Laufs, *Reaktorsicherheit,* esp. 992–97. On the example of Antonio Hurtado, head of the Institute for Energy Technology at the Technical University of Dresden and for a skeptical evaluation of the latest nuclear power research, see "Der wahre Grünstrom?" *WirtschaftsWoche* 35 (26 August 2013): 66, GENIOS, retrieved 10 January 2018 from http://www.genios.de/presse-archiv/artikel/WW/20130826/der-wahre-gruenstrom/C4B4026C-4089-479C-A471-255CFB44EFB2.html.

112. Kristin Shrader-Frechette, "Climate Change, Nuclear Economics, and Conflicts of Interest," *Science & Engineering Ethics* 17, no. 1 (2011): 75–107. Her examples are international in scope.

113. Eduardo Porter, "Embrace of Renewables Has a Hidden Cost," *New York Times* (20 July 2016). This proponent of nuclear power advocates state subsidies for nuclear power. For a critique of calls for subsidies for nuclear power, see Rudolf Balmer, "Nukleare Start-ups," *Tageszeitung-taz* (28 April 2016).

114. Eicke Weber, "Wende nicht Verschlafen," *Süddeutsche Zeitung* (12 June 2016). Also Kate Galbraith, "Filling the Gaps in the Flow of Renewable Energy," *New York Times* (2 October 2013).

115. American Council for an Energy-Efficient Economy, "The International Energy Efficiency Scorecard: Germany," retrieved 31 October 2017 from http://aceee.org/sites/default/files/pdf/country/2016/germany.pdf.

116. International Energy Agency, *Energy Efficiency Market Report 2015* (Paris: OECD/IEA, 2015), 18, 25, 44–45, 60, 129. Retrieved 31 October 2017 from https://www.iea.org/publications/freepublications/publication/MediumTermEnergyefficiencyMarketReport2015.pdf.

117. Bund/Friends of the Earth Germany, "Energiesparen: Der Schlüssel zu Klimaschutz und *Energiewende*," retrieved 22 July 2016 from http://www.bund.net/themen_und_projekte/klima_und_energie/energie_sparen/; Dagmar Dehmer, "Deutschland schlampt beim Umweltschutz," *Die Zeit* (11 August 2015).

118. Detlef Jahn and Sebastian Korolczuk, "German Exceptionalism: The End of Nuclear Energy in Germany!" *Environmental Politics* 21, no. 1 (2012): 163; article on 159–64.

119. "Atomanlage in Gronau blockiert," *Tageszeitung-taz* (11 July 2016); "Suche nach einem Endlager für Atommüll: Stress mit der Bewegung," *Tageszeitung-taz* (5 July 2016); "Anti-AKW-Demo am 23. 04.2016 in Gundremmingen—sofort abschalten und gegen das Zwischenlager," retrieved 31 October 2017 from http://www.landkreiskanal.de/anti-akw-demo-am-23-04-2016-in-gundremmingen-sofort-abschalten-und-gegen-das-zwischenlager/. On a 2015 demonstration against Neckarwestheim, see "2015-03-08 Anti-AKW-Demo bei Neckarwestheim: Start am Bahnhof Kirchheim I," retrieved 31 October 2017 from https://www.youtube.com/watch?v=sc17xs4y35s.

Conclusion

The German turn away from nuclear power represents one of the most successful popular revolts against technocratic thinking of the twentieth and early twenty-first centuries. The impetus came from the mobilization of a broad swath of West Germans who rebelled against authority and set out to create a better society through new social movements. They took on political, economic, and scientific elites that were prepared to defend the development of atomic energy to the hilt. Activists in the GDR criticized party and state nuclear policies, albeit without calling Communist rule into question. Unlike in France, atomic power did not become part of a narrative of national identity or self-assertion in either German state to any significant degree. Rather, in both Germanys, nuclear power meant, above all, prosperity and release from the twentieth-century German "trauma of scarcity."[1]

Elites were guided by what they perceived as a duty to ensure *Sicherheit*. This word, which means both safety and security, encapsulates the historically rooted association of two values that were highly prized in postwar Germany (both East and West): social peace through economic security and high safety standards ensured by engineering and scientific excellence. The German approach to technology, in both the Federal Republic and the GDR, was both state centered and dismissive of lay points of view. This technocratic mindset and its link with a search for stability helps explain the stiff resistance of political, scientific, technical, and economic elites to the questioning of what they saw as an objective need for nuclear power. Activists rejected these patterns, which had persisted across the 1945 divide. The anti–nuclear power movement grew out of a breakdown of trust.

In the early years, East and West German media presented nuclear power in similar ways. Highly emotionalized utopian and dystopian visions abounded in depictions of the Atomic Age. However, journalists, officials, and representatives of industry also tried to win over public support for nuclear power by presenting viewers with dispassionate scientific explanations. There is no way to know how East German television viewers responded to a steady stream of educational programs. However, visitors to the 1955 US "Atoms for Peace" exhibition reacted in a very positive way that reflects the great esteem that science enjoyed in that period.

This optimistic narrative competed with negative coverage by the mid-1950s. Across the globe, aboveground tests of nuclear weapons, particularly the Lucky Dragon incident (when radioactive fallout from a US nuclear bomb test on Bikini killed Japanese fishermen), spurred anxieties about radiation. In West Germany, increased fears of nuclear war during the Berlin crisis evidently also colored perceptions of nonmilitary nuclear technologies. In this period, *Stern* ran articles about accidents at nuclear facilities, the fallout from bomb tests, and possible negative health consequences of radiation exposure through x-rays.

Memories of bombings during World War II and their association with Hiroshima intensified the sense of dread surrounding nuclear technologies. Some of this anxiety seeped across the border into the East German popular press. Germany was the place where transnational exchanges across the Iron Curtain were most intense. However, a great divergence between the two countries began to take shape in the early 1960s, as a more critical, skeptical approach to science and politics took hold in the Federal Republic, characterized by investigative journalism and a politically and ethically oriented critique of science.

The "antiauthoritarian revolt" of the sixties, to use Timothy Brown's phrase, had profound repercussions for the media, civil society, and popular perceptions of science and technology in the Federal Republic. No mere spin-off of the student movement of the 1960s, the anti–nuclear power activism of the 1970s and 1980s mobilized not only the young but also older age groups and not only the Left but also some moderate, religious, and even conservative segments of society. Local citizens' initiatives created the foundations for anti–nuclear power activism. A heterogeneous conglomerate of leftists, environmentalists, feminists, and other groups with national and transnational ties helped build it into a mass movement. Although decentralized and diverse, the movement was highly organized. What united activists of various stripes was their insistence that the leadership did not have the right to force them to accept a technology that they considered too dangerous. Ambitious and politically engaged journalists carried this message to a broader public.

West German political authorities felt under siege. Combating RAF terrorism, Filbinger and others saw public protests as subversive and treated them primarily as a security issue. Chancellor Helmut Schmidt had a more cautious approach to the use of force, but he viewed the anti–nuclear power movement as a challenge to what should be the priorities of policymakers—progress, prosperity, and full employment. Profoundly fearful of popular dissent since the 1953 uprising, the SED tried to quell all manifestations of rebellion in the GDR.

The anti–nuclear power movement emerged from conflict. West German state governments answered demonstrations in Wyhl, Grohnde, Brokdorf, Gorleben, and Kalkar with major shows of force. In Baden-Württemberg, Hans Filbinger's government authorized the police to make use of all means allowed under West German law to clear the nuclear power plant construction site in Wyhl on 19 February 1975. Police used high-powered water cannons and truncheons against peaceful demonstrators. Locals, among them farmers, women, and people in late middle age, cried out against what they saw as police brutality. When protesters returned to the site on 23 February, police showed little resolve to force them to leave.

What seems to have been crucial was that primarily police officers from Baden were called in who shared cultural ties with the protesters. Even the police commissioner displayed a lack of resolution. As a result, activists were able to occupy the construction site and remain there for another seven months. Many sympathizers visited the Wyhl site, which became a place where the Western European anti–atomic power movement coalesced. Disruption through the occupation of construction sites of nuclear facilities became a central strategy of the movement.

Schleswig-Holstein's governor, Gerhard Stoltenberg—determined not to allow a second Wyhl to occur—initiated an unprecedented militarization of the police response to popular protests. To prevent the storming of the Brokdorf nuclear power plant construction site, Stoltenberg ordered the building of a high fence and reinforced barriers and the deployment of national riot police and Federal Border Guard units. Paramilitary units were generally put in charge of the more militarized forms of riot control at Brokdorf, involving the use of tear gas and mace, water cannons, dogs, horses, and helicopters. It is difficult to discern whether this display of state power actually precipitated the pitched battles that took place between demonstrators and police, but it certainly heightened aggression on both sides.

More fundamentally, the Brokdorf conflict brought about a shift from use of local and regional police units with ties to the protesting population to the deployment of quasi-military units and militarization of tactics and strategies. In addition, the Verfassungsschutz, the Federal Republic's domestic security agency, monitored and reported on activists, who could be denied the right to work as state employees if they were found to be radicals. However, all these measures failed to quell the unrest.

The media publicized these conflicts. Opinion tended to split along left–right lines, particularly since states governed by Christian Democrats made the most concerted attempts to expand nuclear power. Policing measures were also the harshest there. However, public broadcasting stations disrupted the usual partisan divides, reaching a mass audience with programs that presented the point of view of the protesters in Wyhl and Brokdorf. Officials, accustomed to dominating radio and television coverage, expressed outrage, particularly when it came to issues to be defined and decided by "experts." In the end, the authorities lost the fight and politically moderate West Germans began—slowly—to listen to the opponents of atomic energy. This expansion of civil society was also qualitative. Activists insisted that their emotions were a valid expression of a sense of alarm and anger over government policies, as well as a justified way of formulating a call for change.

What undermined support for the movement among average, older West Germans was its penchant for violence. After their worst defeats at the hands of the security forces in Brokdorf, activists, including militants, began to think more strategically about violence. The influx of feminists, environmentalists, and less politicized individuals who were part of the "alternative scene" also helped bring about a reduction in hand-to-hand combat with police officers, rock throwing, and the like. The later renewed upsurge of violence in the wake of Chernobyl reflects, in part, the widespread sense of outrage over government handling of the disaster. Another major cause was the rise of the autonomists, whose extremist, revolutionary goals put them at odds with many anti–atomic power activists.

The ranks of those who simply wanted to demonstrate peacefully also grew after Chernobyl. The Hamburg government and police were widely criticized for holding peaceful demonstrators under humiliating circumstances in 1986. In physical conflicts with the police, protesters were generally the losers, at least in terms of pain and fear. However, they found a winning formula when their movement formed a coalition with the peace movement in the 1980s, and they embraced strategies such as human blockades. The police hardly knew how to deal with this radicalized form of passive resistance.

In the GDR, the public was more accepting of nuclear power. In combating dissent on this topic, the SED followed German and Soviet patterns in which state and technical experts colluded to frame technical answers to technological questions as apolitical and inevitable. East German popular culture glorified nuclear energy long after it had come to be seen as more problematic in the West. Although slow to emerge, environmentalism found a home in the Protestant Church in the 1980s. Scientists and counterexperts using scientific arguments played an even more central role in the origins of anti–atomic power activism in the GDR than in the Federal Republic. In the wake of Chernobyl, activists gathered signatures, formed study groups, and wrote and distributed samizdat flyers and booklets on nuclear power.

In response to the Stasi confiscation of printing presses and arrest of members of the staff of the *Ecology Newsletter* at the Zion Church in East Berlin in 1987, activists organized a vigil at the church, held a protest march—a great rarity in the GDR—and even contacted the Western press. As in the West, these activists embraced emotion in a way otherwise not seen or tolerated in public culture. They overcame passivity and fear and took on a more active role, rooted in group solidarity. Stasi informants in their midst made their lives difficult, but activists remained undeterred.

Science—the second major strand in this book—played a central role in the decades-long fight over nuclear power. Proponents of atomic energy had long appeared to have science on their side. After World War II, the Americanization and Sovietization of science and technology in the western and eastern zones of occupation helped lay the foundations for the establishment of atomic energy programs, both in technological and in cultural terms. The US Atoms for Peace campaign provided access to technologies and fissionable material, while at the same time convincing West Germans of the promise of this technology as well as the good intentions of the United States. West German industry adopted the American light water reactor and US safety philosophy, while the West German public was initially receptive to propaganda that hailed nuclear power as the energy source of the future.

Intent on preventing German resurgence, Soviet authorities made use of German nuclear technology and prevented East German nuclear research from developing too much independence. The Soviet Union probably played a leading role in the shut-down of the GDR's nuclear power program in the mid-1960s. The SED looked to the Soviet Union to supply hardware and ensure nuclear power plant safety. Engineers experienced problems on a daily basis that made clear how problematic such an approach to safety was. In monitoring and finding solutions to technical problems, engineers were de facto operating within the paradigm of "engineered safeguards," the safety philosophy that predominated in the West. Here, the German tradition of engineering, with its emphasis on professional autonomy and isolation from society, was of great benefit. It represented the best hope for preventing nuclear accidents in the GDR, since the system, rigged to preserve SED rule at all costs, did not allow the public any insight into the inner workings of the industry.

Safety problems were treated in ways specific to the socialist system. The Stasi investigated nuclear power plant accidents and handled them as highly sensitive matters that could not be disclosed even to plant employees, making it very difficult to learn from them. Human error, on the part of either technical personnel or management, was generally viewed as the root cause of the problem, impeding systematic improvement. The tendency in West Germany was to see safety issues as technical problems with technical solutions. West German industry was well equipped to produce safety equipment, and the Federal Republic had access to technologies from other Western countries.

Nonetheless, the West German nuclear industry tried to conceal weaknesses in its safety systems. The anti–atomic power movement forced industrial watchdogs to publish reports that revealed these failures. In the GDR, activists were not able to gain access to such records. Rather, it was the SAAS (Staatliches Amt für Atomsicherheit und Strahlenschutz, State Office for Nuclear Safety and Radiation Protection of the GDR) that conducted special investigations on nuclear power plant safety; the SAAS pointed to critical failings resulting mainly from general problems of the planned economy, which were virtually insoluble.

Historian Joachim Radkau is partially correct when he asserts, "The origins of the anti-atomic power movement lie in the development of nuclear energy."[2] Alarm over the impact of low-dose radiation, such as that feared to emanate from atomic power plants, and criticism of nuclear safety emerged from the scientific community. US scientists were more likely to speak out on these issues, taking on the role of public intellectuals. In West Germany, those who disseminated this critique were generally experts from the margins of the scientific community or laypersons with some scientific background. Holger Strohm was among the first counterexperts who made scientific arguments against nuclear power accessible to a broad popular audience in West Germany (in 1973). Activist organizations such as the BBU (Bundesverband Bürgerinitiativen Umweltschutz, or Federal Association of Citizens' Initiatives on Environmental Protection) took up the counterexperts' arguments, which various factions of the anti–nuclear power movement absorbed and further disseminated.

The movement's embrace of science was a crucial turnabout. Before Wyhl, anti–atomic power activism had focused on local issues that could have applied to a non–nuclear power plant, such as the impact on the microclimate or the temperature of river water. Science placed their cause in a far larger context and gave it a legitimacy that it had been lacking up until then. Anti-Wyhl citizens' initiatives discovered this when, in 1977, the Freiburg administrative court stopped construction because of concerns over the risk of a reactor explosion. It was expert testimony that led the court to make this ruling. In the same year, activists founded the Eco Institute in Freiburg, where scientifically grounded research on nuclear power was conducted.

In 1979, activists forced open the records on nuclear power plant accidents, revealing a far worse track record than most had expected. Participants in the campaigns in Wyhl, Brokdorf, and Gorleben studied popular science accounts of nuclear power and the effect of radiation on human health. In the GDR, anti–nuclear power activism grew out of scientific arguments about nuclear power.

Ulrich Beck suggested that the way to challenge atomic power was through nonscientific, ethical, and political arguments. Anti-nuclear power activists did so, but they also grounded their critical stance in science. This was a fortuitous choice. Three Mile Island, Chernobyl, and Fukushima made a mockery of experts' claims that atomic power plant accidents would be very rare. The one-two punch of scientific skepticism and distrust of elites was to help bring down the nuclear consensus in Germany.

Essential to the phasing out of nuclear power was the preparing of a concrete, economically viable technological alternative. The Freiburg Eco Institute came up with the dream of a postnuclear, post–fossil fuel future—the "energy transition." With growing awareness of global warming, there has been a revival of enthusiasm for nuclear power among some experts and pundits. In reunited Germany, however, the success of renewable energy sources has opened up the possibility of an energic third way. Wind energy was pioneered by "tinkerers and idealists," solar energy by homeowners and communities, and biomass energy by farmers. For better or for worse, corporations have stepped in to scale up these efforts. There has been some popular resistance, for example to wind turbines, but it has not been overwhelming. As with any process of negotiation, the outcome is not guaranteed.

Political professionalization also aided the antinuclear cause. Under the leadership of Joschka Fischer, the Green Party was part of a coalition with the Social Democratic Party from 1998 to 2005. One can interpret Chancellor Gerhard Schröder's agreement with industry to phase out nuclear power over a 32-year period as either high-handed or realistic. His Christian Democratic successor, Angela Merkel, reversed this decision in 2010. Ironically, she was the chancellor to decide, in the wake of Fukushima, to abandon nuclear power on a tighter schedule than Schröder had envisioned. A pastor's daughter and a physicist by profession, she doubtlessly was receptive to both ethical and scientific arguments. But her reading of the political will of the German people was also a crucial factor in her decision.

The anti–nuclear power movement experienced ebbs and flows over the course of more than forty years. In the period since Germany reunification, it adopted a strategy that greatly strengthened its hand. The Gorleben blockades of roads and railroad lines pushed protest policing to its logical limits and beyond. Forcing through the shipments of nuclear waste with police convoys that deployed hundreds of thousands of officers pushed state finances to the breaking point.

Some historians have argued that the West German anti–nuclear power movement emerged out of an untenable level of risk, while others view it as the product of complex social and cultural forces as well as a deficit of democracy in West Germany. Each of these interpretations has validity but is incomplete. Activists appealed to science in arguing that the state did not have the right to expose citizens to an unacceptable level of risk, that experts should not be the ultimate arbiters of such issues, and that the political system had to be more responsive to the popular will. In attenuated form, this narrative was also found in the GDR. Angela Merkel's 2011 decision was made possible by the convergence of two paths—that of the anti–nuclear power movement and that of science and technology as evolving, contested, deprofessionalizing and reprofessionalizing fields.

Over the past fifty years, the significance of scientific and technical issues has increased exponentially. Germany faced the same question that all modern democracies face: Should some decisions be walled off from common citizens because they do not have the specialized background supposedly needed to understand the issues, or should the public have the right to sit at the table at which major decisions about science and technology are made? The German anti–nuclear power movement forced a path of inclusiveness on their leaders, at least for one moment in history.

Notes

1. Schüring, "Advertising," 384.
2. Radkau, *Aufstieg und Krise,* 14.

Bibliography

I. Archives

1. Die Behörde des Bundesbeauftragten für die Unterlagen des Staatsicherheitsdienstes der ehemaligen DDR [Stasi Archive], Berlin
2. Bundesarchiv [German Federal Archives], Berlin
3. Stiftung Archiv der Parteien und Massenorganisationen der DDR im Bundesarchiv [Foundation of the Archive of the Parties and Mass Organization of the GDR in the Federal German Federal Archives], Berlin
4. Archiv der DDR-Opposition, Robert-Havemann-Gesellschaft e. V. [Archive of the GDR Opposition, Robert Havemann Society], Berlin
5. Deutsche Kinemathek [German Film Library], Berlin
6. Archiv an der Freien Universität Berlin [Extraparliamentary Opposition Archive, Free University Berlin]
7. Archiv Grünes Gedächtnis [Green Party Archive], Berlin
8. National Archives II in College Park, Maryland
9. NDR Archive [Northern German Broadcasting Archive], Hamburg
10. Deutsches Rundfunkarchiv [German Broadcasting Archives], Babelsberg
11. Landesarchiv Baden-Württemberg Hauptstaatsarchiv Stuttgart [Provincial Archives of Baden-Württemberg, Main State Archive, Stuttgart]
12. Landesarchiv Berlin [Provincial Archives of Berlin]
13. Landesarchiv Schleswig-Holstein [Provincial Archives of Schleswig-Holstein], Schleswig
14. Staatsarchiv Hamburg [State Archive of Hamburg]
15. Papier-Tiger-Archiv, Berlin
16. *Stern* magazine, Manhattan bureau
17. WDR Historisches Archiv [West German Broadcasting Historical Archive], Cologne

18. ZDF Unternehmensarchiv [Archive of the Second German Television network], Mainz

II. Newspapers and Popular Periodicals

Arbeiterkampf
Bayern Kurier
Bild Zeitung
Der Spiegel
Der Tag
Deutsche Volkszeitung
Die Tat
Die Zeit
Flensburger Tageblatt
Frankfurter Allgemeine Zeitung
Frankfurter Rundschau
Graswurzelrevolution
Hamburger Abendblatt
Manager Magazin
Neue Berliner Illustrierte
Neues Deutschland
New York Times
Rheinischer Merkur
Spiegel Online
Stern
Süddeutsche Zeitung
Tagesspiegel
Tageszeitung-taz
Zeitung für Kommunalwirtschaft

III. Online Reports

Der Bundesminister des Innern, "Übersicht über besondere Vorkommnisse in Kernkraftwerken der Bundesrepublik Deutschland," 1965–1984, http://www.bfs.de/DE/themen/kt/ereignisse/berichte/jahresberichte/jahresberichte.html.

Der Bundesminister für Umwelt, Naturschutz und Reaktorsicherheit, "Übersicht über besondere Vorkommnisse in Kernkraftwerken der Bundesrepublik Deutschland," 1985–1990, http://www.bfs.de/DE/themen/kt/ereignisse/berichte/jahresberichte/jahresberichte.html.

IAEA, "INES: The International Nuclear and Radiological Event Scale," Information Series 08-26941-E, https://www.iaea.org/sites/default/files/ines.pdf.

IV. Other Works Consulted

Abele, Johannes. *Kernkraft in der DDR: Zwischen nationaler Industriepolitik und sozialistischer Zusammenarbeit 1963–1990.* Dresden: Hannah-Arendt-Institut für Totalitarismusforschung, 2000.

Abele, Johannes and Eckhard Hamper. "Kernenergiepolitik der DDR." In *Zur Geschichte der Kernenergie in der DDR,* edited by Peter Liewers, Johannes Abele, and Gerhard Barkleit, 29–89. Frankfurt am Main: Peter Lang, 2000.

Abrahamson, Dean. "Citizens vs. Atomic Power." *Bulletin of the Atomic Scientists* 29, no. 5 (1973): 43–45.

Adede, A.O. *The IAEA Notification and Assistance Conventions in Case of a Nuclear Accident: Landmarks in the Multi-Lateral Treaty-Making Process.* Boston, MA: M. Nijhoff; London: Graham and Trotman, 1987.

Adorno, Theodor and Max Horkheimer. "The Culture Industry: Enlightenment as Mass Deception." In *Dialectic of Enlightenment: Philosophical Fragment.* Translated by Edmund Jephcott. Stanford, CA: Stanford University Press, 2002. First self-published in 1944 as "Dialektik der Aufklärung" in the form of a hectograph manuscript.

Alers, Kristen and Philip Banse, eds., *Unruhiges Hinterland: Portraits aus dem Widerstand im Wendland (in Text und Bild).* Lüchow: AJB-Verlag, 1997.

Andrews, Kenneth. *Freedom Is a Constant Struggle: The Mississippi Civil Rights Movement and Its Legacy.* Chicago, IL: University of Chicago Press, 2004.

Arbeitsgruppe "Wiederaufarbeitung" (WAA) and der Universität Bremen. *Atommüll oder Der Abschied von einem teuren Traum.* Reinbek bei Hamburg: Rowohlt Verlag, 1977.

Arndt, Melanie. *Tschernobyl: Auswirkungen des Reaktorunfalls auf die Bundesrepublik Deutschland und die DDR.* Erfurt: Landeszentrale für Politische Bildung Thüringen, 2011.

Augustine, Dolores. "The Impact of Two Reunification-Era Debates on the East German Sense of Identity." *German Studies Review* 27 (2004): 563–78.

———. "Learning from War: Media Coverage of the Nuclear Age in the Two Germanies." In *The Nuclear Age in Popular Media: A Transnational History, 1945–1965,* edited by Dick van Lente, 79–116. New York and Houndmills, UK: Palgrave Macmillan, 2012.

———. *Red Prometheus: Engineering and Dictatorship in East Germany, 1945–1990.* Cambridge, MA: MIT Press, 2007.

Augustine, Dolores and Dick van Lente. "Conclusion: One World, Two Worlds, Many Worlds?" In *The Nuclear Age in Popular Media: A Transnational History, 1945–1965,* edited by Dick van Lente, 233–48. New York and Houndmills, UK: Palgrave Macmillan, 2012.

Bahro, Rudolf. *Elemente einer neuen Politik: Zum Verhältnis von Ökologie und Sozialismus.* West Berlin: Olle und Wolter, 1980.

Bailes, Kendall. *Technology and Society under Lenin and Stalin: Origins of the Soviet Technical Intelligentsia, 1917–1941.* Princeton, NJ: Princeton University Press, 1978.

Barck, Simone, Martina Langermann, and Siegfried Lokatis, eds. *"Jedes Buch ein Abenteuer": Zensur-System und literarische Öffentlichkeiten in der DDR bis Ende der Sechziger Jahre.* Berlin: Akademie-Verlag, 1997.

Bayer, Anton. Deutsche Risikostudie Kernkraftwerke. Vol. 8: Unfallfolgenrechnung und Risikoergebnisse. Cologne: Verlag TÜV Rhineland, 1981.

Beaucaire, Dominique. "Atomkonsens, Laufzeitverlängerung und Atomausstieg: Eine Policy-Analyse der deutschen Atompolitik von 2000 bis 2011." Diplom thesis, Free University of Berlin, 2012.

Beck, Ulrich. *Gegengifte: Die organisierte Unverantwortlichkeit.* Frankfurt am Main: Suhrkamp Verlag, 1988.

———. *Risikogesellschaft: Auf dem Weg in eine andere Moderne.* Frankfurt am Main: Suhrkamp Verlag, 1986. First English edition: Ulrich Beck, *Risk Society: Towards a New Modernity.* Translated by Mark Ritter. London and Newbury Park, CA: Sage, 1992.

Behrens, William, Donella Meadows, Dennis Meadows, and Jørgen Randers. *The Limits to Growth: A Report for the Club of Rome's Project on the Predicament of Mankind.* New York: Universe Books, 1972.

Beleites, Michael. *Untergrund: Ein Konflikt mit der Stasi in der Uran-Provinz.* Berlin: BasisDruck, 1991.

Biess, Frank. "Feelings in the Aftermath: Toward a History of Postwar Emotions." In *Histories of the Aftermath: The Legacies of the Second World War in Europe,* edited by Frank Biess and Robert Moeller, 30–48. New York: Berghahn Books, 2010.

———. *German Angst? Fear and Democracy in Postwar West Germany.* New York: Oxford University Press, forthcoming.

Blackbourn, David. *The Conquest of Nature: Water, Landscape, and the Making of Modern Germany.* New York: Norton, 2006.

Bloch, Ernst. *Das Prinzip Hoffnung,* vol. 2. Frankfurt am Main: Suhrkamp, 1973; 1st edition Berlin [GDR]: Aufbau-Verlag, 1955.

Böhm, Karl and Rolf Dörge. *Gigant Atom.* Berlin: Verlag Neues Leben, 1957.

Bonk, Sebastian, Florian Key, and Peer Pasternack. "Die Offene Arbeit in den Evangelischen Kirchen der DDR: Fallbeispiel Halle-Neustadt." In

Wissensregion Sachsen-Anhalt: Hochschule, Bildung und Wissenschaft: Die Expertisen aus Wittenberg, edited by Peer Pasternack, 203–207. Leipzig: Akademische Verlagsanstalt, 2014.

Bösch, Frank. "Geteilt und verbunden: Perspektiven auf die deutsche Geschichte seit den 1970er Jahren." In *Geteilte Geschichte: Ost- und Westdeutschland 1970–2000*, edited by Frank Bösch, 7–38. Göttingen: Vandenhoeck & Ruprecht, 2015.

———. *Mass Media and Historical Change: Germany in International Perspective, 1400 to the Present*. Translated by Freya Buechter. New York: Berghahn Books, 2015.

———. "Politische Macht und gesellschaftliche Gestaltung: Wege zur Einführung des privaten Rundfunks in den 1970er/80er Jahren." In *Wandel des Politischen: Die Bundesrepublik Deutschland während der 1980er Jahre*, edited by Meik Woyke, 195–214. Bonn: Verlag J.H.W. Dietz Nachf., 2013.

———. "Taming Nuclear Power: The Accident Near Harrisburg and the Change in West German and International Nuclear Policy in the 1970s and Early 1980s." *German History* 35, no. 1 (2017): 71–95.

Bösch, Frank and Norbert Frei. "Die Ambivalenz der Medialisierung: Eine Einführung." In *Medialisierung und Demokratie im 20. Jahrhundert*, edited by Frank Bösch und Norbert Frei, 7–23. Göttingen: Wallstein Verlag, 2006.

Bossel, Hartmut, Karl-Friedrich Müller-Reißmann, and Florentin Krause. *Energiewende Wachstum und Wohlstand ohne Erdöl und Uran*. Frankfurt am Main: S. Fischer, 1980.

Bower, Tom. *The Paperclip Conspiracy: The Hunt for the Nazi Scientists*. Boston, MA: Little, Brown, 1987.

Boyer, Paul. *By the Bomb's Early Light: American Thought and Culture at the Dawn of the Atomic Age*. New York: Pantheon Books, 1985.

Brown, Kate. *Plutopia: Nuclear Families, Atomic Cities, and the Great Soviet and American Plutonium Disasters*. Oxford: Oxford University Press, 2012.

Brown, Timothy. *West Germany and the Global Sixties: The Anti-Authoritarian Revolt, 1962–1978*. Cambridge, UK and New York: Cambridge University Press, 2013.

Bruns, Elke, Johann Köppel, Dörte Ohlhorst, and Bernd Wenzel. *Renewable Energies in Germany's Electricity Market: A Biography of the Innovation Process*. Dordrecht: Springer, 2011.

Bues, Andrea and Ludger Gailing. "Energy Transitions and Power: Between Governmentality and Depoliticization." In *Conceptualizing Germany's Energy Transition: Institutions, Materiality, Power, Space*, edited by Ludger Gailing and Timothy Moss, 69–91. New York: Palgrave Macmillan, 2016.

Buiren, Shirley Van. *Die Kernenergie-Kontroverse im Spiegel der Tageszeitungen: Inhaltsanalytische Auswertung eines exemplarischen Teils der*

Informationsmedien: Empirisch Untersuchung im Auftrag des Bundesministers des Inneren. Munich and Vienna: R. Oldenbourg, 1980.

Bundesminister für Forschung und Technologie. *Kernenergie: Eine Bürgerinformation*. Bonn: Der Bundesminister für Forschung und Technologie, 1975.

Bundesverband Bürgerinitiativen Umweltschutz. *Unfälle in deutschen Kernkraftwerken: Veröffentlichungen der vertraulichen Störfallberichte der Bundesregierung*. 2nd ed. Ludwigshafen: Bundesverband Bürgerinitiativen Umweltschutz, 1979.

Busch, Justin. *The Utopian Vision of H.G. Wells*. Jefferson, NC and London: McFarland, 2009.

Carlson, Jennifer. "Unruly Energies: Provocations of Renewable Energy Development in a Northern German Village." Ph.D. diss., University of Texas at Austin, 2014.

Carson, Cathryn. *Heisenberg in the Atomic Age: Science and the Public Sphere*. Cambridge, UK and New York: Cambridge University Press, 2010.

Cassidy, David. *Uncertainty: The Life and Science of Werner Heisenberg*. New York: Freeman, 1992.

Ciesla, Burghard. "Droht der Menschheit Vernichtung? Der Schweigende Stern—First Spaceship on Venus: Ein Vergleich." *Apropos: Film 2002: Das Jahrbuch der DEFA-Stiftung*: 121–36.

Clark, Lee. *Worst Cases: Terror and Catastrophe in the Popular Imagination*. Chicago, IL: University of Chicago Press, 2006.

Collatz, S., D. Falkenberg, and P. Liewers. "Forschungs- und Entwicklungsarbeiten des Zentralinstituts für Kernforschung Rossendorf zur Kernenergienutzung." In *Zur Geschichte der Kernenergie in der DDR*, edited by Peter Liewers, Johannes Abele, and Gerhard Barkleit, 411–74. Frankfurt am Main: Peter Lang, 2000.

Coopersmith, Jonathan. *The Electrification of Russia, 1880–1926*. Ithaca, NY: Cornell University Press, 1992.

Cooter, Roger and Stephen Pumphrey. "Separate Spheres and Public Places: Reflections on the History of Science Popularization and Science in Popular Culture." *History of Science* 32, no. 3 (1994): 237–67.

Cozzens, Susan and Edward Woodhouse. "Science, Government, and the Politics of Knowledge." In *Handbook of Science and Technology Studies*, edited by Sheila Jasanoff, Gerald Markle, James Peterson, and Trevor Pinch, 533–53. Thousand Oaks, CA: Sage, 1995.

Creager, Angela. "Radiation, Cancer, and Mutation in the Atomic Age." *Historical Studies in the Natural Sciences* 45, no. 1 (2015): 14–48.

Danker, Uwe and Astrid Schwabe. *Schleswig-Holstein und der Nationalsozialismus*. Neumünster: Wachholtz, 2005.

Del Sesto, Steven. "Uses of Knowledge and Values in Technical Controversies: The Case of Nuclear Reactor Safety in the US." *Social Studies of Science* 13, no. 3 (1983): 395–416.

Downey, Gary. "Reproducing Cultural Identity in Negotiating Nuclear Power: The Union of Concerned Scientists and Emergency Core Cooling." *Social Studies of Science* 18, no. 2. (1988): 231–64.

Drogan, Mara. "Atoms for Peace, U.S. Foreign Policy and the Globalization of Nuclear Technology, 1953–1960." Ph.D. diss., SUNY Albany, 2011.

Ehmke, Wolfgang. *Zwischenschritte: Die Anti-Atomkraft-Bewegung zwischen Gorleben und Wackersdorf.* Cologne: Kölner Volksblatt Verlag, 1987.

Eith, Ulrich. "Nai hämmer gsait! Stilbildender ziviler Widerstand in Wyhl am Kaiserstuhl." In *Aufbruch, Protest und Provokation: Die bewegten 70er- und 80er-Jahre in Baden-Württemberg,* edited by Reinhold Weber, 35–54. Darmstadt: Theiss Verlag, 2013.

Engels, Jens Ivo. "Gender Roles and German Anti-Nuclear Protest: The Women of Wyhl." In *Le demon moderne: la pollution dans les sociétés urbaines et industrielles d'Europe,* edited by Christoph Bernhardt, 407–424. Clermont-Ferrand: Presses univ. Blaise Pascal, 2002.

_____. "'Inkorporierung' und 'Normalisierung' einer Protestbewegung am Beispiel der westdeutschen Umweltproteste in den 1980er Jahren," *Moving the Social: Journal of Social History and the History of Social Movements* 40 (2008): 81–100. http://moving-the-social.ub.rub.de/index.php/ Moving_the_social/issue/view/53.

Enquete-Kommission des Deutschen Bundestages. *Zukünftige Kernenergie-Politik: Kriterien—Möglichkeiten—Empfehlungen: Bericht der Enquete-Kommission des Deutschen Bundestages.* Bonn: Dt. Bundestag, Presse- und Informationszentrum, 1980.

Fahlenbrach, Kathrin and Laura Stapane. "Mediale und visuelle Strategien der Friedensbewegung." In *"Entrüstet Euch!" Nuklearkrise, NATO-Doppelbeschluss und Friedensbewegung,* edited by Christoph Becker-Schaum, Philipp Gassert, Martin Klimke, Wilfried Mausbach, and Marianne Zepp. Paderborn: Ferdinand Schöningh, 2012.

Fiske, John. "The Popular Economy." In *Cultural Theory and Popular Culture: A Reader,* edited by John Storey. 3rd ed. Harlow, UK: Pearson Education, 2006.

Forman, Paul. "Behind Quantum Electronics: National Security as Basis for Physical Research in the United States, 1940–1960." *Historical Studies in the Physical and Biological Sciences* 18, no. 1 (1987): 149–229.

Frank, Charles. *Operation Epsilon: The Farm Hall Transcripts.* Berkeley, CA: University of California Press, 1993.

Frankland, E. Gene and Donald Schoonmaker. *Between Protest and Power: The Green Party in Germany.* Boulder, CO: Westview, 1992.

Frohman, Larry. "Datenschutz, the Defense of Law, and the Debate over Precautionary Surveillance: The Reform of Police Law and the Changing Parameters of State Action in West Germany," *German Studies Review* 38, no. 2 (2015): 307–32.

Fuller, John. *We Almost Lost Detroit.* New York: Reader's Digest Press, 1975.

Fürmetz, Gerhard, Herbert Reinke, and Klaus Weinhauer, eds., *Nachkriegspolizei: Sicherheit und Ordnung in Ost- und Westdeutschland 1945–1969.* Hamburg: Ergebnisse, 2001.

Gahalaut, Vineet and Harsh Gupta. *Three Great Tsunamis: Lisbon (1755), Sumatra-Andaman (2004) and Japan (2011).* Dordrecht and New York: Springer, 2013.

Geffken, Rolf. *Arbeit und Arbeitskampf im Hafen: Zur Geschichte der Hafenarbeit und der Hafenarbeitergewerkschaft.* Bremen: Edition Falkenberg, 2015.

Giddens, Anthony. *The Third Way: The Renewal of Social Democracy.* Cambridge, UK: Polity, 1998.

Gienow-Hecht, Jessica. *Transmission Impossible: American Journalism as Cultural Diplomacy in Post-War Germany.* Baton Rouge, LA: Louisiana State University Press, 1999.

Gieseke, Jens. *The History of the Stasi: East Germany's Secret Police, 1945–1990.* Translated by David Burnett. New York: Berghahn Books, 2014.

———. "The Stasi and East German Society: Some Remarks on Current Research." *GHI Bulletin* supplement 9 (2014): 59–72.

Goebel, Jan, Christian Krekel, Tim Tiefenbach, and Nicolas R. Ziebarth. "How Natural Disasters Can Affect Environmental Concerns, Risk Aversion, and Even Politics: Evidence from Fukushima and Three European Countries." *Journal of Population Economics* 28 (2015): 1137–80.

Goeckel, Robert. *The Lutheran Church and the East German State: Political Conflict and Change under Ulbricht and Honecker.* Ithaca: Cornell University Press, 1990.

Gofman, John and Arthur Tamplin. *Poisoned Power: The Case against Nuclear Power Plants.* Emmaus, PA: Rodale, 1971.

Goudsmit, Samuel. *Alsos: The Failure of German Science.* New York: Schuman, 1947.

Gouldner, Alvin. *The Two Marxisms.* New York: Seabury Press, 1980.

Graeub, Ralph. *Die sanften Mörder.* Rüschlikon-Zurich, Switzerland: Albert Müller Verlag, 1972.

Graham, Loren. *The Ghost of the Executed Engineer: Technology and the Fall of the Soviet Union.* Cambridge, MA and London: Harvard University Press, 1993.

———. *Science in Russia and the Soviet Union.* Cambridge, UK and New York: Cambridge University Press, 1993.

Grant, Edward. *The Foundations of Modern Science in the Middle Ages: Their Religious, Institutional and Intellectual Contexts.* Cambridge, UK and New York: Cambridge University Press, 1996.

Grasselt, Nico. *Die Entzauberung der Energiewende: Politik- und Diskurswandel unter schwarz-gelben Argumentationsmustern.* Wiesbaden: Springer VS, 2015.

Graßmann, Mathias. "Die Polizei in der Bundesrepublik Deutschland." In *Beiträge zu einer vergleichenden Soziologie der Polizei,* edited by Jonas Grutzpalk, Anja Bruhn, Julia Fatianova, Franziska Harnisch, Christiane Mochan, Björn Schülzke, and Tanja Zischke, 89–107. Potsdam: Universitätsverlag Potsdam, 2009.

Greene, Gayle. *The Woman Who Knew Too Much: Alice Stewart and the Secrets of Radiation.* Ann Arbor, MI: University of Michigan Press, 1999.

Grüning, Juliane and Isabelle Faragallah. "Mein ganzes Herz ist mit Tschernobyl verbunden: Zeitzeuginnengespräch mit Eva Quistorp." In *Frauen aktiv gegen Atomenergie: wenn aus Wut Visionen werden. 20 Jahre Tschernobyl,* edited by Ulrike Röhr, 16–22. Wiesbaden: genanet, 2006.

Grupe, Hans and Winfried Koelzer. *Fragen und Antworten zur Kernenergie.* Bonn: Informationszentrale der Elektrizitätswirtschaft, 1975.

Gumbert, Heather. *Envisioning Socialism: Television and the Cold War in the German Democratic Republic.* Ann Arbor, MI: University of Michigan Press, 2014.

Habermas, Jürgen. *Strukturwandel der Öffentlichkeit: Untersuchungen zu einer Kategorie der bürgerlichen Gesellschaft.* Neuwied: Luchterhand, 1962. English translation: Jürgen Habermas, *The Structural Transformation of the Public Sphere: An Inquiry into a Category of Bourgeois Society.* Translated by Thomas Burger with Frederick Lawrence. Cambridge, MA: MIT Press, 1989.

Hager, Carol. *Technological Democracy: Bureaucracy and Citizenry in the German Energy Debate.* Ann Arbor, MI: University of Michigan Press, 1995.

_____. "The Grassroots Origins of the German Energy Transition." In *Germany's Energy Transition: A Comparative Perspective,* edited by Carol Hager and Christoph Stefes, 1–26. New York: Palgrave Macmillan, 2016.

Hansen, Jan. "Zwischen Staat und Straße: Der Nachrüstungsstreit in der deutschen Sozialdemokratie (1979–1983)." In *Wandel des Politischen. Die Bundesrepublik Deutschland während der 1980er Jahre.* Bonn: Verlag J.H.W. Dietz Nachf, 2013.

Hanshew, Karrin. *Terror and Democracy in West Germany.* Cambridge, UK and New York: Cambridge University Press, 2012.

Häberlen, Joachim and Jake Smith. "Struggling for Feelings: The Politics of Emotions in the Radical New Left in West Germany, c.1968–84." *Contemporary European History* 23, no. 4 (2014): 615–37.

Hargraves, Robert and Ralph Moir. "Liquid Fluoride Thorium Reactors." *American Scientist* 98 (2010): 304–13.

Harms, Erik. *Der kommunikative Stil der Grünen im historischen Wandel. Eine Überblicksdarstellung am Beispiel dreier Bundestagswahlprogramme.* Frankfurt am Main: Peter Lang, 2008.

Haseloff, Otto. *Stern: Strategie und Krise einer Publikumszeitschrift.* Mainz: v. Hase & Koehler, 1977.

Häusler, Jürgen. *Der Traum wird zum Alptraum: Das Dilemma einer Volkspartei— die SPD im Atomkonflikt.* West Berlin: Sigma, 1988.

Havemann, Robert. *Morgen: Die Industriegesellschaft am Scheideweg: Kritik und reale Utopie.* Munich: Piper, 1980.

Hecht, Gabrielle. *The Radiance of France: Nuclear Power and National Identity after World War II.* Cambridge, MA: MIT Press, 1998.

Hegen, Hannes. *Mosaik von Hannes Hegen: Geheimsache Digedanium.* Reprint. Berlin: Buchverlag Junge Welt, 2003.

Heimann, Thomas. "Television in Zeiten des Kalten Krieges: Zum Programmaustausch des DDR-Fernsehens in den sechziger Jahren." In *Massenmedien im Kalten Krieg: Akteure, Bilder, Resonanzen,* edited by Thomas Lindenberger, 235–61. Cologne: Böhlau Verlag, 2006.

Hockenos, Paul. *Joschka Fischer and the Making of the Berlin Republic: An Alternative History of Postwar Germany.* New York: Oxford University Press, 2008.

Hodenberg, Christina von. *Konsens und Krise: Eine Geschichte der westdeutschen Medienöffentlichkeit, 1945–1973.* Göttingen: Wallstein Verlag, 2006.

Hoffmann, Dieter. *Max Planck: Die Entstehung der modernen Physik.* Munich: Verlag C.H. Beck, 2008.

Huff, Tobias. *Natur und Industrie im Sozialismus: eine Umweltgeschichte der DDR.* Göttingen: Vandenhoeck & Ruprecht, 2015.

Hughes, Jeff. *The Manhattan Project: Big Science and the Atom Bomb.* Cambridge, UK: Icon Books, 2002.

Hülsberg, Werner. *The German Greens.* London: Verso, 1988.

Hünemörder, Kai. *Die Frühgeschichte der globalen Umweltkrise und die Formierung der deutschen Umweltpolitik: 1950–1973.* Wiesbaden: Franz Steiner Verlag, 2004.

———. "Zwischen Bewegungsforschung und Historisierungsversuch: Anmerkungen zum Anti-Atomkraft-Protest aus umwelthistorischer Perspektive." In *1968 und die Anti-Atomkraft-Bewegung der 1970er Jahre: Überlieferungsbildung und Forschung im Dialog,* ed. Robert Kretzschmar, Clemens Rehm, and Andreas Pilger, 151–67. Stuttgart: Verlag W. Kohlhammer, 2008.

Institut für Demoskopie. *Kernenergie und Öffentlichkeit: Ergebnisse einer Befragung von Politikern, Journalisten, Experten und der Bevölkerung.* Allensbach: Institut für Demoskopie, 1984.

Hürtgen, Renate. *Ausreise per Antrag: Der lange Weg nach drüben: Eine Studie über Herrschaft und Alltag in der DDR-Provinz.* Göttingen: Vandenhoeck & Ruprecht, 2014.

Joppke, Christian. *Mobilizing against Nuclear Energy: A Comparison of Germany and the United States.* Berkeley, CA: University of California Press, 1993.

Josephson, Paul. "'Projects of the Century' in Soviet History: Large-Scale Technologies from Lenin to Gorbachev." *Technology and Culture* 36, no. 3 (1995): 519–59.

———. *Red Atom: Russia's Nuclear Power Program from Stalin to Today.* New York: Freeman, 2000.

Jordan, Pascual. *Atomkraft: Drohung und Versprechen.* Munich: Wilhelm Heyne, 1954.

Jung, Matthias. *Öffentlichkeit und Sprachwandel: Zur Geschichte des Diskurses über die Atomenergie.* Opladen: Westdeutscher Verlag, 1994.

Jungk, Robert. *Der Atom-Staat: Vom Fortschritt in die Unmenschlichkeit.* Munich: Kindler, 1977. English translation: Robert Jungk. *The New Tyranny: How Nuclear Power Enslaves Us.* Translated by Christopher Trump. New York: F. Jordan Books/Grosset & Dunlap, 1979.

———. *Und Wasser bricht den Stein: Streitbare Beiträge zu drängenden Fragen der Zeit.* Edited by Marianne Oesterreicher-Mollwo. Paperback ed. Munich: Deutscher Taschenbuchverlag, 1988.

Kaarkoski, Miina. "'*Energiemix*' versus '*Energiewende*': Competing Conceptualisations of Nuclear Energy Policy in the German Parliamentary Debates of 1991–2001." Ph.D. diss., University of Jyväskylä [Finland], 2016. https://jyx.jyu.fi/dspace/handle/123456789/50961.

Karapin, Roger. *Protest Politics in Germany: Movements on the Left and Right Since the 1960s.* University Park, PA: Pennsylvania State University Press, 2007.

Karlsch, Rainer and Rudolf Boch. "Die Geschichte des Uranbergbaus der Wismut. Forschungsstand und neue Erkenntnisse." In *Uranbergbau im Kalten Krieg: Die Wismut im sowjetischen Atomkomplex,* edited by Rudolf Boch and Rainer Karlsch, 9–15. Berlin: Ch. Links Verlag, 2011.

Kendall, Henry, Richard Hubbard, and Gregory Minor. *The Risks of Nuclear Power Reactors: A Review of the NRC Reactor Safety Study, WASH-1400: NUREG-75/014.* Cambridge, MA: Union of Concerned Scientists, 1977. German translation: Ökö-Institut Freiburg. *Die Risiken der Atomkraftwerke: Der Anti-Rasmussen-Report der Union of Concerned Scientists.* Translated by Richard Donderer. Fellbach: Adolf Bonz, 1980.

Kershaw, Ian. *Popular Opinion and Political Dissent in the Third Reich, Bavaria 1933–1945.* Oxford: Oxford University Press, 1983.

Kirchhelle, Claas and Frank Uekötter. "Wie Seveso nach Deutschland kam: Umweltskandale und ökologische Debatte von 1976 bis 1986." In *Wandel des Politischen: Die Bundesrepublik Deutschland während der 1980er Jahre,* edited by Meik Woyke, 321–38. Bonn: Verlag J.H.W. Dietz Nachf., 2013.

Kirchhof, Astrid. "Frauen in der Antiatomkraftbewegung: Am Beispiel der Mütter gegen Atomkraft 1986–1990." *Ariadne. Forum für Frauen- und Geschlechtergeschichte* 62 (2013): 48–57.

Kirchhof, Astrid Mignon and Jan-Henrik Meyer. "Global Protest against Nuclear Power. Transfer and Transnational Exchange in the 1970s and 1980s." *Historical Social Research/Historische Sozialforschung* 39, no. 1 (2014): 165–190.

Kirsch, Scott. *Proving Grounds: Project Plowshare and the Unrealized Dream of Nuclear Earthmoving.* New Brunswick: Rutgers University Press, 2005.

Kneale, George and Alice Stewart. "Childhood Cancers in the UK and Their Relation to Background Radiation." *Radiation and Health* 16 (1987): 203–20.

Köhler, Bertram. "Schwerpunkte der Entwicklung im Kraftwerksanlagenbau der DDR." In *Zur Geschichte der Kernenergie in der DDR,* edited by Peter Liewers, Johannes Abele, and Gerhard Barkleit, 115–61. Frankfurt am Main: Peter Lang, 2000.

Kojevnikov, Alexei. *Stalin's Great Science: The Times and Adventures of Soviet Physicists.* London: Imperial College Press, 2004.

Kolb, Felix. "Der Castor-Konflikt: Das Comeback der Anti-AKW Bewegung," *Forschungsjournal Neue SozialBewegungen* 10, no. 3 (1997): 16–29.

Koopmans, Ruud. *Democracy from Below: New Social Movements and the Political System in West Germany.* Boulder, CO: Westview Press, 1995.

Koppe, Johannes. *Zum besseren Verständnis der Kernenergie: 66 Fragen: 66 Antworten.* Hamburg: HEW, 1971.

Köplin, Wilfried. "Die Energiewende erfolgreich gestalten: Mit der Industrie und mit dem Markt." In *Energiewende zwischen Klimaschutz und Atomausstieg— Lösungen in die Umsetzung tragen: 18. Internationale Sommerakademie St. Marienthal,* ed. Fritz Brickwedde and Dirk Schötz, 58–64. Berlin: Erich Schmidt Verlag, 2013.

Korte, Karl-Rudolf. "Solutions for the Decision Dilemma: Political Styles of Germany's Chancellors." *German Politics* 9, no. 1 (2000): 1–22.

Kosing, Alfred, Rolf Dörge, Diedrich Wattenberg, Rudolf Jubelt, Jacob Segal, Werner Rothmaler, Ilselotte Groth, Wolfgang Padberg, R.F. Schmiedt, Heinrich Scheel, Bernhard Bittighöfer, Stefan Doernberg, Karl Böhm, and Fred Müller, *Weltall Erde Mensch: Ein Sammelwerk zur Entwicklungsgeschichte von Natur und Gesellschaft.* 13th ed. Berlin: Verlag Neues Leben, 1965.

Kowalczuk, Ilko-Sascha. *Endspiel: Die Revolution von 1989 in der DDR.* 2nd ed. Munich: Beck, 2009.

Kramer, Thomas. *Micky, Marx und Manitu.* Berlin: Weidler Buchverlag, 2002.

Krementsov, Nikolai. *Stalinist Science.* Princeton: Princeton University Press, 1997.

Krige, John. *American Hegemony and the Postwar Reconstruction of Science in Europe*. Cambridge, MA: MIT Press, 2006.

———. "The Peaceful Atom as Political Weapon: Euratom and American Foreign Policy in the Late 1950s." *Historical Studies in the Natural Sciences* 38, no. 1 (2008): 5–44.

———. *Sharing Knowledge, Shaping Europe: U.S. Technological Collaboration and Nonproliferation*. Cambridge, MA: MIT Press, 2016.

Kühn, Andreas. "Kalkar 1977: Anti-Atomkraft-Bewegung und Polizei im Wandel." *Geschichte im Westen* 22 (2007): 269–89.

Kuhn, Vollrad. "Die Energieproblematik in Synodenbeschlüssen der Ev. Kirchen in der DDR und in den Papieren der 2. Vollversammlung für Gerechtigkeit, Frieden und Bewahrung der Schöpfung." *Arche Nova* (26 April 1989): 39–43, BstU MfS BV Bln Abt. XX Nr. 7192, 40–44.

Kumagai, Yuji and Shinya Mizokami. "Event Sequence of the Fukushima Daiichi Accident." In *Reflections on the Fukushima Daiichi Nuclear Accident,* edited by Joonhong Ahn, Cathryn Carson, Mikael Jensen, Kohta Juraku, Shinya Nagasaki, and Satoru Tanaka, 21–50. Cham: Springer, 2015.

Kupper, Patrick. *Atomenergie und gespaltene Gesellschaft: Die Geschichte gescheiterten Projektes Kernkraftwerk Kaiseraugst*. Zurich: Chronos, 2003.

Lancaster, John. *Engineering Catastrophes: Causes and Effects of Major Accidents*. 3rd ed. Cambridge: Woodhead; Boca Raton, FL: CRC Press, 2005.

Laufs, Paul. *Reaktorsicherheit für Leistungskernkraftwerke: Die Entwicklung im politischen und technischen Umfeld der Bundesrepublik Deutschland*. Berlin: Springer Verlag, 2013.

Lauvhjell, Ingeborg. "Die Atomkraftdebatte in Deutschland. Eine Diskursanalyse über die Repräsentation der Atomkraft in der deutschen Öffentlichkeit." M.A. thesis, University of Oslo, 2012.

Layton, Edwin. *The Revolt of the Engineers: Social Responsibility and the American Engineering Profession*. Cleveland and London: Press of Case Western Reserve University, 1971.

Leistner, Alexander. "Ein Grengänger auf Suche nach Heimat." *Forschungsjournal Soziale Bewegungen* 26, no. 4 (2013): 78–82.

Lengsfeld, Vera. *Ich wollte frei sein: Die Mauer, die Stasi, die Revolution*. Munich: F.A. Herbig, 2011.

Liewers, Peter, Johannes Abele, and Gerhard Barkleit, eds. *Zur Geschichte der Kernenergie in der DDR*. Frankfurt am Main: Peter Lang, 2000.

Lindee, M. Susan. *Suffering Made Real: American Science and the Survivors at Hiroshima*. Chicago, IL: University of Chicago Press, 1994.

Lindenberger, Thomas. "Einleitung." In *Massenmedien im Kalten Krieg: Akteure, Bilder, Resonanzen,* edited by Thomas Lindenberger, 9–23. Cologne: Böhlau Verlag, 2006.

————. "'Havarie': Reading East-German Society through the Violenc/se of Things," *Divinatio* 42-43 (2016): 301–369.

——— "Havarie: Die sozialistische Betriebsgemeinschaft im Ausnahmezustand." In *German Zeitgeschichte: Konturen eines Forschungsfeldes*, edited by Thomas Lindenberger and Martin Sabrow. Wallstein: Göttingen 2016, 242–264.

Linehan, Jennifer Rose. "Fukushima to Fessenheim: France and Germany's Policy Responses to the Fukushima Nuclear Disaster, 11 March to 18 June 2011." Master's thesis, Carleton University, Ottawa, 2013.

Ludwig, Andrea. *Neue oder deutsche Linke?: Nation und Nationalismus im Denken von Linken und Grünen.* Opladen: Westdeutscher Verlag, 1995.

Mählert, Ulrich. Kleine Geschichte der DDR. 7th ed. Munich: Verlag C.H. Beck, 2010.

Markovits, Andrei and Philip Gorski. *The German Left: Red, Green and Beyond.* New York: Oxford University Press, 1993.

Marszolek, Inge. "Radio Zebra in Bremen. Amateur-Radios und soziale Bewegungen in den frühen 80er Jahren." *Rundfunk und Geschichte* 39, nos. 1–2. (2013): 41–55.

Martin, Julia. "Der Berufsverband der Journalisten in der DDR. VDJ." In *Journalisten und Journalismus in der DDR*, edited by Jürgen Wilke, 7–78. Cologne: Böhlau Verlag, 2007.

Matthes, Felix Christian. *Stromwirtschaft und deutsche Einheit: Eine Fallstudie zur Transformation der Elektrizitätswirtschaft in Ost-Deutschland.* Berlin: F.C. Matthes; Norderstedt: Libri, 2000.

Maubach, Klaus-Dieter. *Energiewende: Wege zu einer Bezahlbaren Energieversorgung.* 2nd ed. Wiesbaden: Springer Fachmedien, 2014.

Mende, Silke. *"Nicht rechts, nicht links, sondern vorn": Eine Geschichte der Gründungsgrünen.* Munich: Oldenbourg Verlag, 2011.

Mende, Silke and Birgit Metzger. "Ökopax: Die Umweltbewegung als Erfahrungsraum der Friedensbewegung." In *"Entrüstet Euch!" Nuklearkrise, NATO-Doppelbeschluss und Friedensbewegung,* edited by Christoph Becker-Schaum, Philipp Gassert, Martin Klimke, Wilfried Mausbach, and Marianne Zepp, 118–34. Paderborn: Ferdinand Schöningh, 2012.

Merrit, Anna and Richard Merrit. *Public Opinion in Occupied Germany: The OMGUS Surveys, 1945–1949.* Urbana, IL: University of Illinois Press, 1970.

————. *Public Opinion in Semisovereign Germany: The HICOG Surveys, 1949–1955.* Urbana, IL: University of Illinois Press, 1980. https://archive.org/details/publicopinionins00merr.

Metzger, Birgit. *"Erst stirbt der Wald, dann Du!" Das Waldsterben als westdeutsches Politikum (1978–1986).* Frankfurt am Main: Campus Verlag, 2015.

Mewes, Horst. "A Brief History of the German Green Party." In *The German Greens: Paradox between Movement and Party,* edited by Margit Mayer and John Ely, 29–48. Philadelphia, PA: Temple University Press, 1998.

Meyen, Michael. *Denver Clan und Neues Deutschland*. Berlin: Christoph Links Verlag, 2003.

Meyen, Michael and William Hillman. "Communication Needs and Media Change: The Introduction of Television in East and West Germany." *European Journal of Communication* 18, no. 4 (2003): 455–76.

Mick, Christoph. *Forschen für Stalin: Deutsche Fachleute in der sowjetischen Rüstungsindustrie 1945–1958*. Munich and Vienna: R. Oldenbourg Verlag, 2000.

Milder, Stephen. *Greening Democracy: The Anti-Nuclear Movement and Political Environmentalism in West Germany and Beyond, 1968–1983*. Cambridge, UK and New York: Cambridge University Press, 2017.

———. "'Today the Fish, Tomorrow Us:' Anti-Nuclear Activism in the Rhine Valley and Beyond, 1970–1979." Ph.D. diss., University of North Carolina at Chapel Hill, 2012.

Moniz, Ernest. "Why We Still Need Nuclear Power: Making Clean Energy Safe and Affordable," *Foreign Affairs* 90, no. 6 (2011): 83–94.

Moore, Walter. *Schrödinger: Life and Thought*. Cambridge, UK and New York: Cambridge University Press, 1989.

Mrasek-Robor, Heike. "Technisches Risiko und Gewaltenteilung." Dr. iur. diss., University of Bielefeld, 1997.

Müller, Wolfgang. *Geschichte der Kernenergie in der Bundesrepublik Deutschland*. Vol. 1: *Anfänge und Weichenstellungen*. Stuttgart: Schäffer Verlag für Wirtschaft und Steuern, 1990.

Münkel, Daniela. *Willy Brandt und die 'Vierte Gewalt': Politik und Massenmedien in den 50er bis 70er Jahren*. Frankfurt am Main: Campus Verlag, 2005.

Nehring, Holger. "Cold War, Apocalypse and Peaceful Atoms. Interpretations of Nuclear Energy in the British and West German Anti-Nuclear Weapons Movements, 1955–1964." *Historical Social Research* 29, no. 3 (2004): 150–70.

———. "Transnationale Netzwerke der bundesdeutschen Friedensbewegung." In *"Entrüstet Euch!" Nuklearkrise, NATO-Doppelbeschluss und Friedensbewegung*, edited by Christoph Becker-Schaum, Philipp Gassert, Martin Klimke, Wilfried Mausbach, and Marianne Zepp, 277–93. Paderborn: Ferdinand Schöningh, 2012.

Nehring, Holger and Benjamin Ziemann. "Führen alle Wege nach Moskau? Der NATO-Doppelbeschluss und die Friedensbewegung—eine Kritik." *Vierteljahrshefte für Zeitgeschichte* 59, no. 1 (2011): 81–100.

Neubert, Ehrhart. *Geschichte der Opposition in der DDR 1949–1989*. 2nd ed. Berlin: Christian Links Verlag, 1998.

Niclauß, Karlheinz. *Kanzlerdemokratie: Von Konrad Adenauer bis Angela Merkel*. Wiesbaden: Springer VS, 2015.

Nye, David. *American Technological Sublime*. Cambridge, MA: MIT Press, 1994.

Osgood, Kenneth. *Total Cold War: Eisenhower's Secret Propaganda Battle at Home and Abroad.* Lawrence, KS: University of Kansas, 2006.

Palmowski, Jan. "Citizenship, Identity and Community in the German Democratic Republic." In *Citizenship and National Identity in Twentieth-Century Germany,* edited by Geoff Eley and Jan Palmowski, 73–93. Stanford, CA: Stanford University Press, 2008.

Pauling, Linus. "Nuclear Fission Is Not the Answer," *Bulletin of the Atomic Scientists* 33, no. 3 (1977): 66.

Pence, Katherine. "Showcasing Cold War Germany in Cairo: 1954 and 1957 Industrial Exhibitions and the Competition for Arab Partners." *Journal of Contemporary History* 47, no. 1 (2011): 69–95.

Pensold, Wolfgang. *Eine Geschichte des Fotojournalismus.* Wiesbaden: Springer VS, 2015.

Perlin, John. *Let It Shine: The 6,000-Year Story of Solar Energy.* Novato, CA: New World Library, 2013.

Perrow, Charles. *Normal Accidents: Living with High-Risk Technologies.* New York: Basic Books, 1984.

Pflugbeil, Sebastian. "Preface." In "Pechblende: Der Uranbergbau in der DDR und seine Folgen," by Michael Beleites (hectograph manuscript, Lutherstadt Wittenberg: Kirchliches Forschungsheim Wittenberge, 1988), BStU, HA XVIII, Nr. 18237, 84.

Plamper, Jan. *The History of Emotions: An Introduction.* Translated by Keith Tribe. New York: Oxford University Press, 2015.

Popp, Manfred. "Misinterpreted Documents and Ignored Physical Facts: The History of 'Hitler's Atomic Bomb' Needs to be Corrected." *Berichte zur Wissenschaftsgeschichte* 39, no. 3 (2016). DOI: 10.1002/bewi.201601794.

Port, Andrew. *Conflict and Stability in the German Democratic Republic.* Cambridge and New York: Cambridge University Press, 2007.

Principe, Lawrence. "Alchemy Restored." *Isis* 102, no. 2 (2011): 305–12.

Quistorp, Eva. "Frauen in Bürgerinitiativen." In *Die Neue Frauenbewegung in Deutschland: Abschied vom kleinen Unterschied: Eine Quellensammlung,* edited by Ilse Lenz, 851–53. 2nd ed. Wiesbaden: VS Verlag für Sozialwissenschaften/Springer Verlag, 2010.

Radkau, Joachim. *Aufstieg und Krise der deutschen Atomwirtschaft 1945–1975: Verdrängte Alternativen in der Kerntechnik und der Ursprung der nuklearen Kontroverse.* Reinbek bei Hamburg: Rowohlt Taschenbücher, 1983.

———. "Der Größte Anzunehmende Unfall." In *Ökologische Erinnerungsorte,* edited by Frank Uekötter, 50–60. Göttingen: Vandenhoeck & Ruprecht, 2014.

———. *Die Ära der Ökologie: Eine Weltgeschichte.* Munich: C.H. Beck, 2011.

————. "Mythos German Angst. Zum neusten Aufguss einer alten Denunziation der Umweltbewegung." *Blätter für deutsche und internationale Politik* 5 (2011): 73–82.

————. *Nature and Power: A Global History of the Environment.* Translated by Thomas Dunlap. New York: Cambridge University Press, 2008.

Radkau, Joachim and Lothar Hahn. *Aufstieg und Fall der deutschen Atomwirtschaft.* Munich: oekom, 2013.

Radkau, Joachim and Frank Uekötter, eds., *Naturschutz und Nationalsozialismus.* Frankfurt and New York: Campus Verlag, 2003.

Reichardt, Sven. *Authentizität und Gemeinschaft: Linksalternatives Leben in den siebziger und frühen achtziger Jahren.* Berlin: Suhrkamp Verlag, 2014.

Reichert, Mike. *Kernenergiewirtschaft in der DDR.* St. Katharinen: Scripta Mercaturae Verlag, 1999.

Roesler, Jörg. "Die Produktionsbrigaden in der Industrie der DDR: Zentrum der Arbeitswelt?" In *Sozialgeschichte der DDR,* edited by Hartmut Kaelble, Jürgen Kocka, and Hartmut Zwahr, 144–70. Stuttgart: Klett-Cotta Verlag, 1994.

Röhl, Anja. "Wir lassen uns unsere Angst nicht ausreden!" In *Frauen aktiv gegen Atomenergie: wenn aus Wut Visionen werden: 20 Jahre Tschernobyl,* edited by Ulrike Röhr. 31–35. Wiesbaden: genanet, 2006.

Ross, Corey. *Media and the Making of Modern Germany: Mass Communications, Society, and Politics from the Empire to the Third Reich.* Oxford and New York: Oxford University Press, 2008.

Rüddenklau, Wolfgang. *Störenfried: DDR-Opposition 1986–1989: Mit Texten aus den "Umweltblättern."* Berlin: BasisDruck Verlag, 1992.

Rumiel, Lisa. "'Random Murder by Technology': The Role of Scientific and Biomedical Experts in the Anti-Nuclear Movement, 1969–1992." Ph.D. diss., York University, Canada, 2009.

Rusinek, Bernd-A. "'Kernenergie, Schöner Götterfunken!' Die 'umgekehrte Demontage.' Zur Kontextgeschichte der Atomeuphorie." *Kultur und Technik* 4 (1993): 15–21.

Saliba, George. *A History of Arabic Astronomy: Planetary Theories during the Golden Age of Islam.* New York: New York University Press, 1994.

Scharf, Thomas. *The German Greens: Challenging the Consensus.* Providence: Berg, 1994.

Schattenberg, Susanne. *Stalins Ingenieure: Lebenswelten zwichen Technik und Terror in den 1930er Jahren.* Munich: R. Oldenbourg Verlag, 2002.

Scheer, Hermann. *Energy Autonomy: The Economic, Social & Technological Case for Renewable Energy.* New York: Routledge, 2006.

Schmid, Sonja. "Celebrating Tomorrow Today: The Peaceful Atom on Display in the Soviet Union." *Social Studies of Science* 36, no. 3 (2006): 331–65.

————. *Producing Power: The Pre-Chernobyl History of the Soviet Nuclear Industry.* Cambridge, MA: MIT Press, 2015.

Schöllgen, Gregor. *Gerhard Schröder: Die Biographie.* Munich: Deutsche Verlags-Anstalt, 2015.

Schönfelder, Jan. *Mit Gott gegen Gülle: Die Umweltgruppe Knau/Dittersdorf 1986–1991. Eine regionale Protestbewegung in der DDR.* Rudolstadt and Jena: Hain-Verlag, 2000.

Schönherr, Alexander. "Die ersten vier Blöcke des KKW Gleifswald von der Vorbereitung bis zur Abschaltung." In *Zur Geschichte der Kernenergie in der DDR,* edited by Peter Liewers, Johannes Abele, and Gerhard Barkleit, 221–308. Frankfurt am Main: Peter Lang, 2000.

Schramm, Manuel. "Strahlenschutz im Uranbergbau: DDR und Bundesrepublik Deutschland im Vergleich (1945–1990)." In *Uranbergbau im Kalten Krieg: Die Wismut im sowjetischen Atomkomplex,* edited by Rudolf Boch and Rainer Karlsch, 271–328. Berlin: Ch. Links Verlag, 2011.

Schulz, Jan-Hendrik. "Die Großdemonstration in Brokdorf am 28. Februar 1981—eine empirische Verlaufsstudie mit Blick auf die Fraktionen der Demonstrierenden und der Polizei," B.A. thesis, University of Bielefeld, 2007.

Schüring, Michael. "Advertising the Nuclear Venture: The Rhetorical and Visual Public Relation Strategies of the German Nuclear Industry in the 1970s and 1980s." *History and Technology* 29, no. 4 (2013): 369–98.

————. *'Bekennen gegen den Atomstaat': Die evangelischen Kirchen in der Bundesrepublik Deutschland und die Konflikte um die Atomenergie 1970–1990.* Göttingen: Wallstein Verlag, 2015.

Schwarz, Uta. "Der blockübergreifende Charme dokumentarischer Filme." In *Massenmedien im Kalten Krieg: Akteure, Bilder, Resonanzen,* edited by Thomas Lindenberger, 203–34. Cologne: Böhlau Verlag, 2006.

Schwarzmeier, Jan. "Die Autonomen zwischen Subkultur und sozialer Bewegung." Dr.phil. diss., University of Göttingen, 1999.

Schwerin, Alexander von. "Der gefährdete Organismus: Biologie und Regierung der Gefahren am Übergang vom 'Atomzeitalter' zur Umweltpolitik: 1950–1970." In *Wissenobjekt Mensch: Humanwissenschaftliche Praktiken im 20. Jahrhundert,* edited by Florence Vienne and Christina Brandt, 187–214. Berlin: Kulturverlag Kadmos, 2008.

————. *Strahlenforschung: Bio- und Risikopolitik der DFG, 1920–1970.* Stuttgart: Franz Steiner Verlag, 2015.

Semendeferi, Ioanna. "Legitimating a Nuclear Critic: John Gofman, Radiation Safety, and Cancer Risks." *Historical Studies in the Natural Sciences* 38, no. 2 (2008): 259–301.

Shrader-Frechette, Kristin. "Climate Change, Nuclear Economics, and Conflicts of Interest." *Science & Engineering Ethics* 17, no. 1 (2011): 75–107.

Shapin, Steven. *Never Pure: Historical Studies of Science.* Baltimore: Johns Hopkins Press, 2010.

———. "Science and the Public." In *Companion to the History of Modern Science,* edited by R. Olby G.N. Cantor, J.R.R. Christy, and M.J.S. Hodge, 990–1007. London: Routledge, 1990.

Siddiqi, Asif. *The Soviet Space Race with Apollo.* Gainesville: University Press of Florida, 2003.

Smith, Edward, A.H. Fox, R. Tom Sawyer, and H.R. Austin, *Applied Atomic Power.* New York: Prentice-Hall, 1946.

Solnit, Rebecca. *A Paradise Built in Hell: The Extraordinary Communities That Arise in Disaster.* New York: Viking, 2009.

Sommer, Gisela. *Grenzüberschreitungen: Evangelische Studentengemeinde in der DDR und BRD: Geschichte-Verhältnis-Zusammenarbeit in zwei deutschen Staaten.* Stuttgart: Alektor-Verlag, 1984.

Sposto, Richard, Dale Preston, Yukiko Shimizu, and Kiyohiko Mabuchi. "The Consultant's Forum: The Effect of Diagnostic Misclassification on Non-Cancer and Cancer Mortality Dose Response in A-Bomb Survivors." *Biometrics* 48 (1992): 615–17.

Steinmetz, Rüdiger and Reinhold Viehoff, eds. *Deutsches Fernsehen Ost: Eine Programmgeschichte des DDR-Fernsehens.* Berlin: Verlag für Berlin-Brandenburg, 2008.

Steinmetz, Willibald. "Ungewollte Politisierung durch die Medien: Die Contergan-Affäre." In *Die Politik der Öffentlichkeit—die Öffentlichkeit der Politik,* edited by Bernd Weisbrod, 195–228. Göttingen: Wallstein Verlag, 2003.

Steurer, Reinhard. "Der Wachstumsdiskurs in Wissenschaft und Politik: Von der Wachstumseuphorie über 'Grenzen des Wachstums' zur Nachhaltigkeit." Dr.phil. diss., University of Salzburg, 2001. Retrieved 8 January 2018 from https://www.wiso.boku.ac.at/fileadmin/data/H03000/H73000/H73 200/_TEMP_/Steurer_Wachstumsdiskurs_in_Wissenschaft_und_Politik_ Diss_01.pdf.

Steurer, Reinhard. *Der Wachstumsdiskurs in Wissenschaft und Politik: Von der Wachstumseuphorie über 'Grenzen des Wachstums' zur Nachhaltigkeit.* Berlin: VWF Verlag für Wissenschaft und Forschung, 2002.

Stölken-Fitschen, Ilona. *Atombombe und Geistesgeschichte: Eine Studie der Fünfziger Jahre aus deutscher Sicht.* Baden-Baden: Nomos Verlagsgesellschaft, 1995.

———. "Der verspätete Schock: Hiroshima und der Beginn des atomaren Zeitalters." In *Moderne Zeiten: Technik und Zeitgeist im 19. und 20. Jahrhundert,* edited by Michael Salewski and Ilona Stölken-Fitschen, 139–55. Stuttgart: Franz Steiner Verlag, 1994.

Storey, John. *Cultural Theory and Popular Culture: An Introduction.* 7th ed. New York: Routledge, 2015.

Stötzel, Georg and Martin Wengeler. *Kontroverse Begriffe: Geschichte des öffentlichen Sprachgebrauchs in der Bundesrepublik Deutschland.* Berlin and New York: de Gruyter, 1995.

Strohm, Holger. *Friedlich in die Katastrophe: Eine Dokumentation über Kernkraftwerke.* Hamburg: Verlag Association, 1973.

Sturm, Michael. "Polizei und Friedensbewegung." In *"Entrüstet Euch!" Nuklearkrise, NATO-Doppelbeschluss und Friedensbewegung,* edited by Christoph Becker-Schaum, Philipp Gassert, Martin Klimke, Wilfried Mausbach, and Marianne Zepp, 277–93. Paderborn: Ferdinand Schöningh, 2012.

Szöllösi-Janze, Margit. "National Socialism and the Sciences." In *Science in the Third Reich,* edited by Margit Szöllösi-Janze, 1–35. Oxford and New York: Berg, 2001.

Tarrow, Sidney. *Power in Movement: Social Movements and Contentious Politics.* 3rd rev. ed. Cambridge, UK and New York: Cambridge University Press, 2011.

Tandler, Agnes. "Geplante Zukunft: Wissenschaftler und Wissenschaftspolitik in der DDR 1955–1971." Diss., Europäisches Hochschulinstitut, Florence, Italy, 1997.

Tiggemann, Alselm. *Die "Achillesferse" der Kernenergie in der Bundesrepublik Deutschland Zur Kernenergiekontroverse und Geschichte der nuklearen Entsorgung von den Anfängen bis Gorleben 1955 bis 1985.* Lauf an der Pegnitz: Europaforum-Verlag, 2004.

Tompkins, Andrew. *Better Active Than Radioactive! Anti-Nuclear Protest in 1970s France and West Germany.* Oxford: Oxford University Press, 2016.

Uekötter, Frank. *Deutschland in Grün: Eine zwiespältige Erfolgsgeschichte.* Göttingen: Vandenhoeck & Ruprecht, 2015.

Ukena, Fokko. "Sozialistische Persönlichkeit: Grundlagen, Ziele, Methoden und Resultate der sozialistischen Persönlichkeitskonzeption in der DDR." Diss., University of Osnabrück, 1989.

U.S. Atomic Energy Commission. *Reactor Safety Study: An Assessment of Accident Risks in U.S. Commercial Nuclear Power Plants: Summary Report.* Washington, DC: U.S. Government Printing Office, 1974.

Wagar, W. Warren. "H.G. Wells and the Scientific Imagination." *The Virginia Quarterly Review,* 65, no. 3 (1989): 390–400.

Walker, J. Samuel. *Permissible Dose: A History of Radiation Protection in the Twentieth Century.* Berkeley, CA: University of California Press, 2000.

Walker, Mark. *German National Socialism and the Quest for Nuclear Power 1939–1949.* Cambridge, UK: Cambridge University Press, 1989.

Weart, Spencer. *Nuclear Fear: A History of Images.* Cambridge, MA: Harvard University Press, 1989.

Wehner, Christoph. "Grenzen der Versicherbarkeit—Grenzen der Risikogesellschaft: Atomgefahr, Sicherheitsproduktion und Versicherungsexpertise in der Bundesrepublik und den USA." In *Wandel des Politischen: Die Bundesrepublik Deutschland während der 1980er Jahre,* edited by Meik Woyke, 585–609. Bonn: Verlag J.H.W. Dietz Nachf., 2013.

Weiland, Sabine. *Politik der Ideen: Nachhaltige Entwicklung in Deutschland, Großbritannien und den USA.* Wiesbaden: VS Verlag für Sozialwissenschaften, 2007.

Weingart, Peter. *Die Stunde der Wahrheit? Zum Verhältnis der Wissenschaft zu Politik, Wirtschaft und Medien in der Wissensgesellschaft.* Weilerswist, Germany: Velbrück, 2001.

———. "Verwissenschaftlichung der Gesellschaft—Politisierung der Wissenschaft." *Zeitschrift für Soziologie* 12, no. 3 (1983): 225–41.

Weinhauer, Klaus. "Staatsgewalt, Massen, Männlichkeit: Polizeieinsätze gegen Jugend- und Studentenproteste in der Bundesrepublik der 1960er Jahre." In *Polizei, Gewalt und Staat im 20. Jahrhundert,* edited by Alf Lüdtke, Herbert Reinke, and Michael Sturm, 301–24. Wiesbaden: VS Verlag für Sozialwissenschaften, 2011.

Weisbrod, Bernd. *Die Politik der Öffentlichkeit—die Öffentlichkeit der Politik.* Göttingen: Wallstein Verlag, 2003.

Weish, Peter and Eduard Gruber. *Radioaktivität und Umwelt.* Stuttgart: Gustav Fischer Verlag, 1975.

Weisker, Albrecht. "Expertenvertrauen gegen Zukunftsangst: Zur Risikowahrnehmung der Kernenergie." In *Vertrauen: Historische Annäherungen,* edited by Ute Frevert, 394–421. Göttingen: Vandenhoeck & Ruprecht, 2003.

Weiss, Burghard. "Nuclear Research and Technology in Comparative Perspective." In *Science under Socialism: East Germany in Comparative Perspective,* edited by Kristie Macrakis and Dieter Hoffmann, 212–29. Cambridge, MA and London: Harvard University Press, 1999.

Weiss, Matthias. "Öffentlichkeit als Therapie: Die Medien- und Informationspolitik der Regierung Adenauer zwischen Propaganda und kritischer Aufklärung." In *Medialisierung und Demokratie im 20. Jahrhundert,* edited by Frank Bösch und Norbert Frei, 73–120. Göttingen: Wallstein Verlag, 2006.

Wellock, Thomas. "A Figure of Merit: Quantifying the Probability of a Nuclear Reactor Accident." *Technology and Culture* 58, no. 3 (2017): 678–721.

Wells, H.G. *The World Set Free.* London: Macmillan, 1914.

Werkentin, Falco. *Die Restauration der deutschen Polizei: Innere Rüstung von 1945 bis zur Notstandsgesetzgebung.* Frankfurt am Main and New York: Campus Verlag, 1984.

Widmann, Alexander Christian. *Wandel mit Gewalt?: Der deutsche Protestantismus und die politisch motivierte Gewaltanwendung in den 1960er und 1970er Jahren*. Göttingen: Vandenhoeck & Ruprecht, 2013.

Wiggershaus, Renate and Rolf Wiggershaus, "Beim 'Gewaltschutzparagraphen' geht es nicht nur um Gewalt," *Gewerkschaftliche Monatshefte* 27, no. 10 (1976): 597–602.

Willmann, F.W. "Moralische Gefühle und ihre Bedeutung für die sozialistische Persönlichkeit." Diss., Karl-Marx-Stadt, Technische Hochschule, 1982.

Winter, Martin. "Police Philosophy and Protest Policing in the Federal Republic of Germany, 1960–1990." In *Policing Protest: The Control of Mass Demonstrations in Western Democracies*, edited by Donatella Della Porta and Herbert Retter, 188–212. Minneapolis, MN: University of Minnesota Press, 1998.

Wolf, Christa. *Störfall: Nachrichten eines Tages*. Berlin and Weimar: Aufbau-Verlag, 1987. First English edition: *Accident/A Day's News*. Translated by Heike Schwarzbauer and Rick Takvorian. New York: Farrar, Straus & Giroux, 1989.

Wolfrum, Otfried. *Windkraft: Eine Alternative, die keine ist*. Frankfurt am Main: Zweitausendeins, 1997.

Wynne, Brian. "May the Sheep Safely Graze? A Reflexive View of the Expert-Lay Knowledge Divide." In *Risk, Environment and Modernity: Towards a New Ecology*, edited by Scott Lash, Bronislaw Szerszynski, and Brian Wynne, 44–83. London: Sage, 1996.

Zantop, Susanne. *Colonial Fantasies: Conquest, Family and Nation in Pre-Colonial Germany, 1770–1870*. Lincoln, NE: University of Nebraska Press, 1990.

Index

Protest, Culture, and Society

General editors:
Kathrin Fahlenbrach, Institute for Media and Communication, University of Hamburg
Martin Klimke, New York University, Abu Dhabi
Joachim Scharloth, Waseda University, Japan

Protest movements have been recognized as significant contributors to processes of political participation and transformations of culture and value systems, as well as to the development of both a national and transnational civil society.

This series brings together the various innovative approaches to phenomena of social change, protest, and dissent which have emerged in recent years, from an interdisciplinary perspective. It contextualizes social protest and cultures of dissent in larger political processes and socio-cultural transformations by examining the influence of historical trajectories and the response of various segments of society, political, and legal institutions on a national and international level. In doing so, the series offers a more comprehensive and multi-dimensional view of historical and cultural change in the twentieth and twenty-first century.